NORTH-HOLLAND
MATHEMATICS STUDIES 12

Notas de Matemática (54)

Editor: Leopoldo Nachbin

Universidade Federal do Rio de Janeiro
and University of Rochester

Infinite Dimensional Holomorphy and Applications

Edited by

MARIO C. MATOS

Universidade Estadual de Campinas, Brazil

1977

NORTH-HOLLAND PUBLISHING COMPANY - AMSTERDAM • NEW YORK • OXFORD

North-Holland ISBN: 0 444 85084 8

PUBLISHERS:
NORTH-HOLLAND PUBLISHING COMPANY
AMSTERDAM, NEW YORK, OXFORD

SOLE DISTRIBUTORS FOR THE U.S.A. AND CANADA:
ELSEVIER / NORTH HOLLAND, INC.
52 VANDERBILT AVENUE, NEW YORK, N.Y. 10017

Library of Congress Cataloging in Publication Data

International Symposium on Infinite Dimensional Holomorphy, Universidade Estadual de Campinas, 1975.
Infinite dimensional holomorphy and applications.

(Notas de matemática ; 54) (North-Holland mathematics studies ; 12)
Includes index.
1. Holomorphic functions--Congresses.
2. Domains of holomorphy--Congresses. I. Matos, Mário Carvalho de. II. Title. III. Series.
QA1.N86 no. 54 [QA331] 510'.8s [515'.9]
ISBN 0-444-85084-8 77-20010

PRINTED IN THE NETHERLANDS

FOREWORD

This book contains the Proceedings of the International Symposium on Infinite Dimensional Holomorphy held at the Universidade Estadual de Campinas, Brazil, during August 4-8,1975. It contains twenty five original research articles corresponding to the lectures given at the Symposium and some papers by invited lecturers who could not attend the meeting for one reason or another. The articles include complete proofs and they cover the current most active research lines of Inifinite Dimensional Holomorphy and its applications.

The meeting received support from the "Conselho Nacional de Pesquisas (CNPq)", "Coordenação do Aperfeiçoamento do Pessoal de Nivel Superior (CAPES)", "Fundação de Amparo à Pesquisa do Estado de São Paulo (FAPESP)", "Financiadora de Estudos e Projetos (FINEP)", and "Universidade Estadual de Campinas (UNICAMP)".

The organizing committee for this meeting was formed by J. A. Barroso, G. I. Katz, M. C. Matos (chairman), L. Nachbin and D. Pisanelli.

There were participants from the following countries: Brazil, Chile, France, Germany, Ireland, Sweden, U.S.A. and Yugoslavia.

I would like to register my thanks to the following persons: Professor Ubiratan D'Ambrosio, director of the Institute of Mathematics of UNICAMP, whose support made the meeting possible; Miss Elda Mortari, who typed this volume; Mr. Gilberto Rodrigues Queiroz, who provided all technical facilities for the typing of the manuscript.

Mário C. Matos

TABLE OF CONTENTS

Infinite Dimensional Holomorphy and Applications, Matos (ed.)

APPROXIMATION OF DIFFERENTIABLE FUNCTIONS ON A BANACH SPACE

By *RICHARD M. ARON* (*)

The study of approximation of holomorphic and differentiable functions defined on Banach spaces has received considerable attention in recent years. For example, in the complex case, there has been work done on polynomial approximation of analytic functions, defined on Runge or polynomially convex sets in infinite dimensional spaces (see, for example, [2,5,9,11]). In the real case, there has been interest in the general problem of approximating, in one of several topologies, certain classes of differentiable functions by smoother ones, such as polynomials or real analytic functions. In this note, we will outline some new results relating to approximation of differentiable functions with respect to four topologies.

In Section 1, we examine the approximation of differentiable

(*) Research partially supported by the Instituto de Matemática, Universidade Federal do Rio de Janeiro, Brasil, Conselho Nacional de Desenvolvimento Científico e Tecnologico (CNPq) and Financiadora de Estudos e Projetos (FINEP).

functions, in the fine topology, defined on $C(K)$ where K is a compact Hausdorff space. Our result generalizes slightly work of Wells [15] and Wulbert [17]. In Sections 2-4, our main interest is with polynomial approximation of k-continuously differentiable functions defined on a Banach space E, with respect to topologies studies by Prolla and Guerreiro [13], Lesmes [7], Restrepo [14], Bombal [3], Llavona [8], and others. In these sections, the topologies we consider will be generated by the following seminorms:

$$p(f) = \sup \{ |\hat{d}^j f(x)(y)| : x \in X,\ y \in Y \},$$

where X and Y are subsets of E and $j \in \mathbb{N}$, $j \leq k$. If X and Y range over the compact subsets of E, we get the topology τ_c^k studied in [3,8,12,13]. In Section 2, we show that E has the (Grothendieck) approximation property if and only if $(C^k(E), \tau_c^k)$ has the approximation property for some, and hence for all, $k \geq 1$. If X ranges over the compact subsets of E and Y over the bounded subsets of E, we get the topology τ_u^k (cf [7]), and if both X and Y are allowed to range over the bounded subsets of E, we get the topology τ_b^k which has been studied in the case $k = 1$ in [14]. In Sections 3 and 4, we study the completion of the finite type polynomials on E with respect to τ_u^k and τ_b^k.

For E and F real Banach spaces, $C^k(E;F)$ denotes the space of k-continuously Frechet differentiable F-valued mappings on E. We recall that the space of finite type polynomials from E to F, $\mathcal{P}_f(E;F)$, is the vector space generated by $\mathcal{P}_f({}^nE:F)$ $(n \in \mathbb{N})$, where $\mathcal{P}_f({}^nE;F)$ is the subspace of $\mathcal{P}({}^nE;F)$ (of all continuous n-homogeneous F-valued polynomials on E) spanned by $\{\phi^n \cdot b : \phi \in E',\ b \in F\}$. When F is the scalar field, $C^k(E;F)$ will be denoted $C^k(E)$, $\mathcal{P}_f(E;F)$ by $\mathcal{P}_f(E)$, etc. Throughout, we restrict our attention to scalar valued functions defined on E, although the obvious extensions to the vector-valued case and to functions defined on open subsets of E also hold.

Much of the work outlined here was done while the author was visiting the Instituto de Matemática of the Universidade Federal do Rio de Janeiro, and it is a pleasure to thank the institute for its hospitality. In addition, the author is especially grateful to Professor João Bosco Prolla for many fruitful discussions concerning this area of approximation theory. The author is also grateful to Sean Dineen, Paul Berner, and Ray Ryan for some helpful conversations.

SECTION 1.

In [15], Wells showed that a continuously differentiable function f defined on $c_0 = (\equiv C(N \cup \{\infty\}))$ which has a uniformly continuous derivative must have unbounded support (or be identically zero). In this section, we outline an extension of this result to functions defined on C(K) for some compact Hausdorff space K. As a consequence, we show that it is impossible to approximate an arbitrary C^2 function by a C^3 function in the fine topology of order 2(cf [6,10,17]). In contrast to this, in Section 3, we note that for dispersed compact sets K, one can approximate arbitrary C^k functions on C(K) by finite type polynomials in τ_u^k. Throughout this section, E will be the Banach space C(K), where K is compact, Hausdorff and infinite.

PROPOSITION 1.1: *Let* $G \in C^1(E)$, $G \not\equiv 0$. *If* $\hat{d}G : E \to E'$ *is uniformly continuous, the spt* G *is unbounded.*

The proof of this result is modelled on that of Wells and requires the following lemmas.

LEMMA 1.2. *Let* $h \in E$, $x_1,\ldots,x_k \in K$ *fixed points, such that* $\| h \| = |h(x_i)| = \varepsilon$ $(i = 1,\ldots,k)$. *Then there exists a path* $\gamma : [0,1] \to E$ *such that* $\| \gamma(t) \| = \varepsilon$ *and* $\gamma(t)(x_i) = h(x_i)$ $(i = 2,\ldots,k,\ t \in [0,1])$, $\gamma(0) = h$, *and* $\gamma(1)(x_1) = -h(x_1)$.

LEMMA 1.3. *Let* g *and* $h \in E$, $x_1,\ldots,x_k \in K$ *fixed points such that* $\| g \| = \| h \| = \varepsilon$ *and* $g(x_1) = h(x_i)$ *and of modulus* ε $(i = 1,\ldots,k)$. *Then there exists a path* $\gamma : [0,1] \to E$ *such that* $\gamma(0) = h$, $\gamma(1) = g$, $\|\gamma(t)\| = \varepsilon$ *and* $\gamma(t)(x_i) = g(x_i) = h(x_i)$ $(i = 1,\ldots,k,\ t \in [0,1])$.

PROOF OF LEMMA 1.2: Let V_1 and V_2 be open subsets of K such that $x_1 \in V_1$, $x_2 \in V_2$, and $\overline{V_1} \cap \overline{V_2} = \phi$. We define a function g on $\overline{V_1} \cup \overline{V_2}$ to be 1 on $\overline{V_2}$ and -1 on $\overline{V_1}$, and extend g to K, in such a way a that

$$\| g \| = \sup\{| g(x) | : x \in K\} = 1.$$

Define $\beta : [0,1] \to E$ by $\beta(t) = (1-t)1 + tg$, (where 1 is the constant function), and set $\gamma(t) = \beta(t)h$. It is easy to verify that γ has the required properties. Q.E.D.

The proof of Lemma 1.3 is trivial, and we proceed now to sketch a proof of proposition 1.1. If the proposition is false, then there is a $G \in C^1(E)$ such that $G(0) = 1$, $G(f) = 0$ for $\| f \| \geq 1$, and $\hat{d}G$ is uniformly continuous. Thus for some $\varepsilon = 1/n$, we have $\| \hat{d}G(f) - \hat{d}G(g) \| < 1/2$ whenever $\|f - g\| < \varepsilon$. Let $B_o = \{x_1,\ldots,x_{2^{n+1}}\} \subset K$ be arbitrary, and let $h_o = 0 \in E$.

Let $g_1 \in E$, $g_1(x) \equiv \varepsilon$. If $\hat{d}G(h_o)(g_1) = 0$, set $h_1 = g_1$. Otherwise, $\hat{d}G(h_o)(g_1)$ and $\hat{d}G(h_o)(-g_1)$ are of opposite sign. Applying Lemmas 1.2 and 1.3, there is $h_1 \in E$ and a subset $B_1 \subset B_o$ consisting of all but possibly one point x_i such that $\| h_1 \| = \varepsilon = |h_1(x_i)|$ $(i \in B_1)$ and $\hat{d}G(h_o)(h_1) = 0$. Let $S \subset B_1$, card $S = 2^{n-1}$, and let $g_2 \in E$ be such that $g_2(x_i) = h_1(x_i)$ $(i \in S)$, $g_2(x_i) = -h_1(x_i)$ $(i \notin S)$, and $\| g_2 \| = \varepsilon$. If $\hat{d}G(h_o + h_1)(h_1) = 0$, set $h_2 = h_1$. Otherwise, $\hat{d}G(h_o + h_1)(h_1)$ and either $\hat{d}G(h_o + h_1)(g_2)$ or $\hat{d}G(h_0 + h_1)(-g_2)$ are of opposite sign. Applying Lemmas 1.2 and 1.3, there exists $h_2 \in E$ such that $\| h_2 \| = \varepsilon$, $\hat{d}G(h_o + h_1)(h_2) = 0$, and $h_2(x_i) = h_1(x_i)$ for all i in a set B_2 of cardinality at least 2^{n-1}. Continuing, we get functions $h_o, h_1, \ldots, h_n \in E$ and sets $B_1 \supset B_2 \supset \ldots \supset B_n$, B_i of cardinality $\geq 2^{n-i+1}$, such that $\| h_i \| = \varepsilon$ $(i = 1, \ldots, n)$, $h_i(x) = h_j(x)$ and of modulus $\varepsilon (x \in B_n,\ i,j \geq 1)$, and

$$\hat{d}G(h_o + h_1 + \ldots + h_i)(h_{i+1}) = 0 \quad (i = 0, \ldots, n-1).$$

Thus, setting $f_j = \sum_{i=0}^{j} h_i$ $(j = 0, 1, \ldots, n)$, we see that the finite sequence (f_j) satisfies the following conditions:

$$\| f_j - f_{j-1} \| = \| h_j \| = \varepsilon,$$

$$\| f_n \| = 1, \text{ since } |f_n(x)| = 1 \text{ for } x \in B_n,$$

$$\sum_{j=1}^{n} \| f_j - f_{j-1} \| = 1, \text{ and}$$

$$\hat{d}G(f_{j-1})(f_j - f_{j-1}) = 0 \quad (j = 1, \ldots, n).$$

Therefore, $|G(f_n) - G(f_o)|$

$$\leq \sum_{j=1}^{n} | G(f_j) - G(f_{j-1}) - \hat{d}G(f_{j-1})(f_j - f_{j-1}) |$$

$$\leq \sum_{j=1}^{n} \| f_j - f_{j-1} \| \sup \{ \| \hat{d}G(f_{j-1} + f) - \hat{d}G(f_{j-1}) \| : f \in [0, f_j - f_{j-1}] \}$$

by the Lagrange mean value theorem (where $[a,b]$ denotes all points on the line segment from a to b),

$$\leq \sum_{j=1}^{n} \varepsilon\, 1/2 = 1/2.$$

We conclude that $|G(f_n)| \geq 1/2$, although $\|f_n\| = 1$, which is a contradiction. Q.E.D.

COROLLARY 1.4. (cf. [10,17]). *Let p be a positive, continuous, bounded function on E such that $p(f) \to 0$ as $\|f\| \to \infty$. Let $F \in C^2(E)$, $F \notin C^3(E)$. Then there exists no $G \in C^3(E)$ such that $|F(f) - G(f)| < p(f)$ and $\|\hat{d}^2F(f) - \hat{d}^2G(f)\| < p(f)$, for all $f \in E$. That is, it is impossible to approximate F by a C^3 function in the fine topology of order 2.*

PROOF: Suppose that such a G exists; without loss of generality, we may suppose that $|F(0) - G(0)| = a > 0$. Let $\phi : R \to R$ be a C^∞ function such that

$$\phi(t) = \begin{cases} 0 & \text{if } |t| \leq a/2 \\ 1 & \text{if } |t| \geq a \end{cases},$$

and consider the function $H = \phi \circ (F - G) : E \to R$. Then $\hat{d}H(f) = \phi'[(F - G)(f)]\, \hat{d}(F - G)(f)$, and $\hat{d}^2H(f) = \phi''[(F - G)(f)](\hat{d}(F - G)(f))^2 + \phi'((F - G)(f))\, \hat{d}^2(F - G)(f)$. Hence

$$\|\hat{d}^2H(f)\| \leq MC + |\phi''((F - G)(f))|\,\|\hat{d}(F - G)(f)\|^2$$

where $M = \sup\{\|\hat{d}^2(F - G)(f)\| : f \in E\} < \infty$ and $C = \sup\{|\phi'(t)|, |\phi''(t)| : t \in R\}$. For some $r > 0$, $|(F-G)(f)| < p(f) < a/2$ for all $f \in E$, $\|f\| > r$, so that $\phi''((F - G)(f)) = 0$ for $\|f\| > r$. For $\|f\| \leq r$, an application of the Lagrange mean value theorem shows that $\|\hat{d}(F-G)(f)\| \leq \|\hat{d}(F-G)(0)\| + rM$, so that $\sup\{\|\hat{d}^2H(f)\| : f \in E\} < \infty$. A further application of the Lagrange mean value theorem shows the uniform continuity of $\hat{d}H$.

By Proposition 1.1, H must either have unbounded support or $H \equiv 0$. But since $p(f) \to 0$ as $\|f\| \to \infty$, $|(F - G)(f)| \to 0$ as $\|f\| \to \infty$, so $H \equiv 0$, However, $H(0) = 1$, a contradiction, so that G does not exist. Q.E.D.

We remark that if K is compact, metrizable and uncountable, then the above Proposition and Corollary are trivial and in fact much weaker that what is already known (see, for example, [16,17]). In fact, as Wulbert notes, dens $C(K)' = \text{card } K$. Thus, if K is uncountable, dens $C(K)' >$ dens $C(K)$, so that there does not exist a non-trivial $F \in C^1(C(K))$ with bounded support. Consequently, if $F \in C^1(C(K))$, $F \notin C^2(K))$, there does not exist $G \in C^2(C(K))$ such that $|F(f) - G(f)| \to 0$ as $\|f\| \to \infty$.

SECTION 2.

Throughout this and the subsequent sections, E will denote a real Banach space. In this section, we outline a proof of the following result.

PROPOSITION 2.1: $(C^k(E), \tau_c^k)$ *has the approximation property for some (hence for all)* $k \geq 1$ *if and only if* E *has the approximation property.*

We note that this result complements and, in a sense, completes the analogy first noticed by Llavona [8] and Prolla and Guerreiro [13] with the holomorphic situation [2]. In the complex case, one proves that $(H(E), \tau_c^o)$ has the approximation property if and only if E has the approximation property by using the fact that the ε product $(H(E), \tau_c^o) \varepsilon F \cong (H(E;F), \tau_c^o)$ for any complex Banach space F and then using a characterization of the approximation property in terms of the ε-product (see, for example [2]). However, in the real case, a similar product formula and the completeness of $(C^k(E), \tau_c^k)$ are unknown. Thus, our approach to proving the assertion must be completely different; it will, in fact, be more direct than the ε-product argument. It should be noted that in [4], Bombal and Llavona have studied the topology τ_c^k on spaces of Hadamard differentiable functions, and have proved the completeness of this space as well as various approximation results. Using this, Bombal [3] obtained Proposition 2.1 by a different method.

PROOF OF PROPOSITION 2.1: We prove the result for $k \in N$, only minor modifications being necessary for $k=\infty$. If $(C^k(E), \tau_c^k)$ has the approximation property for some $k \geq 1$, define $\pi : C^k(E) \to E'$ by $\pi(f) = \hat{d}f(0)$. Note that if we regard $E' \subset C^k(E)$ with the induced τ_c^k topology, then π is a continuous projection, since $\sup \{ |\hat{d}^j(\pi f)(x)(y)| : (x,y) \in K \times L, j \leq k\} = \sup\{|\hat{d}f(0)(y)| : y \in L_1 \}$, where K and L are compact subsets of E and $L_1 = K \cup L$. Therefore, E' with the induced topology is a complemented subspace of $C^k(E)$ and hence has the approximation property. However, since the induced topology on E' is just the compact open topology, it follows that E has the approximation property.

To prove the converse, let $\mathcal{K} \subset (C^k(E), \tau_c^k)$ be precompact and let K , L be compact subsets of E . Given $\varepsilon > 0$, we claim that there is a $\delta > 0$ such that if $(x,y) \in K \times L$ and $x',y' \in E$ with $||x' - x|| < \delta$, $||y' - y|| < \delta$, then

$$|\hat{d}^j f(x)(y) - \hat{d}^j f(x')(y')| < \varepsilon \quad (f \in \mathcal{K}, j \leq k) \qquad (*).$$

In fact, if this is false, then for some $j \leq k$, there exist sequences (x'_n) and (y'_n) in E, (x_n) and (y_n) in K and L respectively, and (f_n) in $\mathcal{K}$ such that for all $n \in N$, $\|x_n - x'_n\| < 1/n$, $\|y_n - y'_n\| < 1/n$, and $|\hat{d}^j f_n(x_n)(y_n) - \hat{d}^j f_n(x'_n)(y'_n)| \geq \varepsilon$. Now, $K_1 = \overline{(x_n)}$, $K_2 = \overline{(x'_n)}$, $L_1 = \overline{(y_n)}$, and $L_2 = \overline{(y'_n)}$ are compact in E, so that the seminorm p defined by

$$p(f) = \sup\{|\hat{d}^j f(x_n)(y_n) - \hat{d}^j f(x'_n)(y'_n)| : n \in N\}$$

is τ_c^k continuous. However, for infinitely many m and n, $p(f_n - f_m) \geq \varepsilon/2$, contradicting the precompactness of $\mathcal{K}$. Thus (*) holds.

Let $T \in E' \otimes E$ such that for all $x \in K \cup L$, $\|Tx - x\| < \delta$. Define $\phi : C^k(E) \to C^k(E)$ by $\phi(f) = f \circ T$. Consider now the family $\mathcal{K}|_{T(E)}$ of functions in $\mathcal{K}$ restricted to the finite dimensional space $T(E)$. Since $\mathcal{K}|_{T(E)}$ is a precompact subset of $C^k(T(E))$, which has the approximation property, there exists a continuous linear mapping of finite rank $\psi : C^k(T(E)) \to C^k(T(E))$ such that for all $g \in \mathcal{K}|_{T(E)}$,

$$\sup\{|\hat{d}^j g(x)(y) - \hat{d}^j \psi(g)(x)(y)| : (x,y) \in T(K) \times T(L), j \leq k\} < \varepsilon \qquad (**).$$

Finally, the mapping taking $f \in C^k(E)$ into $\psi(f|_{T(E)}) \circ T \in C^k(E)$ is finite rank, linear and continuous, and if $f \in \mathcal{K}$, $(x,y) \in K \times L$, then

$$|\hat{d}^j f(x)(y) - \hat{d}^j(\psi(f|_{T(E)}) \circ T)(x)(y)| \leq$$
$$|\hat{d}^j f(x)(y) - \hat{d}^j(f \circ T)(x)(y)| + |\hat{d}^j(f \circ T)(x)(y) - \hat{d}^j(f|_{T(E)} \circ T)(x)(y)| +$$
$$+ |\hat{d}^j(f|_{T(E)} \circ T)(x)(y) - \hat{d}^j(\psi(f|_{T(E)} \circ T)(x)(y)|.$$

By (*) and (**), the first and third terms are $< \varepsilon$ while the middle term is 0. Thus, $(C^k(E), \tau_c^k)$ has the approximation property for any k. Q.E.D.

We remark that the proof of the above implication merely employs the fact that if E has the approximation property, then every element of $C^k(E)$ can be approximated in the τ_c^k topology by finite type polynomials (cf [12]), and that the passage of a function in $C^k(E)$ to a polynomial which approximates it can be accomplished in a continuous, linear manner.

SECTION 3.

In this and the next section, we will state, mostly without proof, some results on approximation by finite type polynomials for the topologies τ_u^k and τ_b^k. We will frequently find necessary the following hypothesis on the Banach space E, the first two conditions of which are formally stronger than the assumption that E' has the bounded approximation property.

For some constant $C > 0$, the following holds:

For each $\phi_1,\ldots,\phi_k \in E'$ $(k \in N)$, there exists a sequence $(\pi_n) \in E' \otimes E$ satisfying

(i) $\|\pi_n\| \leq C$ $(n \in N)$.

(ii) $\phi_i \circ \pi_n \to \phi_i$ in E' $(i = 1,\ldots,k)$, and (+)

(iii) $\pi_n(x) \to x$ $(x \in E)$.

Condition (+) is certainly a weaker assumption than the following, which is found in [14] (cf [7]).

There exists a sequence (P_n) of finite rank continuous linear projections satisfying $P_n(x) \to x (x \in E)$ and (++)

$\phi \circ P_n \to \phi$ in E' $(\phi \in E)$.

In fact, condition (++) implies that E' is separable, while ℓ_1 satisfies (+).

PROPOSITION 3.1. $(C^k(E), \tau_u^k)$ *is complete.*

We sketch the proof for $k = 1$, the proof for $k > 1$ being similar. Let $(f_\alpha)_{\alpha \in A}$ be a Cauchy net in $(C^1(E), \tau_u^1)$. It is easy to see that there are continuous functions $f : E \to R$ and $g : E \to E'$ such that $f = \lim f_\alpha$ and $g = \lim \hat{d}f_\alpha$ uniformly on compact subsets of E. It suffices to show that f is differentiable with derivative $\hat{d}f = g$. In fact, if this fails, then there exist $\varepsilon > 0$, $x \in E$, and a sequence (h_n) in E, $h_n \to 0$, such that $|f(x + h_n) - f(x) - g(x)(h_n)| > 3\varepsilon \, \|h_n\|$ $(n \in N)$. Now, for each $\alpha, \beta \in A$ and each $n \in N$, $|f_\alpha(x + h_n) - f_\alpha(x) - f_\beta(x + h_n) - f_\beta(x)| \leq \|h_n\| \sup\{\|\hat{d}f_\alpha(z) - \hat{d}f_\beta(z)\| : z \in \bigcup_n [x, x + h_n]\}$. Noting that $\overline{\bigcup_n [x, x + h_n]}$ is compact, it follows that $|f_\alpha(x + h_n) - f_\alpha(x) - f_\beta(x + h_n) - f_\beta(x)| \leq \varepsilon \|h_n\|$ for all $\alpha, \beta \geq$ some $\alpha_0 \in A$, so that $|f_\alpha(x+h_n) - f_\alpha(x) - f(x+h_n) - f(x)| \leq \varepsilon \|h_n\|$ for all $\alpha \geq \alpha_0$. Also, for $\alpha \geq$ some $\alpha_1 \in A$, $\|g(x) - \hat{d}f_\alpha(x)\| < \varepsilon$. Letting α be any index larger than both α_0 and α_1, there is n_0 such that $|f_\alpha(x + h_n) - f_\alpha(x) - \hat{d}f_\alpha(x)(h_n)| < \varepsilon \, \|h_n\|$ for all $n \geq n_0$. Therefore, for $n \geq n_0$,

$$
\begin{aligned}
&|f(x + h_n) - f(x) - g(x)(h_n)| \leq \\
&\leq |f(x + h_n) - f(x) - f_\alpha(x + h_n) - f_\alpha(x)| + \\
&+ |f_\alpha(x + h_n) - f_\alpha(x) - \hat{d}f_\alpha(x)(h_n)| + |\hat{d}f_\alpha(x)(h_n) - g(x)(h_n)| < \\
&< 3\varepsilon \, \|h_n\| ,
\end{aligned}
$$

a contradiction, which establishes the result. Q.E.D.

We define the space $C_c^k(E;F)$ to be the subspace of $C^k(E;F)$ of all functions g such that for each $j \leq k$ and $x \in E$, $\hat{d}^j g(x) \in \overline{\mathcal{P}_f({}^jE;F)}$; as before, when F is the scalar field,

$C_c^k(E;F)$ is denoted by $C_c^k(E)$. Using Proposition 3.1, it is routine to verify that $C_c^k(E)$ is complete. Note that $C_c^1(E) = C^1(E)$ for every Banach space E; in [1], we note that $C_c^k(E) = C^k(E)$ for $E = C(K)$, where K is a dispersed compact, Hausdorff space, and in particular $C_c^k(c_0) = C^k(c_0)$. When E is a real separable Hilbert space, Lesmes [7] proved that $\mathcal{P}_f(E)$ is τ_u^1 dense in $C^1(E)$ and noted that the result fails for $C^k(E)$, $k \geq 2$. In fact, we have the following.

PROPOSITION 3.2. *Let* E *satisfy condition* (+). *Then for all* $k \geq 1$, $\mathcal{P}_f(E)$ *is* τ_u^k *dense in* $C_c^k(E)$.

Assuming (++), Prolla has noted that for any Banach space F, $(C_c^k(E), \tau_u^k) \varepsilon F = C_c^k(E;F)$. Using the vector valued form of Proposition 3.2 (namely, that $\mathcal{P}_f(E;F)$ is τ_u^k dense in $C_c^k(E;F)$), he concludes that $C_c^k(E) \otimes F$ is dense in $(C_c^k(E;F), \tau_u^k)$, so that $(C_c^k(E), \tau_u^k)$ has the approximation property. In fact, assuming (+), one can show this directly by noting, as in Proposition 2.1, that the passage of a function in $C_c^k(E)$ to a finite type polynomial approximating it can be done in a continuous linear manner. Summarizing, we have the following.

PROPOSITION 3.3. *If* E' *has property* (+), *then* $(C_c^k(E), \tau_u^k)$ *has the approximation property. Conversely, if* $(C_c^k(E), \tau_u^k)$ *has the approximation property, then so does its complemented subspace* E'.

SECTION 4.

In [14], Restrepo, considering a reflexive Banach space E which satisfies condition (++), studied the completion of the finite type polynomials $\mathcal{P}_f(E)$ for the topology τ_b^1. He found that if $g \in C^1(E)$ is weakly continuous on bounded sets and uniformly (one) differentiable on bounded sets, then g is a τ_b^1 limit of elements of $\mathcal{P}_f(E)$. Here, under the assumption (+), we discuss the τ_b^k completion of $\mathcal{P}_f(E)$ for all $k \geq 0$.

A function g between two Banach spaces E and F is said to be weakly continuous on bounded sets if for all $M > 0, \varepsilon > 0$, and each $x \in E$, $\| x \| \leq M$, there exist $\delta > 0$ and $\phi_1,\ldots,\phi_k \in E'$ such that if $y \in E$, $\| y \| < M$, and $|\phi_i(x - y)| < \delta$ $(i=1,\ldots,k)$, then $|g(x) - g(y)| < \varepsilon$. The function is uniformly weakly continuous on bounded sets if, in the above definition, δ and $\phi_1 \ldots \phi_k$ can be chosen independent of x, $\| x \| < M$. Finally, g is uniformly differentiable of order n on bounded sets if for all $M > 0$ and all $\varepsilon > 0$, there is $\delta > 0$ such that for all $x, h \in E$, $\| x \| \leq M$, $\| h \| \leq \delta$,

$$|g(x + h) - g(x) - \hat{d}g(x)(h) - \ldots - \frac{\hat{d}^n g(x)}{n!}(h)| \leq \varepsilon \| h \|^n .$$

If E is reflexive, then the weak compactness of the ball of E implies that a function which is weakly continuous on bounded sets is uniformly weakly continuous on bounded sets. Further, we have the following.

PROPOSITION 4.1. (cf [14, Theorem 6]). *Let* $g : E \to F$ *be such that* $g, \hat{d}g, \ldots, \hat{d}^{n-1}g$ *are uniformly weakly continuous on bounded sets. If* g *is uniformly differentiable of order* n, *then* $\hat{d}^n g$ *is uniformly weakly continuous on bounded sets.*

We do not know if the converse holds, even in the case $n = 1$. That is, if g and $\hat{d}g$ are uniformly weakly continuous on bounded sets, is g uniformly differentiable of order 1 on bounded sets ? Note that the space $C^k_{wu}(E;F)$, of functions $g : E \to F$ such that $g, \hat{d}g, \ldots, \hat{d}^k g$ are uniformly weakly continuous on bounded sets, is complete when endowed with the τ^k_b topology; we do not know if $\{g \in C^k(E;F) : g$ is uniformly weakly continuous and uniformly differentiable of order $\leq k\}$ is complete when endowed with the τ^k_b topology.

PROPOSITION 4.2. (cf [14 , Theorem 8]). *Let* E *satisfy condition* (+). *Then the set of functions* $g \in C^k_{wu}(E)$ *such that* $\hat{d}^j g(x) \in \overline{\mathcal{P}_f({}^jE)}$ $(x \in E,\ j \leq k)$ *is the* τ^k_b *completion of* $\mathcal{P}_f(E)$.

We prove Proposition 4.2 in the case $k = 0$ without assuming condition (+); that is, we show that for any Banach space E, if $g : E \to R$ is uniformly weakly continuous on bounded sets, then g is a uniform limit (on bounded sets) of finite type polynomials (cf [14 , Theorem 3]). The proof of the general case of Proposition 4.2 makes use of following.

LEMMA 4.3. *Let* T *be a (not necessarily linear) mapping from* E *to* F . *If* T *is uniformly weakly continuous on bounded sets, then* T *maps balls in* E *to relatively compact subsets of* F.

To show Proposition 4.2 in the case $k=0$, Let $g : E \to R$ be uniformly weakly continuous on bounded sets, let B be a bounded subset of E , and let $\varepsilon > 0$. For some $\delta > 0$ and $\phi_1, \ldots, \phi_k \in E'$, we have that if $|\phi_i(x - y)| < \delta$ $(x,y \in B,\ i=1,\ldots,k)$, then $|g(x) - g(y)| < \varepsilon$. Consider the subset $\{\phi(x) : x \in B\}$ in R^k, where $\phi(x) = (\phi_1(x), \ldots, \phi_k(x))$. By compactness, there exist $x_1, \ldots, x_n \in B$ such that for all $x \in B$, there is some x_j with $\sup\{|\phi_i(x) - \phi_i(x_j)| : i = 1, \ldots, k\} < \delta/2$. Now,defining

$U_j = \{(y_1,\ldots,y_k) \in R^k : |y_i - \phi_i(x_j)| < \delta \quad (i=1,\ldots,k)\}$ for $j=1,\ldots,n$, let $c_1,\ldots,c_n$ be continuous functions from R^k to R such that

(i) $c_j(y) \geq 0 \quad (y \in R^k, \quad j=1,\ldots,n)$,

(ii) $\sum_{j=1}^{n} c_j(y) = 1 \; (y = \phi(x), \quad$ for some $\quad x \in B)$,

(iii) $\operatorname{spt} c_j \subset U_j \quad (j=1,\ldots,n)$.

Define $h : R^k \to R$ by $h(y) = \sum_{j=1}^{n} c_j(y) g(x_j)$. Then, if $x \in B$,

$$|h \circ \phi(x) - g(x)| = |\sum_{j=1}^{n} c_j(\phi(x)) g(x_j) - g(x)|$$

$$\leq \sum_{j=1}^{n} c_j(\phi(x)) |g(x_j) - g(x)| < \varepsilon .$$

Finally, there exists a polynomial $p : R^k \to R$ such that

$|p(y) - h(y)| < \varepsilon \; (y \in \bigcup_{j=1}^{n} U_j)$. Thus, $|p \circ \phi(x) - g(x)| < 2\varepsilon \; (x \in B)$, and $p \circ \phi$ is a polynomial of finite type. Q.E.D.

REFERENCES

[1] R. M. ARON - Compact polynomials and compact differentiable mappings between Banach spaces, to appear.

[2] R. M. ARON AND R. M. SCHOTTENLOHER - Compact holomorphic mappings on Banach spaces and the approximation property, to appear in J. Funct. Anal.

[3] F. BOMBAL GORDÓN - Differentiable function spaces with the approximation property, to appear.

[4] F. BOMBAL GORDÓN AND J. L. G. LLAVONA - La propiedad de aproximación en espacios de functions diferenciables, to appear.

[5] S. DINEEN - Runge domains in Banach spaces, Proc.R.I.A., 71, Sect. A, nº 7(1971).

[6] J. KURZWEIL - On approximation in real Banach spaces, Studia Math. 14 (1954), 214 - 231.

[7] J. LESMES - On the approximation of continuously differentiable functions in Hilbert spaces, Rev. Colombiana de Matemáticas 8, (1974), 217 - 223.

[8] J. L. G. LLAVONA - Aproximación de funciones diferenciables, Thesis, Universidad Complutense, Madrid.

[9] C. MATYSZCZYK - Approximation of analytic and continuous mappings by polynomials in Frechet spaces, to appear in Studia Math.

[10] N. MOULIS - Approximation de functions differentiables sur certains espaces de Banach, Ann. Inst. Fourier 21 (1971), 293 - 345.

[11] Ph. NOVERRAZ - Pseudo - convexite, convexite polynomiale, et domaines d'holomorphie en dimension infinie, Mathematics Studies 3, North Holland (1973).

[12] J. B. PROLLA - On polynomial algebras of continuously differentiable functions, Rend. dell'Accad. dei Lincei, to appear.

[13] J. B. PROLLA AND C. S. GUERREIRO - An extension of Nachbin's theorem to differentiable functions on Banach spaces with the approximation property, to appear.

[14] G. RESTREPO - An infinite dimensional version of a theorem of Bernstein, Proc. A.M.S. 23 (1969), 193-198.

[15] J. WELLS - Differentiable functions on c_o, Bull, A.M.S. 75(1969), 117 - 118.

[16] J. H. M. WHITFIELD - Differentiable functions with bounded nonempty support on Banach spaces, Bull.A.M.S. 72 (1966), 145 - 146.

[17] D. WULBERT - Approximation by C^k - functions in approximation Theory, Proc. of Intern. Symp., G. G. Lorentz (ed.), Acd. Press (1973), 217 - 239.

School of Mathematics,
39 Trinity College,
Dublin 2, Ireland.

ADDED IN PROOF: The questions raised after Proposition 4.1 all have affirmative answers. In addition, W. B. Johnson has pointed out that condition (+) is equivalent to E' having the bounded approximation property. A much fuller investigation of the completion of spaces of polynomials, containing the material discussed in Sections 3 and 4 of this paper as well as the above points, is contained in a joint paper by the author and J. B. Prolla, "Polynomial Approximation of Differentiable Functions on Banach Spaces", to appear.

Infinite Dimensional Holomorphy and Applications, Matos (ed.)

FONCTIONS MEROMORPHES SUR C^Λ

par *VOLKER AURICH*

INTRODUCTION

Toute fonction holomorphe sur un domaine étalé $p:X \longrightarrow \mathbb{C}^\Lambda$ où Λ est un ensemble arbitraire se factorise à travers un domaine de dimension finie (à savoir son domaine d'existence). Cela reste vrai pour toute fonction méromorphe. En utilisant des résultats en dimensions finies on obtient que toute fonction méromorphe sur un domaine étalé au-dessus de $\mathbb{C}^\Lambda$ est le quotient de deux fonctions holomorphes et se prolonge à l'enveloppe d'holomorphie. Donc il suffit d'étudier les fonctions méromorphes sur un domaine de Stein. On sait qu'un domaine de Stein au-dessus de $\mathbb{C}^\Lambda$ est isomorphe à $S \times \mathbb{C}^{\Lambda-\Psi}$ où $\Psi \subset \Lambda$ est fini et S est un domaine de Stein étalé au-dessus de $\mathbb{C}^\Psi$ ([1],[3]). Nous démontrons que l'espace $\mathcal{M}(S \times \mathbb{C}^{\Lambda-\Psi})$ des fonctions méromorphes sur $S \times \mathbb{C}^{\Lambda-\Psi}$ est la limite inductive des $\mathcal{M}(S \times \mathbb{C}^{\Phi-\Psi})$, $\Phi \supset \Psi$ fini. Un théorème analogue pour les données de Cousin n'est pas vrai. Dans [4] DINEEN a démontré que sur tout ouvert de $\mathbb{C}^{\mathbb{N}}$ il existe une donnée de Cousin I non résoluble. Cependant, une propriété de factorisation pour certaines classes de données de Cousin permet d'énoncer des conditions nécessaires et suffisantes pour qu'une donnée de Cousin sur un domaine de Stein soit résoluble.

La notation est pour la plupart la meme que celle dans [1]. Λ désigne toujours un ensemble et Fin Λ est l'ensemble des sousensembles finis de Λ. Si $p:X \to \mathbb{C}^\Lambda$ est un domaine étalé $\mathcal{O}_X$ ou simplement $\mathcal{O}$ est le faisceau des fonctions holomorphes.

1. PROPRIÉTÉS FONDAMENTALES DES FONCTIONS MÉROMORPHES SUR $\mathbb{C}^\Lambda$

Soit $p:X \to \mathbb{C}^\Lambda$ un domaine étalé. Pour tout ouvert $U \subset X$ soit $M(U)$ l'anneau total des fractions de $\mathcal{O}(U)$. M est un préfaisceau sur X. Le faisceau associé est appelé le faisceau des fonctions méromorphes sur X et noté $\mathcal{M}_X$ ou simplement $\mathcal{M}$. Toute fonction méromorphe sur X, c'est-à-dire toute section $m \in \mathcal{M}(X)$ peut être representée par une famille $(U_i, f_i, \tilde{f}_i)_{i\in I}$ où $(U_i)_{i\in I}$ est un recouvrement ouvert de X et $f_i, \tilde{f}_i \in \mathcal{O}(U_i)$ tels que l'intérieur de $\{\tilde{f}_i = 0\}$ soit vide et $f_i\tilde{f}_j = f_j\tilde{f}_i$ sur $U_i \cap U_j$. Les U_i peuvent être choisis comme des polydisques Inversement, une telle famille détermine une fonction méromorphe. Evidemment, $\mathcal{O}_X$ peut être considéré comme un sousfaisceau de $\mathcal{M}_X$. Pour chaque $x \in X$ $\mathcal{M}_{X,x}$ est le corps des fractions de $\mathcal{O}_{X,x}$.

A chaque $m \in \mathcal{M}(X)$ on associe une fonction:
Soit $D_m := \{y \in X: m_y \in \mathcal{O}_{X,y}\}$. Parce que $x \in D_m$ si et seulement si ils existent une famille $(U_i, f_i, \tilde{f}_i)_{i\in I}$ representant m et un $i \in I$ tels que $\tilde{f}_i \neq 0$ la valeur $F_m(x) := f_i(x)/\tilde{f}_i(x)$ est bien définie. Ainsi on obtient une fonction $F_m: D_m \to \mathbb{C}$.

Comme en dimensions finies on verifie que l'espace étalé de $\mathcal{M}$ est séparé. Donc on a le principe du prolongement méromorphe.

1.1. PROPOSITION *Soient* $U \subset X$ *un ouvert connexe et* $m, n \in \mathcal{M}(X)$. *Si les sections* m *et* n *sont égales en un point elles sont égales dans* U. *Si les fonctions* F_m *et* F_n *sont égales dans un ouvert non vide elles sont égales. Donc* $\mathcal{M}(V)$, $V \subset X$ *ouvert, est un corps si et seulement si* V *est connexe.*

1.2 PROPOSITION $\mathcal{O}_{\mathbb{C}^\Lambda, o}$ *est un anneau factoriel.*

DÉM.: Ou bien comme dans [7] ou bien en utilisant que $\mathcal{O}_{\mathbb{C}^\Phi, o}$ est factoriel pour tout $\Phi \in \text{Fin}\,\Lambda$, que $\mathcal{O}_{\mathbb{C}^\Lambda, o} = \lim\limits_{\Phi \in \text{Fin}\Lambda} \text{ind}\, \mathcal{O}_{\mathbb{C}^\Phi, o}$ ([1]) et que $q \in \mathcal{O}_{\mathbb{C}^\Phi, o}$ est irréductible dans $\mathcal{O}_{\mathbb{C}^\Phi, o}$ si et seulement si q est irréductible dans $\mathcal{O}_{\mathbb{C}^\Lambda, o}$.

1.3 LEMMA *Soit* $m \in \mathcal{M}(X)$. *Il existe une famille* $(U_i, f_i, \tilde{f}_i)_{i \in I}$ *représentant* m *telle que chaque* U_i *soit un polydisque et que pour tout* $i \in I$ *et tout* $x \in U_i$ *les germes* $[f_i]_x$ *et* $[\tilde{f}_i]_x$ *soient premiers entre eux.*

DÉM.: Soit $(V_j, g_j, \tilde{g}_j)_{j \in J}$ une famille représentant m. Pour chaque $x \in X$ choisis $j(x) \in J$ tel que $x \in V_{j(x)}$. Ils existent un polydisque W_x ouvert de centre x et une fonction $h_{j(x)} \in \mathcal{O}(W_x)$ telle que $[h_{j(x)}]_x$ soit le plus grand diviseur commun de $[\tilde{g}_{j(x)}]_x$ et $[g_{j(x)}]_x$, et des fonctions $f_x, \tilde{f}_x \in \mathcal{O}(W_x)$ telles que $g_{j(x)}|W_x = h_{j(x)} \cdot f_x$ et $\tilde{g}_{j(x)}|W_x = h_{j(x)} \cdot \tilde{f}_x$. $[f_x]_x$ et $[\tilde{f}_x]_x$ n'ont pas de diviseurs communs, donc on sait ([5], p.149) qu'il existe un polydisc ouvert U_x de centre $x, U_x \subset W_x$, tel que $[f_x]_y$ et $[\tilde{f}_x]_y$ soient premiers entre eux en tout point $y \in U_x$. Evidemment $(U_x, f_x|U_x, \tilde{f}_x|U_x)_{x \in X}$ représente m.

Nous dirons que $x \in X$ est un *pôle* de $m \in \mathcal{M}(X)$ si $m_x \notin \mathcal{O}_{X,x}$ mais $m_x^{-1} \in \mathcal{O}_{X,x}$ et que x est un *point d'indétermination* si $m_x \notin \mathcal{O}_{X,x}$ et $m_x^{-1} \notin \mathcal{O}_{X,x}$.

1.4 PROPOSITION *Soit* $m \in \mathcal{M}(X)$. $(U_i, f_i, \tilde{f}_i)_{i \in I}$ *soit choisie comme dans* 1.3. *Alors on équivalence entre*

(i) x *est un pôle de* m.

(ii) $x \in U_i$ *entraîne* $f_i(x) \neq 0$ *et* $\tilde{f}_i(x) = 0$.

(iii) *Pour toute suite* $(x_n)_{n \in \mathbb{N}}$ *dans* D_m *qui converge vers* x $F_m(x_n)$ *tend vers* ∞.

On a équivalence entre

(iv) x *est un point d'indétermination de* m.

(v) $x \in U_i$ *entraîne* $f_i(x) = 0 = \tilde{f}_i(x)$.

(vi) *Pour tout voisinage* V *de* x *on a* $F_m(V \cap D_m) = \mathbb{C}$.

DÉM.: Comme en dimensions finies en utilisant [5] 6.2.3.

1.5 COROLLAIRE *L'ensemble des points d'indétermination et l'ensemble des pôles et des points d'indétermination sont des ensembles analytiques.*

Comme dans [7], p. 23 on prouve

1.6 COROLLAIRE *Pour tout* $m \in \mathcal{M}(X)$ D_m *est un ouvert connexe.*

1.7 COROLLAIRE *Pour tout* $m \in \mathcal{M}(X)$ F_m *est une fonction holomorphe.*

Pour une fonction holomorphe f sur une variété $q: Y \longrightarrow \mathbb{C}^\Lambda$ et $x \in Y$ on définit $\operatorname{dep}_x f :=$ l'intersection de tous les sousensembles Φ de Λ tels que f dépend au voisinage de x seulement des variables $q_j, j \in \Phi$. $\operatorname{dep}_x f$ est un ensemble fini. Si Y est connexe $\operatorname{dep}_x f$ ne dépend pas de $x \in Y$, et sa valeur constante sera notée $\operatorname{dep} f$. (voir [1]).

Pour $\Phi \in \operatorname{Fin}\Lambda$ et U ouvert dans X nous définissons

$$\mathcal{M}^\Phi(U) := \{m \in \mathcal{M}(U) : \operatorname{dep}_x F_m \subset \Phi \text{ pour chaque } x \in U\}.$$

$\mathcal{M}^\Phi$ est un faisceau. Il est le faisceau associé au prefaisceau

$U \to$ l'anneau total des fractions de $\mathcal{O}^{\Phi}(U)$.

1.8 PROPOSITION $\mathcal{M}(U) = \bigcup\{\mathcal{M}^{\Phi}(U) : \Phi \in \mathrm{Fin}\,\Lambda\}$ *pour tout ouvert* U *connexe dans* X.

DÉM.: Parce que D_m est connexe $\mathrm{dep}_x F_m$ est constant sur D_m.

2. LE PROBLÈME DE POINCARÉ ET L'ENVELOPPE DE MÉROMORPHIE

2.1 DEFINITION Soient $p:X \to \mathbb{C}^{\Lambda}$ un domaine et $m \in \mathcal{M}(X)$. Un domaine $q:Y \to \mathbb{C}^{\Delta}$, $\Delta \subset \Lambda$, est appelé un domaine d'existence de m s'ils existent une fonction méromorphe $n \in \mathcal{M}(Y)$ et un morphisme $\mu:X \to Y$ tels que les conditions suivantes soient satisfaites:

(i) $m = n \circ \mu$

(ii) Etant donnés un domaine $q':Y' \to \mathbb{C}^{\Delta'}$, $n' \in \mathcal{M}(Y')$ et un morphisme $\mu':X \to Y'$ tels que $m = n' \circ \mu'$ il existe un morphisme $\phi:Y' \to Y$ tel que $\mu = \phi \circ \mu'$.

Le domaine d'existence d'une fonction méromorphe est unique à un isomorphisme près (s'il existe).

2.2 PROPOSITION *Toute fonction méromorphe* m *sur un domaine* $p:X \longrightarrow \mathbb{C}^{\Lambda}$ *admet un domaine d'existence* $p_m:X_m \longrightarrow \mathbb{C}^{\mathrm{dep}\,F_m}$.

DÉM.: Choisis $x \in X$. m induit un germe $g \in \mathcal{M}_{\mathbb{C}^{\Phi},p(x)}$, $\Phi = \mathrm{dep}\,F_m$. X_m soit la composante connexe de g dans l'espace étalé de $\mathcal{M}_{\mathbb{C}^{\Phi}}$.

Elle est un domaine étalé au-dessus de $\mathbb{C}^{\Phi}$. Dans la manière usuelle on démontre qu'elle satisfait (i) et (ii) (voir p.ex. [8]).

2.3 REMARQUE Il est connu qu'un domaine d'existence d'une fonction méromorphe en dimensions finies est pseudoconvexe ([2], p. 86, conséquence du "Kontinuitatssatz" de Hartogs-Kneser dans

[6]), donc il est un domaine de Stein.

2.4 THÉORÈME (Poincaré) *Toute fonction méromorphe sur un domaine étalé au-dessus de* $\mathbb{C}^\Lambda$ *est le quotient de deux fonctions holomorphes.*

DÉM.: Appliquer 2.2, 2.3 et [5] 7.4.6.

2.5. THÉORÈME *Toute fonction méromorphe sur un domaine étalé au-dessus de* $\mathbb{C}^\Lambda$ *se prolonge a l'enveloppe d'holomorphie.*

DÉM.: Conséquence immédiate de 2.4.

2.6. COROLLAIRÉ *L'enveloppe de méromorphie d'un domaine étalé au-dessus de* $\mathbb{C}^\Lambda$ *est égale a l'enveloppe d'holomorphie.*

3. LES FONCTIONS MÉROMORPHES SUR UN DOMAINE DE STEIN

3.1 LEMME *Toute fonction méromorphe* m *sur un domaine* $p : X \to \mathbb{C}^\Lambda$ *admet un représentant* $(U_i, g_i, \tilde{g}_i)_{i \in I}$ *tel que chaque* U_i *soit un polydisc ouvert et que* dep $g_i \subset$ dep F_m *et* dep $\tilde{g}_i \subset$ dep F_m *pour chaque* $i \in I$. *En plus on peut obtenir que* $[g_i]_x$ *et* $[\tilde{g}_i]_x$ *soient premiers entre eux en tout point* $x \in U_i$.

DÉM.: $N := X - D_m$. $\Phi :=$ dep F_m. Choisis un représentant $(U_i, f_i, \tilde{f}_i)$ de m comme dans 1.3. Soit $i \in I$. On peut supposer que $p|U_i \to p(U_i)$ soit topologique. Choisis $\tilde{a} \in U_i - N$ et un polydisque ouvert $U \subset U_i - N$ de centre $\tilde{a}$. $a := \pi_{\Lambda-\Phi}(\tilde{a})$. On définit pour $x \in U_i$ $g_i(x) := f_i \circ (p|U_i)^{-1}(\pi_\Phi \circ p(x), a)$ et $\tilde{g}_i(x) := \tilde{f}_i \circ (p|U_i)^{-1}(\pi_\Phi \circ p(x), a)$. A cause du principe du prolongement analytique l'intérieur de $\{\tilde{g}_i = 0\}$ est vide. Parce que sur U $g_i / \tilde{g}_i = F_m = f_i / \tilde{f}_i$ on obtient sur U_i $g_i \tilde{f}_i = f_i \tilde{g}_i$, donc $(U_i, g_i, \tilde{g}_i)$ détermine m. En procédant maintenant dans la manière de 1.3 on démontre la deuxième partie du lemme.

3.2 PROPOSITION *Etant donnés un domaine* $q:Y \to \mathbb{C}^\Phi$ *tel que* $\Phi \in$ Fin Λ *et un polydisque ouvert* $P \subset C^{\Lambda-\Phi}$, *la projection* σ *de* $X := Y \times P$ *sur* Y *induit un isomorphisme* σ^* *de* $\mathcal{M}(Y)$ *sur* $\mathcal{M}^\Phi(X)$.

DÉM.: Pour tout polydisque $U \subset \mathbb{C}^\Phi$ la projection $U \times P \to U$ induit un isomorphisme de $\mathcal{O}(U)$ sur $\mathcal{O}^\Phi(U\times P)$, donc σ induit un monomorphisme $\sigma^*: \mathcal{M}(Y) \to \mathcal{M}^\Phi(X)$. Il reste à démontrer que σ^* est surjective. Soit $m \in \mathcal{M}^\Phi(X)$. Choisis $(U_i, g_i, \tilde{g}_i)$ comme dans 2.1. $V_i := \sigma(U_i)$. Pour $x \in V_i$ soient $h_i(x) := g_i(\sigma^{-1}(x) \cap U_i)$ et $\tilde{h}_i(x) := \tilde{g}_i(\sigma^{-1}(x) \cap U_i)$. h_i et $\tilde{h}_i$ sont bien définies et holomorphes et l'intérieur de $\{\tilde{h}_i = 0\}$ est vide. Soit $N :=$ l'ensemble des pôles et des points d'indétermination de $m = \{x \in X: \tilde{g}_i(x) = 0$ si $x \in U_i\}$. D'après le lemme 2.3 ci-dessous $N = \sigma(N) \times P$, donc $D_m = \sigma(D_m) \times P$. Par conséquent F_m se factorise à travers $\sigma(D_m)$. Pour $x \in V_i \cap V_j \cap \sigma(D_m)$ on obtient $h_i/\tilde{h}_i(x) = g_i/\tilde{g}_i(\sigma^{-1}(x) \cap U_i) = F_m(\sigma^{-1}(x) \cap U_i) = g_j/\tilde{g}_j(\sigma^{-1}(x) \cap U_j) = h_j/\tilde{h}_j(x)$, donc par continuité $h_i\tilde{h}_j = h_j\tilde{h}_i$ sur $V_i \cap V_j$. Cela prouve que $(V_i, h_i, \tilde{h}_i)$ représente une fonction méromorphe $\tilde{m}$ sur Y. Evidemment $\sigma^*(\tilde{m}) = m$.

3.3 LEMME *Soient* $q:Y \to \mathbb{C}^\Phi$ *un domaine,* $\Phi \in$ Fin Λ, P *un polydisque ouvert dans* $\mathbb{C}^{\Lambda-\Phi}$ *et* A *un ensemble analytique dans* $Y \times P$ *qui soit localement définissable par une fonction* f *telle que* dep $f \subset \Phi$. *Alors* $A = \sigma(A) \times P$ *où* σ *est la projection de* $Y \times P \to Y$.

DÉM.: Soient $a = (a', a'') \in A$ et $b = (a', b'') \in \sigma^{-1}(\sigma(a))$. Parce que le segment $[a,b] := \{(a', a'' + t(b'' - a'')): t \in [0,1]\}$ est compact il existe un recouvrement fini par des polydisques ouverts $U_1, \ldots, U_n$ avec les propriétés suivantes: $a \in U_1$, $b \in U_n$, $U_i \cap U_{i+1} \neq \emptyset$ et $\sigma(U_i) = \sigma(U_{i+1})$ pour tout i, ils existent des $f_i \in \mathcal{O}^\Phi(U_i)$ tels que $A \cap U_i = f_i^{-1}(0)$. Par conséquent $A \cap U_i = \sigma^{-1}(\sigma(A \cap U_i)) \cap U_i$ pour tout i. Cela entraîne

$\sigma^{-1}(\sigma(A \cap U_i)) \cap U_i \cap U_{i+1} = A \cap U_i \cap U_{i+1} = \sigma^{-1}((A \cap U_{i+1})) \cap U_i \cap$ $\cap U_{i+1}$. Parce que $U_i \cap U_{i+1}$ est cylindrique et non vide $\sigma(A \cap U_i) \cap \sigma(U_i \cap U_{i+1}) = \sigma(A \cap U_{i+1}) \cap \sigma(U_i \cap U_{i+1})$. En obsérvant que $\sigma(U_i) = \sigma(U_i \cap U_{i+1}) = \emptyset$ on obtient $\sigma(A \quad U_i) = \sigma(A \cap U_{i+1})$. Cela implique $a' \in \sigma(A \cap U_n)$ donc $b \in A$.

3.4 REMARQUE $p: X \to \mathbb{C}^\Lambda$ *soit un domaine de Stein. On sait* ([1], [3]) *qu'ils existent* $\Psi \in \mathrm{Fin}\,\Lambda$ *et un domaine de Stein* $q: S \to \mathbb{C}^\Psi$ *tels que* X *soit isomorphe a* $S \times \mathbb{C}^{\Lambda-\Psi}$. *Pour* $\Phi \in \mathrm{Fin}\,\Lambda$, $\Phi \supset \Psi$, *soient* $X_\Phi := S \times \mathbb{C}^{\Phi-\Psi}$ *et* $\sigma_\Phi: X \to X_\Phi$ *la projection.*

3.2 et 3.4 entraîne le corollaire suivant.

3.5 COROLLAIRE $p: X \to \mathbb{C}^\Lambda$ *soit un domaine de Stein. Pour tout* $\Phi \in \mathrm{Fin}\,\Lambda$, $\Phi \supset \Psi$, σ_Φ *induit un isomorphisme* σ_Φ^* *de* $\mathcal{M}(X_\Phi)$ *sur* $\mathcal{M}^\Phi(X)$.

Pour $\Psi \subset \Phi \subset \Phi' \in \mathrm{Fin}\,\Lambda$ on a des morphismes canoniques $\sigma_\Phi^{\Phi'}$ de $X_{\Phi'}$ sur X_Φ tels que $(\mathcal{M}(X_\Phi), (\sigma_\Phi^{\Phi'})^*)$ soit un système inductive.

3.6 COROLLAIRE *Sur un domaine de Stein* $p: X \to \mathbb{C}^\Lambda$ $\mathcal{M}(X)$ *est la limite inductive de* $(\mathcal{M}(X_\Phi), (\sigma_\Phi^{\Phi'})^*)$, $\Psi \subset \Phi \subset \Phi' \in \mathrm{Fin}\,\Lambda$.

DÉM.: Conséquence de 3.5 et 1.8.

4. LES PROBLÈMES DE COUSIN SUR UN DOMAINE DE STEIN

$p: X \to \mathbb{C}^\Lambda$ soit un domaine étalé. On a les suites exactes de faisceaux

$$0 \to \mathcal{O} \to \mathcal{M} \xrightarrow{\nu} \mathcal{M}/\mathcal{O} \to 0$$

$$0 \to \mathcal{O}^* \to \mathcal{M}^* \xrightarrow{\nu^*} \mathcal{M}^*/\mathcal{O}^* \to 0$$

$\mathcal{O}^*$ ($\mathcal{M}^*$) est le faisceau des fonctions holomorphes (fonctions méromorphes) ne s'annulant dans aucun point de X (dans aucun ouvert non vide de X). Une section dans $\Gamma(X, \mathcal{M}/\mathcal{O})$ (dans

$\Gamma(X,\mathcal{M}^*/\mathcal{O}^*))$ est appelée *une donnée de Cousin I* (Cousin II) sur X. Elle peut être representée par une famille $(U_i,m_i)_{i\in I}$ où $(U_i)_{i\in I}$ est un recouvrement de X par des polydisques ouverts et $m_i - m_j \in \mathcal{O}(U_i \cap U_j)$ $(m_i/m_j \in \mathcal{O}^*(U_i \cap U_j))$. Inversement, une telle famille détermine toujours une section.

Une donnée de Cousin I (II) est dite *résoluble* si elle est contenue dans l'image de $\mathcal{V}(\mathcal{V}^*)$. Nous appelons une donnée de Cousin *de dimension finie* si elle admet un représentant (U_i,m_i) tel que les U_i soient des polydisques et $\bigcup\{\text{dep } F_{m_i},\ i\in I\}$ soit fini. A cause de 1.8 on a le lemme suivant.

4.1 LEMME *Une donnée de Cousin résoluble est de dimension finie.*

Pour $\Phi\subset\Lambda$, $\Gamma_\Phi(X,\mathcal{M}/\mathcal{O})(\Gamma_\Phi(X,\mathcal{M}^*/\mathcal{O}^*))$ désigne l'ensemble des données de Cousin I (II) qui admettent un représentant $(U_i,m_i)_{i\in I}$ tel que dep $F_{m_i}\subset\Phi$ pour tout i. Evidemment $\Gamma_\Phi(X,\mathcal{M}/\mathcal{O}) \cong \Gamma(X,\mathcal{M}^\Phi/\mathcal{O}^\Phi)$ et $\Gamma_\Phi(X,\mathcal{M}^*/\mathcal{O}^*) \cong \Gamma(X,(\mathcal{M}^\Phi)^*/(\mathcal{O}^\Phi)^*)$.

4.2 LEMME *Soit* $\Phi\in$ Fin Λ. $q: Y\to\mathbb{C}^\Phi$ *soit un domaine étalé,* $X := Y\times\mathbb{C}^{\Lambda-\Phi}$ *et* $\sigma: X\to Y$ *soit la projection. Alors toute donnée de Cousin dans* $\Gamma_\Phi(X,\mathcal{M}/\mathcal{O})$ *ou* $\Gamma_\Phi(X,\mathcal{M}^*/\mathcal{O}^*)$ *admet un représentant* $(U_i,m_i)_{i\in I}$ *tel que* $U_i = \sigma(V_i)\times\mathbb{C}^{\Lambda-\Phi}$ *et* dep $F_{m_i}\subset\Phi$ *pour tout* i.

DÉM.: Cousin I : $\tau: X\to\mathbb{C}^{\Lambda-\Phi}$ soit la projection. Il existe un représentant $(V_i,n_i)_{i\in I}$ tel que les V_i soient des polydisques ouverts et $n_i\in\mathcal{M}^\Phi(V_i)$. D'après 3.2 chaque n_i définit une fonction méromorphe m_i sur $U_i := \sigma(V_i)\times\mathbb{C}^{\Lambda-\Phi}$. Il faut démontrer que $m_i - m_j$ soit holomorphe sur $U_i\cap U_j$. Soit $x\in U_i\cap U_j$. Choisis $x_i\in V_i$ et $x_j\in V_j$ tels que $\sigma(x_i) = \sigma(x_j) = \sigma(x)$. Le segment $\{(\sigma(x),\ \tau(x_i) + t(\tau(x_j) - \tau(x_i)) : t\in[0,1]\}$ peut être recouvert d'un nombre fini des V_k, disons $V_i = V_{k_1}, \dots, V_{k_r} = V_j$, tel que $V_{k_\nu} \cap V_{k_{\nu+1}} \neq \emptyset$. Parce que $n_{k_\nu} - n_{k_{\nu+1}}$ est holomorphe sur $V_{k_\nu} \cap V_{k_{\nu+1}}$, $f_\nu := m_{k_\nu} - m_{k_{\nu+1}}$ l'est aussi sur $U_{k_\nu} \cap U_{k_{\nu+1}}$. Par conséquent $m_i - m_j = \sum_{\nu=1}^{r-1} f_\nu$ est holomorphe sur $\bigcap_{\nu=1}^{r} U_{k_\nu} = \bigcap_{\nu=1}^{r} \sigma(V_{k_\nu}) \times C^{\Lambda-\Phi}$.

Cousin II: analogue.

4.3 COROLLAIRE *σ induit des isomorphismes*

$\Gamma(Y, \mathcal{M}/\mathcal{O}) \to \Gamma_\Phi(X, \mathcal{M}/\mathcal{O})$ *et* $\Gamma(Y, \mathcal{M}^*/\mathcal{O}^*) \to \Gamma_\Phi(X, \mathcal{M}^*/\mathcal{O}^*)$.

4.4 THÉORÈME *Une donnée de Cousin I sur un domaine de Stein* $p: X \to \mathbb{C}^\Lambda$ *est résoluble si et seulement si elle est de dimension finie.*

DÉM.: $\Longrightarrow$: 4.1. Nous prouvons $\Longleftarrow$. D'apres 3.4 et 4.2 on a pour tout $\Phi \in \mathrm{Fin}\,\Lambda$, $\Phi \supset \Psi$, un diagramme commutative

$$\begin{array}{ccccc} \Gamma(X_\Phi, \mathcal{M}) & \to & \Gamma(X_\Phi, \mathcal{M}/\mathcal{O}) & \to & H^1(X_\Phi, \mathcal{O}) \\ & & & & \\ \Gamma_\Phi(X, \mathcal{M}) & \to & \Gamma_\Phi(X, \mathcal{M}/\mathcal{O}) & & \end{array}$$

Il en résulte le théorème parce que $H^1(X_\Phi, \mathcal{O}) = 0$, X_Φ étant une variété de Stein.

4.5 THÉORÈME *Soit* $p: X \to \mathbb{C}^\Lambda$ *un domaine de Stein. On sait qu'il existe un domaine de Stein* $q: S \to \mathbb{C}^\Psi$, $\Psi \subset \Lambda$ *fini, tel que* X *soit isomorphe a* $S \times C^{\Lambda-\Psi}$. *Supposons que* $H^2(S \times \mathbb{C}^n, Z) = 0$ *pour tout* $n \in \mathbb{N}$. *Alors une donnée de Cousin II sur* X *est résoluble si et seulement si elle est de dimension finie.*

DÉM.: Analogue 4.4, car $H^1(X_\Phi, \mathcal{O}^*) \cong H^2(X_\Phi, Z) = 0$ ([5]).

BIBLIOGRAPHIE

[1] V. AURICH: The spectrum as envelope of holomorphy of a domain over an arbitrary product of complex lines. Proceedings on infinite dimensional holomorphy, p. 109, Springer Lecture Notes 364.

[2] H. BEHNKE, P. THULLEN: Theorie der Funktionen mehrerer komplexer Veränderlichen. Springer 1970.

[3] G. COEURÉ: Analytic functions and manifolds in infinite dimensional spaces. North-Holland 1974.

[4] S. DINEEN: Cousin's first problem on certain locally convex topological vector spaces. Mathematics Research Report No. 75-2, January 1975, University of Maryland.

[5] L. HÖRMANDER: An introduction of complex analysis in several variables. Van Nostrand.

[6] H. KNESER: Ein Satz über die Meromorphiebereiche analytischer Funktionen von mehreren Veränderlichen. Math. Ann. 106, p. 648-655.

[7] J.P. RAMIS: Sous-ensembles analytiques d'une variété banachique complexe. Springer 1970.

[8] M. SCHOTTENLOHER: Das Leviproblem in unendlichdimensionalen Räumen mit Schauderzerlegung. Habilitationsschrift München 1974.

Infinite Dimensional Holomorphy and Applications, Matos (ed.)

ON HOLOMORPHY VERSUS LINEARITY IN CLASSIFYING LOCALLY CONVEX SPACES

By JORGE ALBERTO BARROSO,
MARIO C. MATOS and
LEOPOLDO NACHBIN

1. INTRODUCTION

In the linear theory of locally convex spaces, it is classical to study bornological, barreled, infrabarreled and Mackey spaces. In the holomorphic approach, the corresponding concepts have been introduced recently as holomorphically bornological, holomorphically barreled, holomorphically infrabarreled and holomorphically Mackey spaces, that are more restricted classes than the corresponding linear ones. In this reasonably self-contained, expository paper, we present some basic results in such a study.

Let us introduce the following abbreviations for properties of a complex locally convex space: B = Baire, S = Silva, sm = semimetrizable, hba = holomorphically barreled, hbo = holomorphically bornological, hib = holomorphically infrabarreled, hM = holomorphically Mackey. We have the following implications for the named properties:

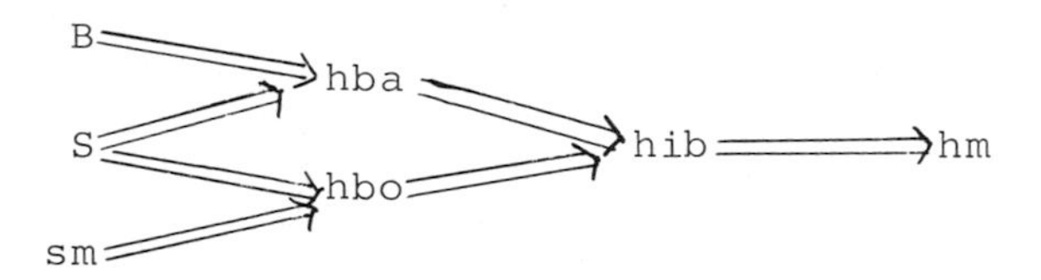

that correspond to classical ones dealing with continuous linear mappings, in place of holomorphic mappings. An interesting highlight is the holomorphic Banach-Steinhaus theorem on a Fréchet space, that contains as a particular case the classical linear Banach-Steinhaus theorem on such a space.

We shall use freely the notation and terminology of [8]; see also the references given there. Let us make a brief review of what will be needed here. Unless stated otherwise, we shall adhere to the following conventions. E and F denote complex locally convex spaces; and U is a nonvoid open subset of E. The set of all continuous seminorms on E is denoted by CS(E). We denote by E_α the vector space E seminormed by α. We represent by wF the weakened space F, that is, the vector space F endowed with the weak topology $\sigma(F,F')$ defined on F by F'. If I is a set and F is a seminormed space, we denote by $\ell^\infty(I;F)$ the seminormed space of all bounded mappings of I into F; and by $c_o(I;F)$ the seminormed subspace of all mappings of I into F tending to 0 at infinity. A mapping $f: U \to F$ is amply bounded if $\beta \circ f$ is locally bounded for every $\beta \in CS(F)$; more generally, a collection of mappings of U into F is amply bounded if the collection $\beta \circ \mathfrak{X}$ is locally bounded for every $\beta \in CS(F)$. We denote by $\mathcal{H}(U;F)$ the vector space of all holomorphic map pings of U into F; and by H(U;F) the vector space of all mappings of U into F which are holomorphic when considered as mappings of U into a fixed completion $\hat{F}$ of F. Unspecified use of the adjective holomorphic refers to $\mathcal{H}$, not to H. We say that f: U → F is algebraically holomorphic if the restriction $f \mid (U \cap S)$ is holomorphic, for every finite dimensional vector space S of E meeting U, where S carries its natural topology. On function spaces from U into F, we represent by τ_o the topology of uniform convergence on compact subsets; and by τ_{of} that for finite dimensional compact subsets only. When F = $\mathbb{C}$, it is not included in the notation for function spaces; thus $\mathcal{H}(U)$ stands for $\mathcal{H}(U;\mathbb{C})$.

2. HOLOMORPHICALLY BORNOLOGICAL SPACES

DEFINITION 1. *A given* E *is a "holomorphically bornological space" if, for every* U *and every* F, *we have that each mapping* f: U → F *belongs to* $\mathcal{H}(U;F)$ *if (and always only if)* f *is algebraically holomorphic, and* f *is bounded on every compact subset of* U.

Remark 4 bellow motivates the above definition, but we need the following preliminary material which is known.

LEMMA 2. *For a given* E, *the following conditions are equivalent:*

(1b) *For every* F, *we have that each mapping* f: E → F *belongs to* $\mathcal{L}(E;F)$ *if (and always only if)* f *is linear, and* f *is bound ed on every bounded subset of* E.

(1c) *For every* F, *we have that each mapping* $f: E \to F$ *belongs to* $\mathcal{L}(E;F)$ *if (and always only if)* f *is linear, and* f *is bounded on every compact subset of* E.

(2b) *Each seminorm* α *on* E *is continuous if (and always only if)* α *is bounded on every bounded subset of* E.

(2c) *Each seminorm* α *on* E *is continuous if (and always only if)* α *is bounded on every compact subset of* E.

PROOF. We shall prove the following implications

$$\begin{array}{ccc} (1b) & \Longrightarrow & (2b) \\ \Uparrow & & \Downarrow \\ (1c) & \Longrightarrow & (2c) \end{array}$$

(1c) $\Longrightarrow$ (1b). This is clear.

(1b) $\Longrightarrow$ (2b). Let α be a seminorm on E that is bounded on every bounded subset of E. Put $F = E_\alpha$. The identity mapping $f = I: E \to F$ is linear, and f is bounded on every bounded subset of E. By (1b), f is continuous. Thus, α is continuous.

(2b) $\Longrightarrow$ (2c). Let α be a seminorm on E that is bounded on every compact subset of E. We claim that α is bounded on every bounded subset X of E. In fact, let $x_m \in X$ ($m \in \mathbb{N}$) be arbitrary. For any $\lambda_m \in \mathbb{C}$ ($m \in \mathbb{N}$) such that $\lambda_m \to 0$, we have that $\lambda_m x_m \to 0$, as $m \to \infty$. Then, α is bounded on $\{\lambda_m x_m ; m \in \mathbb{N}\}$, since this subset together with 0 is compact; that is, $\{\lambda_m \alpha(x_m); m \in \mathbb{N}\}$ is bounded. We deduce that $\{\alpha(x_m); m \in \mathbb{N}\}$ is also bounded, since (λ_m) is arbitrary. Thus $\alpha(X)$ is bounded, because α is bounded on every denumerable subset of X. By (2b), α is continuous.

(2c) $\Longrightarrow$ (1c). Let $f: E \to F$ be linear and bounded on every compact subset of E. If $\beta \in CS(F)$, then $\beta \circ f$ is a seminorm on E that is bounded on every compact subset of E. By (2c), $\beta \circ f$ is continuous. Thus f is continuous, since β is arbitrary.

The proof can also be carried on with the same reasoning, by reversing the arrows. QED

The following definition is classical, particularly in terms of (1b) or (2b).

DEFINITION 3. *A given* E *is a "bornological space" if it satisfies the equivalent conditions of Lemma 2.*

REMARK 4. Definition 1 was formulated in analogy to Definition 3 through (1c), rather than (1b), of Lemma 2. The reason is that each $f \in \mathcal{H}(U;F)$ is always bounded on every compact subset of U; whereas it may occur that some $f \in \mathcal{H}(E)$ is unbounded on some bounded subset of E (see [7], p.28). Actually, as a consequence of the Josefson - Nissenzweig theorem [5], [10], it is known that, if E is an infinite dimensional normed space, and $X \subset E$ has a non void interior, there is some $f \in \mathcal{H}(E)$ which is unbounded on X (see [5]).

PROPOSITION 5. *A holomorphically bornological space* E *is also a bornological space.*

PROOF. It suffices to compare Definitions 1 and 3, by using (1c) of Lemma 2, and by remarking that a linear mapping is algebraically holomorphic. QED

PROPOSITION 6. *A semimetrizable space* E *is a holomorphically bornological space.*

PROOF. Let $f: U \to F$ be algebraically holomorphic, and bounded on every compact subset of U. Since E is semimetrizable, it follows that f is amply bounded. Hence $f \in \mathcal{H}(U;F)$, because f is algebraically holomorphic and amply bounded. QED

REMARK 7. Propositions 5 and 6 imply the known fact that a semimetrizable space E is a bornological space.

The following is a by now known definition.

DEFINITION 8. E *is a said to be a "Silva space" if there is a sequence* E_m *of Banach spaces, where* E_m *is a vector subspace of* E, *and* $E_m \subset E_{m+1}$, *this inclusion mapping being compact (that is, some neighborhood of* 0 *in* E_m *has a compact closure in* E_{m+1}), *for every* $m \in \mathbb{N}$; *moreover, it is assumed that*

$$E = \bigcup_{m \in \mathbb{N}} E_m$$

and that E *carries the inductive limit topology.*

REMARK 9. A Silva space is known to be essentially the same thing as the dual of a Fréchet-Schwartz space, or FS-space for short; thus it is also known as a DFS-space. More explicitly, the strong dual space of a Fréchet-Schwartz space a Silva space; the strong dual space of a Silva space is a Fréchet-Schwartz space; and both Silva spaces and Fréchet-Schwartz spaces are reflexive.

PROPOSITION 10. *A Silva space* E *is a holomorphically bornological space.*

The proof will rest on the following lemma.

LEMMA 11. *Let* E *be a complex vector space,* E_m *a complex locally convex space,* $\rho_m : E_m \to E$ *a linear mapping, and* $\sigma_m : E_m \to E_{m+1}$ *a compact linear mapping such that* $\rho_m = \rho_{m+1} \circ \sigma_m$ for $m \in \mathbb{N}$. *Assume that*

$$E = \bigcup_{m \in \mathbb{N}} \rho_m(E_m)$$

and endow E *with the inductive limit topology. Let* $U \subset E$ *be open. Put* $U_m = \rho_m^{-1}(U)$, *and assume that* U_o *is non-void; hence* U *and* U_m *are nonvoid for* $m \in \mathbb{N}$. *If* F *is a complex locally convex space and* $f: U \to F$, *then* $f \in \mathcal{H}(U;F)$ *if and only if* $f_m \equiv f \circ \rho_m \in \mathcal{H}(U_m;F)$ *for every* $m \in \mathbb{N}$.

PROOF. Necessity being clear, let us prove sufficiency. Assume that $f_m \in \mathcal{H}(U_m;F)$ for every $m \in \mathbb{N}$. We claim that f is algebraically holomorphic. In fact, let S be a finite dimensional vector subspace of E, with $U \cap S \neq \emptyset$. There are $m \in \mathbb{N}$ and a vector subspace S_m of E_m, of same dimension as S, such that $\rho_m(S_m) = S$. Thus, ρ_m is a vector space isomorphism between S_m and S. We have $\rho_m(U_m \cap S_m) = U \cap S$. In particular, $U_m \cap S_m \neq \emptyset$. Since $f_m | (U_m \cap S_m) \in \mathcal{H}(U_m \cap S_m; F)$, it follows that $f | (U \cap S) \in \mathcal{H}(U \cap S; F)$ because ρ_m is a homeomorphism between S_m and S, where S_m and S carry their natural topologies. Thus, the first claim is true. We next claim that f is amply bounded. It is enough to treat F as being seminormed. We may assume that $0 \in U$, and it suffices to show that f is locally bounded at 0. Since f_0 is locally bounded at 0, choose a convex neighborhood V_0 of 0 in U_0 such that $\sigma_0(V_0)$ has a compact closure in E_1 contained in U_1, hence $\rho_0(V_0) \subset U$, and such that

$$(1) \qquad \sup\{\|f_0(x)\| ; x \in V_0\} < M$$

for some $M \in \mathbb{R}$. Assume that, for some $m \in \mathbb{N}$, we have defined a convex neighborhood V_m of 0 in U_m such that $\sigma_m(V_m)$ has a compact closure in E_{m+1} contained in U_{m+1}, hence $\rho_m(V_m) \subset U$, and such that

$$(2) \qquad \sup\{\|f_m(x)\| ; x \in V_m\} < M ;$$

this is indeed the case for $m = 0$, by (1). Since f_{m+1} is locally bounded at the closure of $\sigma_m(V_m)$ in E_{m+1}, hence uniformly continuous there, and such a closure in convex, use (2) to choose a convex neighborhood V_{m+1} of that closure, hence of 0, in U_{m+1} such that $\sigma_{m+1}(V_{m+1})$ has a compact closure in E_{m+2} contained in U_{m+2}, hence $\rho_{m+1}(V_{m+1}) \subset U$, and such that

$$\sup\{\|f_{m+1}(x)\| ; x \in V_{m+1}\} < M ;$$

we also have $\rho_m(V_m) \subset \rho_{m+1}(V_{m+1})$. Proceeding in this way and letting

$$V = \bigcup_{m \in \mathbb{N}} \rho_m(V_m),$$

we get a neighborhood V of 0 in U such that $\|f(x)\| < M$ for every $x \in V$. Hence the second claim is true. It follows that $f \in \mathcal{H}(U;F)$. QED

REMARK 12. It is known that Lemma 11 is true if we replace f and f_m being holomorphic by them being continuous; but Lemma 11 is false if we replace f and f_m being holomorphic by them being amply bounded, as we see even when $E = \mathbb{C}^{(\mathbb{N})}$ and $F = \mathbb{C}$.

REMARK 13. It can be seen that, in Lemma 11, E is necessarily a Silva space.

PROOF OF PROPOSITION 10. Consider the sequence (E_m) of Definition 8. Let $f: U \to F$ be algebraically holomorphic, and bounded on every compact subset of U. Set $U_m = U \cap E_m$; we may assume that $U_o \neq \emptyset$, hence $U_m \neq \emptyset$, for all $m \in \mathbb{N}$. Then $f \mid U_m$ is algebraically holomorphic, and bounded on every compact subset of U_m. By Proposition 6, $f \mid U_m$ is holomorphic for every $m \in \mathbb{N}$. By Lemma 11, f is holomorphic. QED

REMARK 14. Lemma 11 is a reminicense of the known fact that, if E_i $(i \in I)$ is any family of locally convex spaces, E is a vector space, $\rho_i: E_i \to E$ $(i \in I)$ is a linear mapping, E is endowed with the inductive limit topology, and F is a locally convex space, then a linear mapping $f: E \to F$ is continuous if and only if $f \circ \rho_i: E_i \to F$ is continuous for every $i \in I$. Lemma 11 may break down in obsence of compactness (see Example 18 below) or denumerability (see Example 20 below) conditions.

REMARK 15. Proposition 10 is a reminiscense of the known fact that any inductive limit of bornological spaces is a bornological space. A denumerable inductive limit whose connecting mappings σ are not compact (see Example 18 below), or a non-denumerable inductive limit with compact connecting mappings σ (see Example 20 below), of holomorphically bornological spaces may fail to be a holomorphically bornological space.

PROPOSITION 16. *If E is a holomorphically bornological space, then* $\mathcal{H}(U;F)$ *is complete for the compact-open topology* τ_0 *if F is complete.*

PROOF. Let $\tilde{\mathcal{H}}(U;F)$ be the vector space of all mappings $f: U \to F$ which are algebraically holomorphically holomorphic and bounded on the compact subsets of U. Then $\tilde{\mathcal{H}}(U;F)$ is complete for the compact-open topology τ_0, since F is complete. Since E is a holomorphically bornological space, then $\mathcal{H}(U;F) = \tilde{\mathcal{H}}(U;F)$ algebraically and topologically. QED

REMARK 17. It is known that, if F is a bornological space , then $\mathcal{L}(E;F)$ is complete for the strong, or compact-open, topology , if F is complete. Proposition 16 corresponds to the second half of this remark.

EXAMPLE 18. Let X_0 be a separated infinite dimensional complex locally convex space. It is known that an $\mathcal{H}$-bounding subset of X_0 (that is, a subset of X_0 on which every member of $\mathcal{H}(X_0)$ is bounded) has an empty interior [5]. Therefore, if X_0 is metrizable, there is a sequence $g_m \in \mathcal{H}(X_0)$ $(m = 1,2,\ldots)$ such that, given any neighborhood V of 0 in X_0, then some g_m is unbounded on V. Consider the topological direct sum

$$E = \sum_{m=0}^{\infty} X_m ,$$

where $X_m = \mathbb{C}$ $(m = 1,2,\ldots)$. Define $f: E \to \mathbb{C}$ by

$$f(x) = \sum_{m=1}^{\infty} g_m(x_0)x_m$$

if $x = (x_m)_{m\in\mathbb{N}} \in E$. If we let

$$E_m = X_o \oplus \ldots \oplus X_m$$

and consider it as a vector subspace of E, then $f|E_m \in \mathcal{H}(E_m)$ for $m \in \mathbb{N}$. Notice that each E_m is metrizable, and even normable if X_o is normable. We claim that f is not locally bounded at 0. In fact, if V is a neighborhood of 0 in X_o and $\varepsilon_m > 0$ for $m = 1,2,\ldots$, define W as the set of all $x = (x_m)_{m\in\mathbb{N}} \in E$ such that $x_o \in V$ and $|x_m| \leq \varepsilon_m (m=1,2,\ldots)$. If we choose k so that g_k is unbounded on V, then f is unbounded on the set of all $x \in E$ with $x_o \in V$, $x_k = \varepsilon_k$, and $x_m = 0$ for $m \geq 1$, $m \neq k$; thus f is unbounded on W. Since all such W form a basis of neighborhoods of 0 in E, our claim is proved. Hence $f \notin \mathcal{H}(E)$. This shows that Lemma 11 breaks down if the σ_m are assumed to be linear and continuous, but not compact, although denumerability of the family is preserved. Such an example also shows that a denumerable inductive limit E of holomorphically bornological spaces E_m $(m \in \mathbb{N})$ may fail to be a holomorphically bornological space. In fact, f is algebraically holomorphic on E; and it is bounded on the compact subsets of E, since each such subset is contained in some E_m. However, $f \notin \mathcal{H}(E)$. Thus, E is not a holomorphically bornological space. Actually, $\mathcal{H}(E)$ is not complete, or even sequentially complete, for the compact-open topology $\mathcal{T}_o$. To see this, it is enough to introduce the truncated function $f_k \in \mathcal{H}(E)$ defined by

$$f_k(x) = \sum_{m=1}^{k} g_m(x_o) x_m$$

for $k = 1,2,\ldots$; since $f_k \to f$ uniformly on every compact subset of E as $k \to \infty$, but $f \notin \mathcal{H}(E)$, we conclude that $\mathcal{H}(E)$ is not sequentially complete for $\mathcal{T}_o$. Actually, if we look at E as

$$E = X_o \times \mathbb{C}^{(\mathbb{N})},$$

we see that E is a bornological space, as the cartesian product of two bornological spaces. Hence, a bornological space may fail to be a holomorphically bornological space. We also see that a cartesian product of two holomorphically bornological spaces may fail to be a holomorphically bornological space.

We shall need the following lemma in Example 20 below.

LEMMA 19. *Let* E *be a complex vector space whose (algebraic) dimension is at least equal to the continuum. Endow* E *with its largest locally convex topology. Then there is a 2-homogeneous polynomial* $P: E \to \mathbb{C}$ *which is not continuous.*

FIRST PROOF. Let B be a basis for E. Since B is infinite, there is a set S of functions $s: B \to \mathbb{N}$ such that S has the power of the continuum; and such that, for every function $t: B \to \mathbb{R}_+$, there is some $s \in S$ for which $s \leq ct$ is false for all $c \in \mathbb{R}_+$. In fact, fix an infinite denumerable subset I of B, and call S the set of all functions $s: B \to \mathbb{N}$ vanishing off that subset. Then S has the power of $\mathbb{N}^{\mathbb{N}}$, that is, of the continuum. For every $t: B \to \mathbb{R}_+$, we can find $s \in S$ such that $s \leq ct$ is false on I, hence on B, for all $c \in \mathbb{R}_+$. Since the power of B is at least equal to the continuum, there is a surjective mapping $b \in B \to s_b \in S$. Define

$$r(b_1,b_2) = \max\left\{s_{b_1}(b_2),\ s_{b_2}(b_1)\right\}$$

for $b_1, b_2 \in B$; then

$$r(b_2,b_1) = r(b_1,b_2) \geq s_{b_1}(b_2) \geq 0.$$

We claim that there is no $t: B \to \mathbb{R}_+$ such that $r(b_1,b_2) \leq t(b_1).t(b_2)$ for all $b_1,b_2 \in B$. In fact, if t did exist, we could choose $s \in S$ such that $s \leq ct$ is false for all $c \in \mathbb{R}_+$. Let $b_1 \in B$ so that $s = s_{b_1}$. Then $s(b_2) = s_{b_1}(b_2) \leq r(b_1,b_2) \leq t(b_1).t(b_2)$ for all $b_2 \in B$; thus $s \leq ct$ if $c = t(b_1)$, a contradiction. Now, define the symmetric bilinear form $A: E^2 \to \mathbb{C}$ by

$$A(x_1,x_2) = \sum_{b_1,b_2 \in B} r(b_1,b_2)\, b_1^*(x_1)\, b_2^*(x_2)$$

for $x_1,x_2 \in E$, where b^* is the linear form on E which to every $x \in E$ associates its b-component by B, if $b \in B$; the above sum is finite. Let the 2-homogeneous polynomial $P: E \to \mathbb{C}$ be given by $P(x) = A(x,x)$ for $x \in E$. We claim that P is not continuous. Otherwise, A would be continuous too, that is, we would have a seminorm α on E such that $|A(x_1,x_2)| \leq \alpha(x_1).\alpha(x_2)$ for all $x_1,x_2 \in E$. Then, letting $x_1 = b_1 \in B$ and $x_2 = b_2 \in B$, we would get $r(b_1,b_2) \leq \alpha(b_1).\alpha(b_2)$ for all $b_1,b_2 \in B$, a contradiction. QED

SECOND PROOF. Let X be an infinite dimensional complex vector space, and Y be its (algebraic) dual space. Assume firstly that $E = X \times Y$. Let $P: E \to \mathbb{C}$ be the 2-homogeneous polynomial defined by $P(x,y) = y(x)$ for all $x \in X$, $y \in Y$. We claim that P is not continuous if E is given its largest locally convex topology. In fact, assume that P is continuous. Now, the largest locally convex topology on E is the cartesian product of the largest locally convex topologies on X and Y; and P is a bilinear form on $X \times Y$. Then, there are seminorms α on X and β on Y such that $|P(x,y)| \leq \alpha(x).\beta(y)$ for all $x \in X$, $y \in Y$. Once the seminorm α is given, and X is infinite dimensional, there is a linear form b on X which is not continuous for α. However $|b(x)| = |P(x,b)| \leq c.\alpha(x)$ for all $x \in X$, where $c = \beta(b)$, showing that b is continuous for α, a contradiction. Hence, P is not continuous. Coming back to any E, in order to finish the proof, we argue that it is enough to prove the lemma when the dimension of E is equal to the continuum; in fact, the general case reduces to this one because E is a direct sum of two vector subspaces, one of which has dimension equal to the continuum. Now, if the above X has an infinite denumerable dimension, the corresponding Y has dimension equal to the continuum; hence $X \times Y$ has dimension equal to the continuum too. QED

EXAMPLE 20. Let E be a complex vector space whose dimension is at least equal to the continuum. Endow E with its largest locally convex topology; E is the inductive limit of its finite dimensional vector subspaces. By Lemma 19, let $f: E \to \mathbb{C}$ be a 2-homogeneous polynomial which is not continuous. For every finite dimensional vector subspace X of E, it is clear that $f|X \in \mathcal{H}(X)$. However, $f \notin \mathcal{H}(E)$. This shows that Lemma 11 breaks down in absence of denumerability of the family, although compactness of the connecting mappings σ is preserved. Such an example also shows that a non-denumerable inductive limit of holomorphically bornological spaces may fail to be a holomorphically bornological space, even if the connecting mappings σ are compact.

REMARK 21. In Example 18, if X_o is not normable, then every bounded subset of X has an empty interior. In this case, g_m may be chosen to be a continuous linear form. Thus, f is a 2-homogeneous polynomial. This is to be compared to Example 20, where f is also a 2-homogeneous polynomial. The following example is then in order (but X could not be a metrizable locally convex space which is not a normable space, in it).

EXAMPLE 22. We now show that E may fail to be a holomorphically bornological space, and yet have the following property: for every U and every F, we have that each polynomial $f: E \to F$ is continuous if (and always only if) f is bounded on every compact subset of U. In fact, let X be an infinite dimensional complex normed space $Y = \mathbb{C}^{(\mathbb{N})}$ and $E = X \times Y$. Let $f: E \to F$ be a polynomial that is bounded on every compact subset of U. It is enough to treat F as being seminormed, and U as being $V \times W$, where $V \subset X$, $W \subset Y$ are open and non-void. We write, for $x \in X$, $y \in Y$,

$$f(x,y) = \sum_{\alpha} g_{\alpha}(x)y^{\alpha}$$

uniquely, where α is any sequence of positive integers all but finitely many of which are zero, and each $g_{\alpha}: X \to F$ is a polynomial. If $y \in Y$, define $f_y: X \to F$ by by $f_y(x) = f(x,y)$ for $x \in X$. Each g_{α} is a finite linear combination of the f_y. Hence g_{α} is bounded on every compact subset of V. Since X is a normed space, then g_{α} is continuous on X, hence bounded on every bounded subset of X. Set

$$c_{\alpha} = \sup\{\|g_{\alpha}(x)\| \ ; \ \|x\| \leq 1\}$$

and ε be a sequence $\varepsilon_m > 0$ $(m \in \mathbb{N})$ such that

$$\sum_{\alpha} c_{\alpha} \varepsilon^{\alpha} \leq 1 \quad ;$$

then $\|x\| \leq 1$ and $|y_m| \leq \varepsilon_m$ for every $m \in \mathbb{N}$ imply that $\|f(x,y)\| \leq 1$, that is, f is bounded on a neighborhood for 0 in E. Hence, f is continuous.

We now show in Example 26 below that it is not enough to use only $F = \mathbb{C}$ in Definition 1; see however Propositions 54 and 76 below. To this end, we shall need the following results.

LEMMA 23. *Let* $E = c_o(I)$ *be endowed with the supremum norm, and* $(e_i)_{i \in I}$ *be its natural basis. To every* $A \in \mathcal{L}(^mE)$ *associate* $A_*: I^m \to \mathbb{C}$ *defined by*

$$A_*(i_1,\ldots,i_m) = A(e_{i_1},\ldots,e_{i_m})$$

for $i_1,\ldots,i_m \in I$, *where* $m \in \mathbb{N}$. *Then* $A_* \in c_o(I^m)$; *in particular,* A_* *vanishes off a denumerable subset of* I^m.

PROOF. For $m = 0$, the lemma is true, subject to the convention that $c_o(I^o)$ is reduced to 0. Let $m \geq 1$. In case $m \geq 2$, assumed that the lemma is true for $m-1$ to argue by induction. If $x_1,\ldots,x_m \in E$, then

(1) $A(x_1,\ldots,x_m) =$

$$\sum_{i_1,\ldots,i_m \in I} A_*(i_1,\ldots,i_m)\, x_{1i_1} \ldots x_{mi_m}$$

where the series is convergent by partial summation over all finite subsets of I^m that are cartesian products. We must prove that, for every $\varepsilon > 0$, the set of the $(i_1,\ldots,i_m) \in I^m$ for which $|A_*(i_1,\ldots,i_m)| \geq \varepsilon$ has to be finite. Assume that this set is infinite for some ε. Let then $(i_{1n},\ldots,i_{mn}) \in I^m$ be pairwise distinct, and such that $|A_*(i_{1n},\ldots,i_{mn})| \geq \varepsilon$ for $n=1,2,\ldots$. For each fixed $h = 1,\ldots,m$, we must have that $i_{hn} \to \infty$ as $n \to \infty$, meaning that every finite subset of I contains i_{hn} for only finitely many values of n; this is clear if $m = 1$, and if $m \geq 2$ this follows from the assumption that the lemma holds for m-1. By passing to subsequences, we may assume, for each fixed $h = 1,\ldots,m$, that the i_{hn} are pairwise distinct. In case $m \geq 2$, in view of the assumption that the lemma holds for $1,\ldots,m-1$, and by passing to subsequences, we may also assume inductively that

$$(2) \qquad \sum \frac{|A_*(i_{1k_1},\ldots,i_{mk_m}|}{\sqrt[m]{k_1\ldots k_m}} \leq \frac{1}{2^n}$$

for $n \geq 2$, where summation is over all $k_1,\ldots,k_m \in \{1,\ldots,n\}$, one at least but not all of them being equal to n. In case $m = 1$, the above step of the reasoning is to be abolished. Define $x_1,\ldots,x_m \in E$ inductively as follows. Set

$$t_{k_1\ldots k_m} =$$

$$A_*(i_{1k_1},\ldots,i_{mk_m}) \; x_{1i_{1k_1}} \ldots x_{mi_{mk_m}}$$

for $k_1,\ldots,k_m = 1,2,\ldots,$ and

$$s_n = \sum t_{k_1\ldots k_m}$$

where summation is over all $k_1,\ldots,k_m \in \{1,\ldots,n\}$ for $n \geq 1$. Then require:

1) for $h = 1,\dots,m$ and $k = 1,2,\dots$

$$\left| x_{hi_{hk}} \right| = \frac{1}{\sqrt[m]{k}} ,$$

2) $t_{n\dots n}$ has the same argument as s_{n-1} for $n \geq 2$,

3) $x_{hi} = 0$ for $h = 1,\dots,m$ and the remaining $i \in I$.

We have

$$|t_{n\dots n}| \geq \varepsilon/n, \quad |s_1| \geq \varepsilon ,$$

and, by using (2), we get

$$|s_{n-1}| + |t_{n\dots n}| = |s_{n-1} + t_n| \leq |s_n| + 1/2^n$$

for $n \geq 2$, hence

$$|s_n| \geq \varepsilon \sum_{h=1}^{n} 1/h - \sum_{h=2}^{n} 1/2^h ,$$

proving that $s_n \to \infty$, against $s_n \to A(x_1,\dots,x_m)$ as $n \to \infty$, by (1).

QED

DEFINITION 24. *Set* $E = c_o(I)$. *Let* $\mathcal{T}$ *be the topology on* E *defined by the full supremum norm*

$$x \in E \to \|x\| = \sup_{i \in I} |x_i| \in \mathbb{R}$$

whereas let $\mathcal{S}$ *be the topology on* E *defined by the family of the denumerable supremum seminorms*

$$x \in E \to \|x\|_J = \sup_{i \in J} |x_i| \in \mathbb{R}$$

for all denumerable $J \subset I$. *Obviously* $\mathcal{S} \subset \mathcal{T}$; *and* $\mathcal{S} = \mathcal{T}$ *if and only if* I *is denumerable.*

The following result is due to Josefson [4].

LEMMA 25. *If* $E = c_o(I)$ *and* $U \subset E$ *is nonvoid and open for* $\mathcal{S}$, *hence for* $\mathcal{T}$, *then* $\mathcal{H}(U)$ *is the same regardless of the fact that we endow* E *with* $\mathcal{S}$ *or* $\mathcal{T}$.

PROOF. In the following, an index J denotes that we are taking a concept with respect to the seminorm on E defined by the denumerable subset J of I; whereas lack of that index means that we are using the full supremum norm (see Definition 24). It is enough to consider $f \in \mathcal{H}(U)$ for $\mathcal{T}$ and conclude that $f \in \mathcal{H}(U)$ for $\mathcal{S}$. Fix $\xi \in U$. Put $A_m = d^m f(\xi) \in \mathcal{L}(^mE)$ for $\mathcal{T}$ $(m \in \mathbb{N})$. There is $\varepsilon > 0$ such that $B_\varepsilon(\xi) \subset U$ and

$$(1) \qquad f(x) = \sum_{m \in \mathbb{N}} A_m (x - \xi)^m$$

uniformly for $x \in B_\varepsilon(\xi)$. Moreover, (1) holds true pointwisely on the largest ξ-balanced subset U_ξ of U. From the Cauchy-Hadamard formula, it follows that

$$\| A_m \|^{1/m} \qquad (m = 1,2,\ldots)$$

is bounded. By Lemma 23, there is a denumerable subset J of I such that $\| A_m \|_J = \| A_m \|$ and, in particular, A_m is continuous for $\mathcal{S}$, for all $m \in \mathbb{N}$. It follows that

$$\| A_m \|_J^{1/m} \qquad (m = 1,2,\ldots)$$

is bounded. Since U is open for $\mathcal{S}$, we may assume that J is large enough and ε is sufficiently small so that $B_{J\varepsilon}(\xi) \subset U$ and $\varepsilon^m . \| A_m \|_J$ $(m \in \mathbb{N})$ is bounded. Then, (1) holds not only pointwisely on $B_{J\varepsilon}(\xi) \subset U_\xi$ but also uniformly on $B_{J\varepsilon/2}(\xi)$, proving the claim. QED

EXAMPLE 26. Let E be a complex vector space. Assume that $\mathcal{S}$ and $\mathcal{T}$ are two locally convex topologies on E such that:

1. $\mathcal{S} \subset \mathcal{T}$ and $\mathcal{S} \neq \mathcal{T}$;

2. $\mathcal{S}$ and $\mathcal{T}$ have the same compact subsets of E, hence the same bounded subsets of E.

3. for every nonvoid subset $U \subset E$ that is open for $\mathcal{S}$, hence for $\mathcal{T}$, then $\mathcal{H}(U)$ is the same regardless of the fact that we endow E with $\mathcal{S}$ or $\mathcal{T}$.

4. E is holomorphically bornological when it is endowed with $\mathcal{T}$.

Then, if E is endowed with $\mathcal{S}$, we claim that E is not bornological, hence not holomorphically bornological. However, for every nonvoid subset $U \subset E$ that is open for $\mathcal{S}$, each function $f: U \to \mathbb{C}$ belongs to $\mathcal{H}(U)$ if f is algebraically holomorphic, and f is bounded on every compact subset of U. In fact, E endowed with $\mathcal{S}$ is not bornological since the identity mapping $I: (E, \mathcal{S}) \to (E, \mathcal{T})$ is linear, and it maps bounded subsets into bounded subsets, but it is not continuous. Now let $U \subset E$ be nonvoid and open for $\mathcal{S}$, and let $f: U \to \mathbb{C}$ be algebraically holomorphic and bounded on every subset of U which is compact for $\mathcal{S}$. Since U is open for $\mathcal{T}$, and f is algebraically holomorphic and bounded on every subset of U which is compact for $\mathcal{T}$, then $f \in \mathcal{H}(U)$ if E is endowed with $\mathcal{T}$. It follows that $f \in \mathcal{H}(U)$ if E is endowed with $\mathcal{S}$. An instance of this situation is the following. Take a nondenumerable set I, and use $\mathcal{S}$ and $\mathcal{T}$ of Definition 24 on $E = c_o(I)$. Then, all the above four conditions can be checked; the third condition follows from Lemma 25.

REMARK 27. Example 26 also shows that it is not enough to use only $F = \mathbb{C}$ in Definition 3 via (1b) or (1c) of Lemma 2. However, in this case there are simpler classical conditions.

3. HOLOMORPHICALLY BARRELED SPACES

DEFINITION 28. *A given* E *is a "holomorphically barreled space" if, for every* U *and every* F, *we have that each collection* $\mathcal{X} \subset \mathcal{H}(U;F)$ *is amply bounded if (and always only if)* $\mathcal{X}$ *is bounded on every finite dimensional compact subset of* U.

REMARK 29. It will follow from Proposition 38 below that, in Definition 28 and in similar situations, it is equivalent to consider only the affine one dimensional compact subsets of U.

REMARK 30. $\mathcal{X} \subset \mathcal{H}(U;F)$ is amply bounded if and only if $\mathcal{X}$ is equicontinuous and bounded at every point of U. Thus Definition 28 may be rephrased by requiring that $\mathcal{X}$ is equicontinuous if it is bounded on every finite dimensional compact subset of U; then there is no "and always only if".

REMARK 33. below motivates the above definition, but we need the following preliminary material which is known.

LEMMA 31. *For a given* E, *the following conditions are equiva-*

lent:

(1p) For every F, *we have that each collection* $\mathfrak{X} \subset \mathcal{L}(E;F)$ *is amply bounded, or equivalently equicontinuous, if (and always only if)* $\mathfrak{X}$ *is bounded at every point of* E.

(1c) For every F, *we have that each collection* $\mathfrak{X} \subset \mathcal{L}(E;F)$ *is amply bounded, or equivalently equicontinuous, if (and always only if)* $\mathfrak{X}$ *is bounded on every finite dimensional compact subset of* E

(2) Each seminorm α *on* E *is continuous if (and always only if)* α *is lowersemicontinuous.*

PROOF. We shall prove the following implications

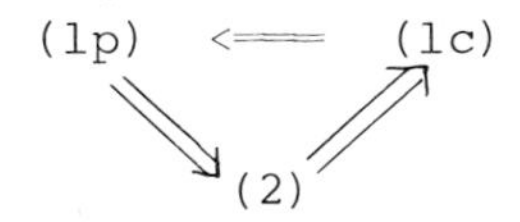

(1p) $\Longrightarrow$ (2). Let α be a seminorm on E that is lowersemicontinuous. Call $\mathfrak{X}$ the collection of the continuous linear forms f on E such that $|f(x)| \leq \alpha(x)$ for all $x \in E$. By the Hahn-Banach theorem, we have $\alpha(x) = \sup\{|f(x)|;\ f \in \mathfrak{X}\}$ for all $x \in E$. Since $\mathfrak{X}$ is bounded at every point of E, it is equicontinuous, by (1p). It follows that α is continuous.

(2) $\Longrightarrow$ (1c). Let $\mathfrak{X} \subset \mathcal{L}(E;F)$ be bounded on every finite dimensional compact subset of E, hence at every point of E. If $\beta \in CS(F)$, then $\alpha(x) = \sup\{\beta[f(x)];\ f \in \mathfrak{X}\}$ for $x \in E$ defines a lowersemicontinuous seminorm α on E. By (2), α is continuous. It follows that $\mathfrak{X}$ is equicontinuous as β is arbitrary.

(1c) $\Longrightarrow$ (1p). Let $\mathfrak{X} \subset \mathcal{L}(E;F)$ be bounded at every point of E. Thus $\mathfrak{X}$ is bounded on every finite dimensional simplex, hence on every finite dimensional compact subset, of E. By (1c), $\mathfrak{X}$ is equicontinuous.

The proof can also be carried on with the same reasoning, by reversing the arrows. QED

The following definition is classical, particularly in terms of (1b) or (2).

DEFINITION 32. *A given* E *is a "barreled space" if it satisfies the equivalent conditions of Lemma 31.*

REMARK 33. Definition 28 was formulated in analogy to Definition 32 trough (1c), rather than (1p), of Lemma 31. The reason is that, by a classical example, it can occur that a sequence $f_m \in \mathcal{H}(\mathbb{C})$ $(m \in \mathbb{N})$ is bounded at every point of $\mathbb{C}$, and yet it fails to be bounded on some compact subset of $\mathbb{C}$, that is, it is not locally bounded.

PROPOSITION 34. *A holomorphically barreled space is also a barreled space.*

PROOF. It suffices to compare Definitions 28 and 32, by using (1c) of Lemma 31, and by remarking that $\mathcal{L}(E;F) \subset \mathcal{H}(E;F)$. QED

PROPOSITION 35. *For a given* E *to be a holomorphically barreled space, it is necessary and sufficient that, for every* U, *we have that each collection* $\mathfrak{X} \subset \mathcal{H}(U)$ *is locally bounded if (and always only if)* $\mathfrak{X}$ *is bounded on every finite dimensional compact subset of* U.

PROOF. Necessity being clear, let us prove sufficiency. Let $\mathfrak{X} \subset \mathcal{H}(U;F)$ be bounded on every finite dimensional compact subset of U. Given any $\beta \in CS(F)$, let $\mathcal{Y}$ be the collection of the linear forms ψ on F such that $|\psi(y)| \leq \beta(y)$ for all $y \in F$. By the Hahn-Banach theorem, we have that $\beta(y) = \sup\{|\psi(y)|; \psi \in \mathcal{Y}\}$ for all $y \in F$. Since the collection $\mathcal{Y} \circ \mathfrak{X}$ of all $\psi \circ f$, where $\psi \in \mathcal{Y}$ and $f \in \mathfrak{X}$, is bounded on every finite dimensional compact subset of U, there results that $\mathcal{Y} \circ \mathfrak{X}$ is locally bounded, for every β. It follows that $\mathfrak{X}$ is amply bounded. QED

REMARK 36. It is known that it is enough to take $F = \mathbb{C}$ in (lp) or (lc) of Lemma 31, when using them in Definition 32. For the case of (lc), Proposition 35 corresponds to this remark.

PROPOSITION 37. *A Baire space* E *is a holomorphically barreled space.*

PROOF. It is enough to treat F as being a seminormed space. We start with two classical remarks.

If X is a nonvoid Baire space, and $\mathcal{U}$ is a pointwise bounded set of continuous mappings of X to F, there is at least a point of X where $\mathcal{U}$ is locally bounded.

If $p: E \to F$ is an m-homogeneous polynomial $(m \in \mathbb{N})$ and $a, b \in E$, then

$$\sup\{\|p(\lambda a+b)\| \;;\; \lambda \in \mathbb{C}, |\lambda| \leq 1\} =$$

$$\sup\{\|p(a+\lambda b)\| \;;\; \lambda \in \mathbb{C}, |\lambda| \leq 1\} \;;$$

in fact, by the maximum principle, we may replace $|\lambda| \leq 1$ by

$|\lambda| = 1$, and then equality is clear via $\lambda \to 1/\lambda$, by m-homogeneity. In particular

$$\|p(b)\| \leq \sup\{\|p(a+\lambda b)\| \; ; \; \lambda \in \mathbb{C}, \; |\lambda| \leq 1\}.$$

Now, let $\mathfrak{X} \subset \mathcal{H}(U;F)$ be bounded on every affine one dimensional compact subset of U, which is the case if $\mathfrak{X}$ is bounded on every finite dimensional compact subset of U. Fix $\xi \in U$. Take a balanced open neighborhood V of 0 in E such that $\xi + V \subset U$. By the Cauchy integral, the set

$$\mathcal{U} = \left\{\frac{1}{m!} \hat{d}^m f(\xi); \; f \in \mathfrak{X} \;, \;\; m \in N\right\}$$

is pointwise bounded on V, because $\mathfrak{X}$ is bounded on every affine one dimensional compact subset $\{\xi + \lambda x \; ; \; \lambda \in \mathbb{C}, \; |\lambda| \leq 1\}$ of U, where $x \in V$. By the first remark above, there is a $a \in V$ where $\mathcal{U}$ is locally bounded, since V is a nonvoid Baire space. Let W be a balanced neighborhood of 0 in E such that $a + W \subset V$ and $\mathcal{U}$ is bounded on $a + W$. By the second remark above, $\mathcal{U}$ is bounded on W. Then Taylor series expansion at ξ shows that $\mathfrak{X}$ is bounded on $\xi + W/2$. Hence $\mathfrak{X}$ is locally bounded. QED

PROPOSITION 38. $\mathfrak{X} \subset \mathcal{H}(U;F)$ *is bounded on every finite dimensional compact subset of* U *if and only if* $\mathfrak{X}$ *is bounded on every affine one dimensional compact subset of* U.

PROOF. Only sufficiency requires justification. It is enough to restrict attention to the case when E is finite dimensional, hence a Baire space. Then, an inspection of the proof of Proposition 35 gives the argument for the present proof. QED

REMARK 39. Propositions 34 and 37 imply the known fact that a Baire space E is a barreled space. Propositions 37 and 38 contain as a particular case the following generalization of the classical Banach-Steinhaus Theorem.

PROPOSITION 40. *(Holomorphic Banach-Steinhaus Theorem). If* E *is a Fréchet space, each collection* $\mathfrak{X} \subset \mathcal{H}(U;F)$ *is equicontinuous if* $\mathfrak{X}$ *is bounded on every affine one dimensional compact subset of* U.

PROPOSITION 41. *A Silva space is a holomorphically barreled space.*

The proof will rest on the following lemma.

LEMMA 42. *In the notation of Lemma 11, then* $\mathfrak{X} \subset \mathcal{H}(U;F)$ *is amply bounded if and only if* $\mathfrak{X}_m \equiv \mathfrak{X} \circ \rho_m \subset \mathcal{H}(U_m;F)$ *is amply bounded for every* $m \in \mathbb{N}$.

PROOF. Necessity being clear, let us prove sufficiency. It is enough to treat F as being a seminormed space. Since each $\mathfrak{X}_m$ is pointwise bounded, it follows that $\mathfrak{X}$ is pointwise bounded too. Consider $g: U \to \ell^\infty(\mathfrak{X};F)$ defined by $g(x)(f) = f(x)$ for $x \in U$ and $f \in \mathfrak{X}$. Since each $\mathfrak{X}_m \subset \mathcal{H}(U_m;F)$ is locally bounded, we see that $g \circ \rho_m: U_m \to \ell^\infty(\mathfrak{X};F)$ is holomorphic for every $m \in \mathbb{N}$. By Lemma 11, we conclude that g is holomorphic. Thus, g is locally bounded, that is, $\mathfrak{X}$ is locally bounded. QED

REMARK 43. Lemma 42 may be proved directly, by a reasoning quite close to that of the proof of Lemma 11, see Lemma 3,

[1]. Notice that Lemma 11 is not the particular case of Lemma 42 when $\mathfrak{X}$ is reduced to one element, as then Lemma 42 is trivial.

PROOF OF PROPOSITION 41. Consider the sequence (E_m) of Definition 8, and use notation of Lemma 42. Let $\mathfrak{X} \subset \mathcal{H}(U;F)$ be bounded on every finite dimensional compact subset of U. Then $\mathfrak{X}_m \subset \mathcal{H}(U_m;F)$ is bounded on every finite dimensional compact subset of U_m. By Proposition 37 (or else 40), $\mathfrak{X}_m$ is amply bounded for every $m \in \mathbb{N}$. By Lemma 42, $\mathfrak{X}$ is amply bounded. QED

REMARK 44. Lemma 42 is a reminiscense of the known fact that, if $E_i (i \in I)$ is any family of locally convex spaces, E is a vector space, $\rho_i : E_i \to E$ $(i \in I)$ is a linear mapping, E is endowed with the inductive limit topology, and F is a locally convex space, then a collection $\mathfrak{X} \subset \mathcal{L}(E,F)$ is amply bounded, or equivalently equicontinuous, if and only if $\mathfrak{X}_i \equiv \mathfrak{X} \circ \rho_i \subset \mathcal{L}(E_i;F)$ is amply bounded, or equivalently equicontinuous, for every $i \in I$. Lemma 42 may break down in absence of compactness (see Example 65 below) or denumerability (see Example 66 below) conditions.

REMARK 45. Proposition 41 is a reminiscense of the known fact that any inductive limit of barreled spaces is a barreled space. A denumerable inductive limit whose connecting mappings σ are not compact (see Example 65 below), or a non-denumerable inductive limit with compact connecting mappings σ (see Example 66 below), of holomorphically barreled spaces may fail to be a holomorphically barreled space, or even to be a holomorphically infrabarreled space in the sense of the next section.

4. HOLOMORPHICALLY INFRABARRELED SPACES

DEFINITION 46. *A given* E *is a "holomorphically infrabarreled space" if, for every* U *and every* F, *we have that each collection* $\mathfrak{X} \subset \mathcal{H}(U;F)$ *is amply bounded if (and always only if) is bounded on every compact subset of* U.

REMARK 47. For the reason given in Remark 30, Definition 46 may be rephrased by requiring that $\mathfrak{X}$ is equicontinuous if it is bounded on every compact subset of U; then there is no "and always only if".

Remark 50 below motivates the above definition, but we need the following preliminary material which is known.

LEMMA 48. *For a given* E, *the following conditions are equivalent:*

(1b) *For every* F, *we have that each collection* $\mathfrak{X} \subset \mathcal{L}(E;F)$ *is amply bounded, or equivalently equicontinuous, if (and always only if)* $\mathfrak{X}$ *is bounded on every bounded subset of* E

(1c) *For every* F, *we have that each collection* $\mathfrak{X} \subset \mathcal{L}(E;F)$ *is amply bounded or equivalently equicontinuous, if (and always only if)* $\mathfrak{X}$ *is bounded on every compact subset of* E.

(2b) *Each seminorm* α *on* E *is continuous if (and always only if)* α *is lowersemicontinuous and bounded on every bounded subset of* E.

(2c) *Each seminorm* α *on* E *is continuous if (and always only if)* α *is lowersemicontinuous and bounded on every compact subset of* E.

PROOF. We shall prove the following implications

$$\begin{array}{ccc} (1c) & \Longrightarrow & (2b) \\ \Uparrow & & \Downarrow \\ (1c) & \Longleftarrow & (2c) \end{array}$$

(1c) $\Longrightarrow$ (1b). This is clear.

(1b) $\Longrightarrow$ (2b). Let α be a seminorm on E that is lowersemicontinous and bounded on every bounded subset of E. Call $\mathcal{X}$ the collection of the continuous linear forms f on E such that $|f(x)| \leq \alpha(x)$ for all $x \in E$. By the Hahn-Banach theorem, we have $\alpha(x) = \sup\{|f(x)|\ ;\ f \in \mathcal{X}\}$ for all $x \in E$. Since $\mathcal{X}$ is bounded on every bounded subset of E, it is equicontinuous, by (1b). It follows that α is continuous.

(2b) $\Longrightarrow$ (2c). Let α be a seminorm on E that is lowersemicontinuous and bounded on every compact subset of E; then α is also bounded on every bounded subset of E (see the same step in the proof of Lemma 2). By (2b), α is continuous.

(2c) $\Longrightarrow$ (1c). Let $\mathcal{X} \subset \mathcal{L}(E;F)$ be bounded on every compact subset of E. If $\beta \in CS(F)$, then $\alpha(x) = \sup\{\beta[f(x)]\ ;\ f \in \mathcal{X}\}$ for $x \in E$ defines a lowersemicontinuous seminorm α on E that is bounded on every compact subset of E. By (2c), α is continuous. It follows that $\mathcal{X}$ is equicontinuous as β is arbitrary.

The proof can be also carried on with the same reasoning, by reversing the arrows. QED

The following definition is classical, particularly in terms of (1b) or (2b).

DEFINITION 49. *A given* E *is an "infrabarreled space" if it satisfies the equivalent conditions of Lemma 48.*

REMARK 50. Definition 46 was formulated in analogy to Definition 49 through (1c), rather than (1b), of Lemma 48. The reason is the same given in Remark 4.

PROPOSITION 51. *A holomorphically infrabarreled space is also an infrabarreled space.*

PROOF. It suffices to compare Definitions 46 and 49, using (1c) of Lemma 48, and by remarking that $\mathcal{L}(E;F) \subset \mathcal{H}(E;F)$. QED

PROPOSITION 52. *For a given* E *to be a holomorphically infrabarreled space, it is necessary and sufficient that, for every* U, *we have that each collection* $\mathfrak{X} \subset \mathcal{H}(U)$ *is locally bounded if (and always only if)* $\mathfrak{X}$ *is bounded on every compact subset of* U.

PROOF. The argument is similar to that of the proof of Proposition 35. QED

REMARK 53. It is known that it is enough to take $F = \mathbb{C}$ in (1b) or (1c) of Lemma 48, when using them in Definition 49. For the case of (1c), Proposition 52 corresponds to this remark.

PROPOSITION 54. *For* E *to be a holomorphically bornological space it is necessary and sufficient that* E *be a holomorphically infrabarreled space, and moreover that, for every* U, *we have that each function* $f: U \to \mathbb{C}$ *belongs to* $\mathcal{H}(U)$ *if (and always only if)* f *is algebraically holomorphic, and* f *is bounded on every compact subset of* U.

PROOF. Let us prove necessity, and assume that E is a holomorphically bornological space. Let $\mathfrak{X} \subset \mathcal{H}(U;F)$ be bounded on the compact subsets of U. It follows that $\mathfrak{X}$ is pointwise bounded too. It is enough to treat F as being a seminormed space. Consider $g: U \to \ell^{\infty}(\mathfrak{X};F)$ defined by $g(x)(f) = f(x)$ for $x \in U$ and $f \in \mathfrak{X}$. Since $\mathfrak{X}$ is bounded on the compact subsets of U, it follows that g is bounded on the compact subsets of U. In particular, $g|(U \cap S)$ is locally bounded for every finite dimensional vector subspace S of E meeting U; hence g is algebraically holomorphic. Since E is a holomorphically bornological space, then g is holomorphic, hence locally bounded. It follows that $\mathfrak{X}$ is locally bounded. This shows that E is a holomorphically infrabarreled space. The rest of necessity is clear. Let us prove sufficiency, and assume that $f: U \to F$ is algebraically holomorphic and bounded on every compact subset of U. For any fixed $\beta \in CS(F)$, let $\mathcal{Y}$ be the collection of the linear forms ψ on F such that $|\psi(y)| \leq \beta(y)$ for all $y \in F$. Each such $\psi \circ f$ is algebraically holomorphic and bounded on the compact subsets of U; thus it is holomorphic. Moreover, $\mathfrak{X} \equiv \mathcal{Y} \circ f \subset \mathcal{H}(U)$ is bounded on every compact subset of U. Thus $\mathfrak{X}$ is locally bounded, since E is a holomorphically infrabarreled space. By the Hahn-

-Banach theorem, we have $\beta(y) = \sup\{|\psi(y)| ; \psi \in \mathcal{Y}\}$ for all $y \in F$. It follows that $\beta \circ f$ is locally bounded. Thus f is amply bounded. Since f is also algebraically holomorphic, it is holomorphic. Thus E is a holomorphically bornological space. QED

REMARK 55. It is known that, for E to be a bornological space it is necessary and sufficient that E be an infrabarreled space, and moreover that each function $f: E \to \mathbb{C}$ belongs to E' if (and always only if) f is linear, and f is bounded on every bounded, or compact, subset of E. Proposition 54 corresponds to the second half of this remark.

DEFINITION 56. *A given* E *has the "Montel property" if, for every* U *and every* F, *we have that each collection* $\mathfrak{X} \subset H(U;F)$ *is relatively compact for* τ_0 *if (and always only if)* $\mathfrak{X}$ *is bounded on every finite dimensional compact subset of* U, *and* $\mathfrak{X}(x) \subset F$ *is relatively compact for every* $x \in U$.

REMARK 57. The terminology in Definition 56 comes, of course, from the classical Montel theorem saying that, if E is finite dimensional and $F = \mathbb{C}$, then $\mathfrak{X} \subset \mathcal{H}(U)$ is relatively compact for τ_0 if and only $\mathfrak{X}$ is bounded on every compact subset of U. We should distinguish between Montel property of Definition 56 and by now classical Montel property of E requiring that every bounded subset of E be relatively compact (see Example 67 below).

PROPOSITION 58. *For* E *to be a holomorphically barreled space it is necessary and sufficient that* E *be a holomorphically infrabarreled space, and moreover that* E *has the Montel property.*

PROOF. Let us prove necessity, and assume that E is holomorphically barreld. Then, comparison of Definitions 28 and 46 shows that E is holomorphically infrabarreled. Moreover, let $\mathfrak{X} \subset H(U;F) \subset \mathcal{H}(U;\hat{F})$ be bounded on the finite dimensional compact subsets of U. Then, $\mathfrak{X}$ is amply bounded, hence equicontinuous . If, in addition, $\mathfrak{X}(x) \subset F$ is relatively compact for every $x \in U$, then Ascoli's theorem implies that $\mathfrak{X} \subset H(U;F)$ is relatively compact. Thus E has Montel property. Let us turn to sufficiency, and assume that E is holomorphically infrabarreled having Montel property. If $\mathfrak{X} \subset \mathcal{H}(U)$ is bounded on every finite dimensional compact subset of U, then $\mathfrak{X} \subset \mathcal{H}(U)$ is relatively compact for $\mathcal{T}_0$ by Montel property; it follows that $\mathfrak{X}$ is bounded for $\mathcal{T}_0$, that is, bounded on every compact subset of U. Since E is holomorphically infrabarreled, there results that $\mathfrak{X}$ is locally bounded. By Proposition 35, E is holomorphically barreld. QED

REMARK 59. We may think of a variation of the Montel property with just $F = \mathbb{C}$; namely that each $\mathfrak{X} \subset \mathcal{H}(U)$ is relatively compact for $\mathcal{T}_0$ if (and always only if) $\mathfrak{X}$ is bounded on every finite dimensional compact subset of U. The proof of Proposition 58 shows that the Montel property with arbitrary F is equivalent to such a variation of it with just $F = \mathbb{C}$ when E is holomorphically infrabarreled. However, they are not equivalent by themselves (see Example 68 below).

REMARK 60. Proposition 58 and Remark 59 correspond to the following known facts. For E to be a barreled space it is necessary and sufficient that E be an infrabarreled space, and moreover that, for every F, we have that each collection $\mathfrak{X} \subset \mathcal{L}(E;F)$ is relatively compact for $\mathcal{T}_o$ if (and always only if) $\mathfrak{X}$ is pointwise bounded, and $\mathfrak{X}(x) \subset F$ is relatively compact for every $x \in E$. It is enough to take $F = \mathbb{C}$ in the above statement: the two conditions on $\mathfrak{X}$ are equivalent when E is infrabarreled; however, they are not equivalent by themselves.

DEFINITION 61. *A given* E *has the "infra-Montel property" if, for every* U *and every* F, *we have that each collection* $\mathfrak{X} \subset H(U;F)$ *is relatively compact for* $\mathcal{T}_o$ *if (and always only if)* $\mathfrak{X}$ *is bounded on every compact subset of* U, *and* $\mathfrak{X}(x) \subset F$ *is relatively compact for every* $x \in U$.

REMARK 62. The terminology in Definition 61 is motivated as in Remark 57, an by comparison between Propositions 58 and 63 below. It is clear that E has the infra-Montel property if it has the Montel property. Except for that, we should distinguish between Montel property, infra-Montel property and classical Montel property (see Example 67 below)

PROPOSITION 63. *A holomorphically infrabarreled space* E *has the infra-Montel property.*

PROOF. The argument is a minor modification of the proof of the corresponding assertion of Proposition 58. QED

REMARK 64. We may think of a variation of the infra-Montel property with just $F = \mathbb{C}$; namely that each $\mathfrak{X} \subset \mathcal{H}(U)$ is relatively compact for $\mathcal{T}_o$ if (and always only if) $\mathfrak{X}$ is bounded on every compact subset of U. This amounts to saying that each $\mathcal{H}(U)$ has the classical Montel property for $\mathcal{T}_o$. The infra-Montel property with arbitrary F is not equivalent to such a variation of it with just $F = \mathbb{C}$ (see Example 68 below).

EXAMPLE 65. Consider Example 18. Call $\mathfrak{X} \subset \mathcal{H}(E)$ the collection of the f_k for all $k = 1,2,\dots$. Then $\mathfrak{X}$ is bounded on every compact subset of E. Hence $\mathfrak{X}|E_m$ is locally bounded for $m \in \mathbb{N}$. However, $\mathfrak{X}$ is not locally bounded at 0, because f is not locally bounded at 0 and $f_k \to f$ pointwisely as $k \to \infty$. This shows that Lemma 42 breaks down if the σ_m are assumed to be linear continuous, but not compact, although denumerability of the family is preserved. Such an example also shows that a denumerable inductive limit E of holomorphically barreled spaces E_m $(m \in \mathbb{N})$ may fail to be a holomorphically infrabarreled space. In fact, if X_o is a Fréchet space, then each E_m is a Fréchet space, hence holomorphically barreled (by Propositions 37 or 40). However, E is not holomorphically infrabarreled.

EXAMPLE 66. Consider Example 20. Fix a basis B for E. For every finite subset I of B, call p_I the projection defined by B, of E onto the vector subspace of E generated by I. Call $\mathfrak{X} \subset \mathcal{H}(E)$ the collection of the $f_I \equiv f \circ p_I$ for all such I. Then $\mathfrak{X}$ is bounded on every compact subset of E. However, is not locally bounded at 0, because f is not locally bounded at 0 and $f_I \to f$ pointwisely as I increases. This shows that

Lemma 42 breaks down in absence of denumerability of the family, although compactness of the connecting mappings σ is preserved. Such an example also shows that a non-denumerable inductive limit of holomorphically barreled spaces may fail to be a holomorphically infrabarreled space, even if the connecting mappings σ are compact.

EXAMPLE 67. An infinite dimensional Banach space E has the Montel property, by Propositions 37 or 40, and 58. However, E fails to have the classical Montel property, by a theorem of Riesz. Conversely, assume that the locally convex space E has the classical Montel property. It may occur that there is some $\mathfrak{X} \subset \mathcal{H}(E)$ which is bounded for $\mathcal{T}_o$, but is not relatively compact for $\mathcal{T}_o$; then E does not have the infra-Montel property. An instance of this situation is described in Example 65, if X_o is assumed to have the classical Montel property. Another instance of the same situation is described in Example 66. Finally, let the locally convex space E be metrizable, but not barreled. Then E is a holomorphically infrabarreled space, by Propositions 6 and 54; thus E has the infra-Montel property, by Proposition 63. However, E does not have the Montel property, by Propositions 34 and 58; in fact, E is holomorphically infrabarreled, but E is not holomorphically barreled because it is not barreled. An instance of this situation is $E = \mathbb{C}^{(\mathbb{N})}$ with the supremum norm.

We now show in Example 68 below that it is not enough to use $F = \mathbb{C}$ in Definitions 56 and 61.

EXAMPLE 68. Let E be a complex vector space. Assume that $\mathcal{S}$ and $\mathcal{T}$ are two locally convex topologies on E such that conditions 1, 2 and 3 of Example 26 are satisfied, and moreover:

4. E is holomorphically barreled when it is endowed with $\mathcal{T}$.

5. There are a Banach space F and a collection $\mathcal{X} \subset \mathcal{L}((E,\mathcal{S});F)$ that is bounded on every compact subset of E and is such that $\mathcal{X}(x) \subset F$ is relatively compact for every $x \in E$; and yet $\mathcal{X} \subset \mathcal{L}(E,\mathcal{S});F)$ is not relatively compact for the pointwise topology.

Then, if E is endowed with $\mathcal{S}$, we claim that E satisfies Definition 56 with $F = \mathbb{C}$ (see Remark 59), hence Definition 61 with $F = \mathbb{C}$ (see Remark 64); but E does not have the infra-Montel property of Definition 61, hence does not have the Montel property of Definition 56, with arbitrary F. In fact, let U be nonvoid and open for $\mathcal{S}$, hence open for $\mathcal{T}$. If $\mathcal{X} \subset \mathcal{H}((U,\mathcal{S}))$ is bounded on every finite dimensional compact subset of U, then $\mathcal{X} \subset \mathcal{H}((U,\mathcal{T}))$ is bounded on every finite dimensional compact subset of U. Hence $\mathcal{X}$ is locally bounded for $(U,\mathcal{T})$. It follows that $\mathcal{X}$ is equicontinuous for $(U,\mathcal{T})$, and also pointwise bounded. By Ascoli's theorem, $\mathcal{X} \subset \mathcal{C}((U,\mathcal{T}))$ is relatively compact for $\mathcal{T}_0$; hence $\mathcal{X} \subset \mathcal{H}((U,\mathcal{T}))$ is relatively compact for $\mathcal{T}_0$ because $\mathcal{H}((U,\mathcal{T}))$ is closed in $\mathcal{C}((U,\mathcal{T}))$ for $\mathcal{T}_0$. We then see that $\mathcal{X} \subset \mathcal{H}((U,\mathcal{S}))$ is relatively compact for $\mathcal{T}_0$. This proves the first half of the claim. Consider now F and $\mathcal{X}$ quoted in condition 5. Then $\mathcal{X} \subset \mathcal{H}((E,\mathcal{S});F)$ is bounded on every compact subset of E, and $\mathcal{X}(x) \subset F$ is relatively compact for every $x \in E$. However, $\mathfrak{X} \subset \mathcal{H}((E,\mathcal{S});F)$ is not relatively compact for $\mathcal{T}_o$, as $\mathfrak{X} \subset \mathcal{H}((E,\mathcal{S});F)$ is not relatively compact for the pointwise topology; in fact, $\mathfrak{X} \subset \mathcal{L}((E,\mathcal{S});F)$ is not relatively compact for the pointwise topology, and $\mathcal{L}((E,\mathcal{S});F)$ is closed in $\mathcal{H}((E,\mathcal{S});F)$ for that topology. This proves the second half of the claim. An instance of this situation is the same $E = c_o(I)$ with the topologies $\mathcal{S}$ and $\mathcal{T}$ of Example 26. Then, all the above five conditions can be checked. Let us verify 5, as the other four conditions are clear by now. We take $F = c_o(I)$ with the full supremum norm. For every denumerable $J \subset I$, let $f_J : E \to F$ be defined by $f_J(x) = y$, where $y_i = x_i$ if $i \in J$ and $y_i = 0$ if $i \in I - J$, for every $x \in E$. Then $f_J \in \mathcal{L}((E,\mathcal{S});F)$. Call $\mathfrak{X}$ the collection of the f_J for all such J. Then $\mathfrak{X}(K)$ is compact for every compact $K \subset E$, since K is based on a denumerable subset of I. Yet $\mathfrak{X} \subset \mathcal{L}((E,\mathcal{S});F)$ is not relatively compact for the pointwise topology. In fact, the identity mapping $I: E \to F$ does not belong to $\mathcal{L}((E,\mathcal{S});F)$; but it belongs to the closure of $\mathfrak{X}$ for the pointwise topology in the vector space F^E of all mappings from E to F.

5. HOLOMORPHICALLY MACKEY SPACE

DEFINITION 69. *A given* E *is a "holomorphically Mackey space" if, for every* U *and every* F, *we have that each mapping* $f: U \to F$ *belongs to* $H(U;F)$ *if (and always only if)* f *is weakly holomorphic, that is,* $\psi \circ f \in \mathcal{H}(U)$ *for every* $\psi \in F'$; *in other notation:* $H(U;F) = H(U;\omega F)$.

Remark 72 below motivates the above definition, but

we need the following preliminary material which is known.

LEMMA 70. *For a given* E, *the following conditions are equivalent:*

(1) For every F, *we have that each mapping* $f: E \to F$ *belongs to* $\mathcal{L}(E;F)$ *if (and always only if)* f *is linear, and* f *is weakly continuous, that is,* $\psi \circ f \in E'$ *for every* $\psi \in F'$; *in other notation:* $\mathcal{L}(E;F) = \mathcal{L}(E;\omega F)$.

(2) A locally convex topology $\mathcal{S}$ *on* E *is smaller than the given topology* $\mathcal{T}$ *on* E *if (and always only if)* $\mathcal{S}$ *has fewer continuous linear forms than* $\mathcal{T}$.

(2g) The given topology $\mathcal{T}$ *on* E *is the greatest locally convex topology on* E *among those defining the same dual space* E'.

(2m) The given topology $\mathcal{T}$ *on* E *is maximal among the locally convex topologies on* E *defining the same dual space* E'.

(3) The given topology $\mathcal{T}$ *on* E *is the topology of uniform convergence on the* $\sigma(E',E)$*-compact convex subsets of* E'.

PROOF. We shall prove the following implications

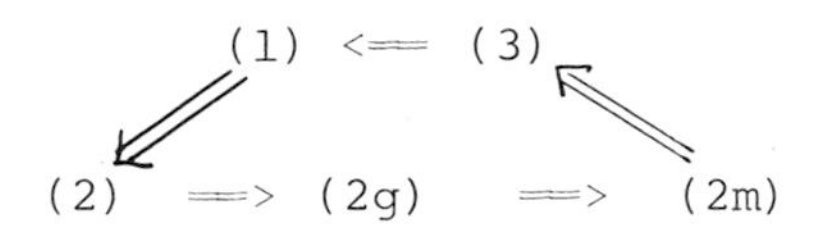

(2) $\Longrightarrow$ (2g) $\Longrightarrow$ (2m). This is clear.

(2m) $\Longrightarrow$ (3). Call $\mathcal{S}$ the topology of uniform convergence on the $\sigma(E',E)$-compact convex subsets of E'; we may restrict attention to such subsets that are also balanced. We claim that $\mathcal{T} \subset \mathcal{S}$. In fact, if V is a $\mathcal{T}$-closed convex balanced neighborhood of 0 for $\mathcal{T}$, then its polar V^{o} in E' is $\sigma(E',E)$-compact, by the Alaoglu-Bourbaki theorem, and also convex. Since V is the polar of V^{o} in E, then V is a neighborhood of 0 for $\mathcal{S}$. This proves our claim. We next claim that a linear form ϕ on E that is continuous for $\mathcal{S}$ is also continuous for $\mathcal{T}$. In fact, there is a $\sigma(E',E)$-compact convex balanced subset $K \subset E'$ such that $|\phi(x)| \leq 1$ if $x \in K^{o}$, where K^{o} is the polar of K in E; thus $\phi \in K^{oo}$, where K^{oo} denotes the polar of K^{o} in the algebraic dual space E^{*} of E. However, K is balanced, convex and $\sigma(E',E)$-compact, hence $\sigma(E',E)$-closed in E^{*}; thus $K^{oo} = K$ and $\phi \in K \subset E'$, showing that ϕ is continuous for $\mathcal{T}$. This proves our claim. Hence $\mathcal{T}$ and $\mathcal{S}$ define the same dual space E'. By (2m), we have $\mathcal{T} = \mathcal{S}$.

(3) $\Longrightarrow$ (1). Let $f: E \to F$ be linear and weakly continuous. We have the transposed linear mapping ${}^{t}f: \psi \in F' \to \psi \circ f \in E'$ that is continuous from $\sigma(F',F)$ to $\sigma(E',E)$. Let W be any closed convex balanced neighborhood of 0 in F. Its polar W^{o} in F' is convex and $\sigma(F',F)$-compact, by the Alaoglu-Bourbaki theorem. Thus $K \equiv {}^{t}f(W^{o})$ is convex and $\sigma(E',E)$-compact. Hence, the polar $V \equiv K^{o}$ of K in E is a neighborhood of 0 in E. Now, $x \in V$ implies $|\psi[f(x)]| \leq 1$ for every $\psi \in W^{o}$, that is $f(x) \in W^{oo} = W$, where W^{oo} is the polar of W^{o} in F. Thus f is continuous.

(1) $\Longrightarrow$ (2). Let $\mathcal{S}$ have fewer continuous linear forms than $\mathcal{T}$. Put $F = (E, \mathcal{S})$. Then the identity mapping $I: E \to F$ is weakly continuous. By (1), it is continuous. Thus $\mathcal{S} \subset \mathcal{T}$.

The proof can also be carried on with the same reasoning, by reversing the arrows. QED

The following definition is classical, particularly in terms of (2g).

DEFINITION 71. *A given E is a "Mackey space" if it satisfies the equivalent conditions of Lemma 66.*

REMARK 72. Definition 69 was formulated in analogy to Definition 71 through (1).

PROPOSITION 73. *A holomorphically Mackey space is also a Mackey space.*

PROOF. It suffices to compare Definition 69 and 71, by using (1) of Lemma 70, and by remarking that $\mathcal{L}(E;F) \subset \mathcal{H}(E;F)$. QED

PROPOSITION 74. *A holomorphically infrabarreled space E is a holomorphically Mackey space.*

PROOF. Let $f: U \to F$ be weakly holomorphic, that is, $\psi \circ f \in \mathcal{H}(U)$ for every $\psi \in F'$. It follows that f is algebraically holomorphic in the H-sense (not necessarily in the $\mathcal{H}$-sense); in other words, we are using here the fact that, if E is finite dimensional, then it is a holomorphically Mackey space, as it is known. We next prove that f is amply bounded. Now, clearly f(K) is weakly bounded, hence bounded, in F for every compact subset K of U. Thus $\mathfrak{X} = \{\psi \circ f;\ \psi \in \mathcal{Y}\}$ is bounded on all compact subsets of U, for every strongly bounded subset $\mathcal{Y}$ of F'. There results that $\mathfrak{X}$ is locally bounded, because E is holomorphically infrabarreled. It follows that, if $\beta \in CS(F)$ and $\mathcal{Y}$ is the set of all linear forms ψ on F satisfying

$|\psi(y)| \leq \beta(y)$ for every $y \in F$, then $\mathfrak{X}$ is locally bounded. By the Hahn-Banach theorem, we have $\beta(y) = \sup\{|\psi(y)|;\ \psi \in \mathcal{Y}\}$ for all $y \in F$. Thus $\beta \circ f$ is locally bounded for every such β. Hence f is amply bounded. It follows that $f \in H(U;F)$. QED

REMARK 75. It is known that an infrabarreled space is a Mackey space. Proposition 73 corresponds to this remark.

PROPOSITION 76. *For* E *to be a holomorphically bornological space it is necessary and sufficient that* E *be a holomorphically Mackey space, and moreover that, for every* U, *we have that each function* $f: U \to \mathbb{C}$ *belongs to* $\mathcal{H}(U)$ *if (and always only if)* f *is algebraically holomorphic, and* f *is bounded on every compact subset of* U.

PROOF. Let us prove necessity. If E is a holomorphically bornological space, then it follows from Propositions 54 and 74 that E is a holomorphically Mackey space. The rest of necessity is clear. Let us prove sufficiency, and assume that $f: U \to F$ is algebraically holomorphic and bounded on every compact subset of U. Then $\psi \circ f$ is algebraically holomorphic and bounded on every compact subset of U, for every $\psi \in F'$. It follows that $\psi \circ f \in \mathcal{H}(U)$ for every such ψ. Hence $f \in H(U;F)$. QED

REMARK 77. It is known that, for E to be a bornological space it is necessary and sufficient that E be a Mackey space, and moreover that each function $f: E \to \mathbb{C}$ belongs to E' if (and always only if) f is linear, and f is bounded on every bounded, or compact, subset of E. Proposition 76 corresponds to the sec ond half of this remark.

6. ACKNOWLEDGEMENTS

The authors gratefully acknowledge partial financial support from FAPESP and FINEP.

BIBLIOGRAPHY

1. J. A. BARROSO, M. C. MATOS & L. NACHBIN, On bounded sets of holomorphic mappings, Proceedings on Infinite Dimensional Holomorphy (Editors: T.L. Hayden & T.J. Suffridge), Lecture Notes in Mathematics 364 (1974), 123-134.

2. S. DINEEN, Holomorphic functions on locally convex spaces, Annales de l'Institut Fourier 23(1973), 19-54, 153-185.

3. S. DINEEN, Holomorphic Functions on Strong Duals of Fréchet-Montel spaces Infinite Dimensional Holomorphy and Applications (Editor: M.C. Matos), North-Holland Mathematics Studies (1977).

4. B. JOSEFSON, A counterexample in the Levi problem, Proceedings on Infinite Dimensional Holomorphy (Editors: T. L. Hayden & T. J. Suffridge), Lecture Notes in Mathematics 364 (1974), 168-177.

5. B: JOSEFSON, Weak sequential convergence in the dual of a Banach space does not imply norm convergence, Arkiv för Mathematik 13 (1975), 79-89.

6. M. C. Matos, On Locally Convex Spaces with the Montel Property, Functional Analysis (Editor: D. de Figueiredo) , Marcel Dekker (1976).

7. L. NACHBIN, Topology on spaces of holomorphic mappings , Springer-Verlag (1969).

8. L. NACHBIN, A glimpse at Infinite Dimensional Holomorphy, Proceedings on Infinite Dimensional Holomorphy (Editors: T.L. Hayden & T.J. Suffridge), Lecture Notes in Mathematics 364(1974), 69-79.

9. L. NACHBIN, Some holomorphically significant properties of locally convex spaces, Functional Analysis (Editor: D. de Figueiredo), Marcel Dekker (1976).

10. A. NISSENZWEIG, W* sequential convergence, Israel Journal of Mathematics 22(1975), 266-272.

Departamento de Matemática Pura
Universidade Federal do Rio de Janeiro
Rio de Janeiro - RJ ZC-32 Brasil

Departamento de Matemática
Universidade Estadual de Campinas
Campinas SP Brasil

Department of Mathematics
University of Rochester
Rochester NY 14627 USA

Infinite Dimensional Holomorphy and Applications, Matos (ed.)

TOPOLOGIES ON SPACES OF HOLOMORPHIC FUNCTIONS OF CERTAIN SURJECTIVE LIMITS

By *PAUL BERNER*

In this paper we study topologies on spaces of holomorphic functions defined in an open surjective limit of locally convex spaces, especially such spaces as $\mathcal{D}'$ (Schwartz's distributions) which are open compact countable surjective limits of Dual Frechet Nuclear spaces. To do so we introduce an inductive limit topology as follows:

If U is a convex balanced subset of an open surjective limit $E = \text{surj lim}_{\alpha\in A}(E,\pi_\alpha)$ (see definition 2.1) then $H(U) = \bigcup_{\alpha\in A} {}^t\pi_\alpha(H(\pi_\alpha(U)))$ where ${}^t\pi_\alpha$ is the map $f \in H(\pi_\alpha(U)) \to f \circ \pi_\alpha \in H(U)$. So we may define an inductive limit topology on H(U) by the formula $(H(U),\tau_I) \equiv \text{ind lim}_{\alpha\in A}((H(\pi_\alpha(U)),\tau_o),{}^t\pi_\alpha)$. If $U \subset E$ is open and connected but not convex or balanced, then we may have that $H(U) \neq \bigcup_{\alpha\in A} {}^t\pi_\alpha(H(\pi_\alpha(U)))$. For this reason we are forced to consider domains spread over the spaces E_α instead of the sets $\pi_\alpha(U)$ (see Theorem 2.1) in order to obtain a good definition of τ_I on H(U) for all open connected sets U.

When E is a non-trivial open compact countable surjective limit of $\mathcal{DFN}$ spaces, we show that τ_I is a strict (LF)-Montel space and coincides with the $\tau_{\omega b}$ and τ_{δ} topologies. We then use this fact to show, for example, that τ_{ω} is quasi-complete, but τ_{ob} is not quasi-complete.

In Section 1, we give some preliminary results concerning domains spread. In Section 2 we define directed surjective limits and the topology τ_I. The τ_o topology on holomorphic functions defined on a domain spread over a $\mathcal{DFM}$ space is studied in Section 3 and the results are applied to give our main theorem concerning the τ_I topology. Section 4 deals with all the various topologies for holomorphic functions on a non-trivial compact countable surjective limit of $\mathcal{DFM}$ spaces and we conclude this final section with a discussion of further results.

We shall use the standard notation of infinite dimensional holomorphy, and l.c.s. will always mean complex Hausdorff locally convex linear space(s).

Some of these results appeared in the author's University of Rochester Ph.D thesis (1974). The author wishes to thank Drs. S. Dineen and R. Aron for their helpful comments and to acknowledge the financial support of a Department of Education (Ireland) Post-Doctoral Fellowship.

1. DOMAINS SPREAD - PRELIMINARY RESULTS

DEFINITION 1.1 *A connected Hausdorff space Ω together with a local homeomorphism of Ω into a* l.c.s. E, $\phi : \Omega \to E$, *is called a domain spread (over* E*), and denoted by* (Ω, ϕ, E) *or just* Ω.

A connected non-empty open subset $W \subset \Omega$ of a domain spread (Ω, ϕ, E) is called a *chart* if $\phi|_W : W \to \phi(W)$ is a homeomorphism.

Let (Ω,ϕ,E) and (Σ,ψ,F) be two domains spread over l.c.s. E and F respectively, and let $\pi : E \to F$ be a continuous open linear (and consequently surjective) map of E onto F. A continuous map $J : \Omega \to \Sigma$ is called a *π-morphism* iff $\psi \circ J = \pi \circ \phi$. If $J : \Omega \to \Sigma$ is a *π-morphism*, ${}^tJ$ will denote the map ${}^tJ : f \in H(\Sigma) \to f \circ J \in H(\Omega)$.

REMARK Since a π-morphism J is "locally the same as" the continuous linear map π, it is easy to see that ${}^tJ$ is well defined. Since π is open, J is also open, so by the uniqueness of analytic continuation it follows that ${}^tJ$ is injective.

If Ω is a domain spread over an l.c.s., then τ_0, τ_ω, and τ_δ will denote, respectively, on $H(\Omega)$, the compact-open topology, the topology generated by semi-norms ported by compact sets, and the topology generated by semi-norms ported by countable covers (see [6], [9]), and τ_{0b} and $\tau_{\omega b}$ will denote the bornological topology associated to τ_0 and τ_ω respectively.

For the remainder of this section, E and F will denote fixed l.c.s., $\pi : E \to F$ will be a continuous open linear map of E onto F, and (Ω,ϕ,E) will be a fixed domain spread over E.

LEMMA 1.1 *If* $J : \Omega \to (\Sigma,\psi,F)$ *is a π-morphism, then the map* ${}^tJ : H(\Sigma) \to H(\Omega)$ *is continuous for the* τ_0 *(respectively the* τ_δ*) topologies on* $H(\Omega)$ *and* $H(\Sigma)$.

PROOF If $K \subset \Omega$ is compact, then $J(K) \subset \Sigma$ is compact and $\|{}^tJ(f)\|_K = \|f \circ J\|_K = \|f\|_{J(K)}$. It is clear from this that ${}^tJ$ is τ_0-continuous.

Now suppose p is a τ_δ-continuous semi-norm on $H(\Omega)$, it suffices to show that $p \circ {}^tJ$ is τ_δ-continuous on $H(\Sigma)$. Let $\{V_n\}_n$

be any increasing countable open cover of Σ, then $\{J^{-1}(V_n)\}_n$ is an increasing countable cover of Ω so there exists a $C > 0$ and and $N \in \mathbb{N}$ such that $p(h) \leq C\|h\|_{J^{-1}(V_N)}$ for all $h \in H(\Omega)$. But this implies that $p \circ {}^tJ(f) \leq C\|f \circ J\|_{J^{-1}(V_n)} = C\|f\|_{V_N}$ for all $f \in H(\Sigma)$ so $p \circ {}^tJ$ is τ_δ-continuous on $H(\Sigma)$.

REMARK It is easy to show that ${}^tJ$ is also continuous for the τ_{ob}, τ_ω, $\tau_{\omega b}$ topologies.

DEFINITION 1.2 $L(\pi,\Omega) = \{f \in H(\Omega) \mid \exists$ *a chart* $W \subset \Omega$ *and* $g \in H(\pi \circ \phi(W))$ $\ni . f_{|W} = g \circ \pi \circ \phi_{|W}\}$ *will denote the set of holomorphic functions on* Ω *which factor locally through* π.

A π-morphism, $J : \Omega \to (\Sigma,\psi,F)$, is called a *$\pi$-domain of factorization (for Ω)* iff ${}^tJ(H(\Sigma)) \supset L(\pi,\Omega)$.

A π-domain of factorization, $J : \Omega \to \Sigma$ is called *the minimal π-domain of factorization (for Ω)* iff J is surjective and satisfies the following universal property:

If $K : \Omega \to (\Gamma,\eta,F)$ is any other π-domain of factorization such that K is surjective, then there exists a unique Id_F-morphism, $\tilde{K} : \Gamma \to \Sigma$ such that $\tilde{K} \circ K = J$.

LEMMA 1.2 *Let* $x \in \Omega$, *let* W *be a chart in* Ω *containing* x, *and let* $f \in H(\Omega)$. *If* $D_a f_{|W} \equiv 0$ *for each* $a \in \pi^{-1}(0)$, *where* $D_a f(x) \equiv \hat{d}^1 f(X)(a)$, *then* $f \in L(\pi,\Omega)$

PROOF By shrinking, we may assume that $\phi(W)$ is convex. Since $D_a f_{|W} \equiv 0$ for all $a \in \pi^{-1}(0)$, we have that $f \circ (\phi_{|W})^{-1}$ is locally constant on each set of the form $(\phi(y) + \pi^{-1}(0)) \cap \phi(W)$, where $y \in W$. But, by the convexity of $\phi(W)$, each such set is connected, so the function

$$g : z \in \pi \circ \phi(W) \to f \circ (\phi_{|W})^{-1} \circ (\pi_{|\phi(W)})^{-1}(z) \in \mathbb{C}$$

is well defined and evidently G-holomorphic. π is an open mapping, so g is also continuous, hence $g \in H(\pi \circ \phi(W))$. Now $f_{|W} = g \circ \pi \circ \phi_{|W}$ so $f \in L(\pi,\Omega)$.

PROPOSITION 1.1 *If* $J : \Omega \to (\Sigma,\psi,F)$ *is a* π*-domain of factorization for* Ω*, then* ${}^tJ(H(\Sigma))$ *is a closed subspace of* $(H(\Omega),\tau_o)$.

PROOF Suppose $\{{}^tJ(f_\lambda)\}_\lambda$ is a net in ${}^tJ(H(\Sigma))$ which converges to $h \in H(\Omega)$ in the τ_o topology. We may assume that there is a chart $W \subset \Omega$ such that $\phi(W)$ is a balanced neighbourhood of o. Let $W_{1/2} = (\phi_{|W})^{-1}(1/2\,(\phi(W)))$. For each $x \in W_{1/2}$, $o < \rho < 1/2$, and $a \in \phi(W)$, we have $D_a h_\lambda(x) \equiv \hat{d}^1 h_\lambda(x)(a) =$

$$= \int_{|\zeta|=\rho} \zeta^{-2} h_\lambda \circ (\phi_{|W})^{-1}(\phi(x) + \zeta a)\ (2\pi i)^{-1} d\zeta, \text{ where } h_\lambda = {}^tJ(f_\lambda) - h.$$

Since $h_\lambda \to o$ uniformly on the compact set $\{(\phi_{|W})^{-1}(\phi(x)+\zeta a)\ |\ |\zeta|=\rho\}$, and $\phi(W)$ is absorbing, and $\hat{d}h_\lambda(x)(a)$ is linear in a, we conclude that $D_a{}^tJ(f_\lambda)(x) \underset{\lambda}{\to} D_a h(x)$ for all $x \in W_{1/2}$ and all $a \in E$. Since $\psi \circ J \circ (\phi_{|W})^{-1}(\phi(x) + \zeta a) = \pi \circ \phi(x)$ for all $x \in W_{1/2}$ and $a \in \pi^{-1}(o) \cap \phi(W)$, it follows that $D_a{}^tJ(f_\lambda)(x) = o = D_a h(x)$ for all $x \in W_{1/2}$ and $a \in \pi^{-1}(o)$. Hence, by Lemma 1.2, $h \in L(\pi,\Omega) \subset {}^tJ(H(\Sigma))$, so ${}^tJ(H(\Sigma))$ is τ_o-closed in $H(\Omega)$.

DEFINITION 1.3 *Let* X *and* Y *be two topological spaces. A map* $\rho : X \to Y$ *will be called compactly proper iff for each* $K \subset Y$ *compact, there exists an* $L \subset X$ *compact, satisfying* $\rho(L) = K$.

PROPOSITION 1.2 *If* $J : \Omega \to (\Sigma,\psi,F)$ *is a surjective* π*-morphism and* π *is compactly proper, then* J *is also compactly proper.*

PROOF Let $\mathcal{N} = \{W \subset \Omega |$ W is a chart and both $\phi|_{\overline{W}}$ and $\psi|_{\overline{J(W)}}$ are injective$\}$ and let $\mathcal{C} = \{(W,V) \in \mathcal{N} \times \mathcal{N} |\ W \subset V$ and $\phi(W) + U \subset \phi(V)$ for some o-neighbourhood U$\}$. Since J is open

and surjective, $\{J(W)\}_{(W,V)\in\mathcal{E}}$ is an open cover of Σ. Now if $K\subset\Sigma$ is compact, then there exists $(W_1,V_1),\ldots,(W_n,V_n)\in\mathcal{E}$ such that $K\subset\bigcup_{i=1}^n J(W_i)$. For each $i = 1,\ldots,n$, let $L_i=\psi(K\cap\overline{J(W_i)})$. By hypothesis, there exists a compact set $\tilde{L}_i\subset E$ such that $\pi(\tilde{L}_i) = L_i$. Since
$\psi\circ\overline{J(W_i)} = \overline{\pi\circ\phi(W_i)}$, $\tilde{L}_i\subset\pi^{-1}(L_i)\subset\pi^{-1}\circ\overline{\pi\circ\phi(W_i)}\subset\phi(V_i)+\pi^{-1}(o)$.
$\tilde{L}_i$ is compact so there exist $a_1,\ldots,a_m\in\pi^{-1}(o)$ such that $\tilde{L}_i\subset\bigcup_{j=1}^m\phi(V_i)+a_j$. Let $\tilde{K}_i = (\phi|_{\overline{V}_i})^{-1}(\bigcup_{j=1}^m(\tilde{L}_i - a_j)\cap\phi(\overline{V}_i))$, and let $\tilde{K} = \bigcup_{i=1}^n\tilde{K}_i$. Then $\tilde{K}\subset\Omega$ is compact and $J(\tilde{K}) = \bigcup_{i=1}^n J(\tilde{K}_i) = \bigcup_{i=1}^n(\phi|_{J(V_i)})^{-1}(L_i) = K$, hence J is compactly proper.

COROLLARY 1.1 *If* $J:\Omega\to(\Sigma,\psi,F)$ *is a* π*-morphism and* π *is compactly proper, then the relative topology on* $H(\Sigma)$ *induced by the inclusion* ${}^tJ: H(\Sigma)\to(H(\Omega),\tau_o)$ *coincides with the compact open topology* $(H(\Sigma),\tau_o)$.

PROOF In view of Lemma 1.1, it suffices to show that each τ_o-continuous semi-norm, p, on $H(\Sigma)$ can be extended to a τ_o-continuous semi-norm on $H(\Omega)$. Since p is continuous, there are $C>0$ and $K\subset\Sigma$ compact such that $p(f)\leq\|f\|_K$ all $f\in H(\Sigma)$. By Proposition 1.2, there is a $\tilde{K}\subset\Omega$ compact such that $J(\tilde{K}) = K$. The semi-norm $h\in H(\Omega)\to\|h\|_{\tilde{K}}$ is τ_o-continuous on $H(\Omega)$ $p(f)\leq C\|f\|_{J(\tilde{K})} = \|f\circ J\|_{\tilde{K}} = \|{}^tJ(f)\|_{\tilde{K}}$, hence p can be continuously extended.

2. SURJECTIVE LIMITS AND THE τ_I-TOPOLOGY

DEFINITION 2.1 A l.c.s. E *is called a directed surjective limit of* l.c.s. $\{E_\alpha\}_{\alpha\in A}$ *if there is a directed preorder* $\geq$ *on* A *and for all* $\alpha,\beta\in A$ *such that* $\beta\geq\alpha$ *there are continuous sur-*

jective maps $\pi_\alpha : E \to E_\alpha$ *and* $\pi_{\alpha\beta} : E_\beta \to E_\alpha$ *satisfying* $\pi_\alpha = \pi_{\alpha\beta} \circ \pi_\beta$, *and* E *has the projective limit topology determined by the maps* $\{\pi_\alpha\}_{\alpha\varepsilon A}$. *We denote this situation by writing* $E = \text{surj lim}_{\alpha\varepsilon A}(E_\alpha, \pi_\alpha, \pi_{\alpha\beta})_{\beta\geq\alpha}$ (*or* $E = \text{surj lim}_{\alpha\varepsilon A} E_\alpha$ *when the* π_α's *and the* $\pi_{\alpha\beta}$'s *are understood*). *Furthermore we say a directed surjective limit* E *is:*

(a) *open if* π_α *is an open map, all* $\alpha \varepsilon A$ (*equivalently: if* $\pi_{\alpha\beta}$ *is an open map all* $\alpha, \beta \varepsilon A$, $\beta \geq \alpha$).

(b) *compact if* π_α *is compactly proper, all* $\alpha \varepsilon A$.

(c) *countable if* $(A, \geq) = \mathbb{N}$ *with its usual ordering.*

(d) *non-trivial if for each* $\alpha \varepsilon A$, π_α *is not a homeomorphism.*

The strong dual of a Frechet Montel space will be called a $\mathcal{DFM}$ space, and if it is also Nuclear, a $\mathcal{DFN}$ space.

REMARK Surjective limits are extensively studied in [7], where the following result is proved:

PROPOSITION 2.1 *If* F *is a strict inductive limit of a sequence of Frechet Montel spaces* $\{F_n\}_n$, *then the strong dual of* F, F'_β *is an open compact countable surjective limit of the* $\mathcal{DFM}$ *spaces* $(F_n)'_\beta$.

EXAMPLE 2.1 Let $U \subset \mathbb{R}^m$ be open and let $\{V_n\}$ be a fundamental sequence of relatively compact open subsets of U satisfying $\overline{V}_n \subset V_{n+1}$, $n \varepsilon \mathbb{N}$, then the space of distributions $\mathcal{D}'(U)$ is a non-trivial open compact countable surjective limit of the $\mathcal{DFN}$ spaces $\{\mathcal{E}'(V_n)\}_n$.

EXAMPLE 2.2 $\Sigma_{j=o}^{\infty} \mathbb{C} \times \Pi_{i=o}^{\infty} \mathbb{C}$ is a non-trivial open compact countable surjective limit of the $\mathcal{DFN}$ spaces. $\{\Sigma_{j=o}^{\infty} \mathbb{C} \times \Pi_{i=o}^{n} \mathbb{C}\}_n$.

NOTE Every directed surjective limit of $\mathcal{DFM}$ spaces is necessarily open by the open mapping theorem.

For the remainder of this section, $E = \text{surj lim}_{\alpha \in A} (E_\alpha, \pi_\alpha, \pi_{\alpha\beta})_{\beta \geq \alpha}$ will be a fixed open directed surjective limit.

DEFINITION 2.2 *If* (Ω, ϕ, E) *is a domain spread over* E, *let* $A_\Omega = \{\alpha \in A \mid \exists$ *a chart* $W \subset \Omega$ *such that* $\phi(W) = \phi(W) + \pi_\alpha^{-1}(o)\}$.

REMARK By definition of the topology of a directed surjective limit, every neighbourhood in E contains a neighbourhood V satisfying $V = V + \pi_\alpha^{-1}(o)$ for some $\alpha \in A$. It is obvious that : $\alpha \in A_\Omega$, $\beta \in A$, and $\beta \geq \alpha \Longrightarrow \beta \in A_\Omega$. Hence $(A_\Omega, \geq)$ is cofinal in $(A, \geq)$.

The following result is proved in [4]:

THEOREM 2.1 *Let* (Ω, ϕ, E) *be a domain spread over an open directed surjective limit* $E = \text{surj lim}_{\alpha \in A} (E_\alpha, \pi_\alpha, \pi_{\alpha\beta})_{\beta \geq \alpha}$ *Then:*

(1) *For each* $\alpha \in A_\Omega$, *the minimal* π_α*-domain of factorization for* Ω, $J_\alpha : \Omega \to (\Omega_\alpha, \phi_\alpha, E_\alpha)$, *exists.*

(2) *For each* $\alpha, \beta \in A_\Omega$ *such that* $\beta \geq \alpha$, *there exists a (necessarily unique)* $\pi_{\alpha\beta}$*-morphism,* $J_{\alpha\beta} : \Omega_\beta \to \Omega_\alpha$, *such that* $J_\alpha = J_{\alpha\beta} \circ J_\beta$; *furthermore,* $J_{\alpha\beta} : \Omega_\beta \to \Omega_\alpha$ *is the minimal* $\pi_{\alpha\beta}$*-domain of factorization for* Ω_β.

(3) Ω *has the projective limit topology determined by the maps* $\{J_\alpha : \Omega \to \Omega_\alpha\}_{\alpha \in A_\Omega}$. *And*

(4) $H(\Omega) = \bigcup_{\alpha \in A_\Omega} {}^t J_\alpha (H(\Omega_\alpha))$.

With the notation of Theorem 2.1 we make the following definition:

DEFINITION 2.3 *We define the topology* τ_I *on* $H(\Omega)$ *to be the (locally convex) inductive limit topology on* $H(\Omega)$ *determined by the maps* $\{{}^tJ_\alpha : (H(\Omega_\alpha),\tau_o) \to H(\Omega)\}_{\alpha\in A_\Omega}$. *Clearly:* $(H(\Omega),\tau_I) \equiv \text{ind}\lim_{\alpha\in A_\Omega}((H(\Omega_\alpha),\tau_o),{}^\alpha J,{}^{\beta\alpha}J)_{\beta\geq\alpha}$ *in the category of* l.c.s., *where* ${}^\alpha J \equiv {}^tJ_\alpha$ *and* ${}^{\beta\alpha}J \equiv {}^tJ_{\alpha\beta}$.

PROPOSITION 2.2 *Let* Ω *be a domain spread over* E. *Then on* $H(\Omega) : \tau_o \leq \tau_I$.

In particular τ_I is a Hausdorff topology.

PROOF BY Lemma 1.1, ${}^\alpha J : (H(\Omega_\alpha),\tau_o) \to (H(\Omega),\tau_o)$ is continuous for each $\alpha \in A_\Omega$, hence the result.

THEOREM 2.2 *If* E *is also a compact directed surjective limit and* Ω *is a domain spread over* E, *then* $(H(\Omega),\tau_I)$ *is a strict inductive limit of closed subspaces* $\{{}^\alpha J(H(\Omega_\alpha))\}_{\alpha\in A_\Omega}$

PROOF Suppose $\alpha,\beta \in A_\Omega$ and $\beta \geq \alpha$. If $K \subset E_\alpha$ is compact, then since π_α is compactly proper, there exists a $\tilde{K} \subset E$, compact, such that $\pi_\alpha(\tilde{K}) = K$. $\pi_\beta(\tilde{K}) \subset E_\beta$ is compact and $\pi_{\alpha\beta}(\pi_\beta(\tilde{K})) = K$ so Corollary 1.1 applies to $J_{\alpha\beta} : \Omega_\beta \to \Omega_\alpha$ showing that τ_I is strict. Proposition 1.1 and Theorem 2.1 (part 2) show that ${}^\alpha J(H(\Omega_\alpha))$ is closed in ${}^\beta J(H(\Omega_\beta),\tau_o)$.

DEFINITION 2.4 *Let* $\mathfrak{X}$ *be a subset of* $H(\Sigma)$, *where* Σ *is a domain spread over a* l.c.s. $\mathfrak{X}$ *is said to be equi-bounded (or amply-bounded) iff for each* $x \in \Sigma$ *there exists a neighbourhood* V *of* x *in* Σ *such that* $\sup_{f\in\mathfrak{X}} \|f\|_V < \infty$.

PROPOSITION 2.3 *If* (Ω,ϕ,E) *is a domain spread over* E *and* $\mathfrak{X} \subset H(\Omega)$ *is equi-bounded, then there exists an* $\alpha \in A_\Omega$ *such that* $\mathfrak{X} \subset {}^\alpha J(H(\Omega_\alpha))$ *and is equi-bounded in* $H(\Omega_\alpha)$.

PROOF Suppose $\mathfrak{X} \subset H(\Omega)$ is equi-bounded, then there exists a chart V in Ω such that $\sup_{f\in\mathfrak{X}} \|f\|_V < \infty$. By Theorem 2.1 (3) there exists an $\alpha \in A_\Omega$ and a chart $W \subset \Omega_\alpha$ such that $V \supset J_\alpha^{-1}(W)$. Hence for each $f \in \mathfrak{X}$, $a \in \pi_\alpha^{-1}(o)$, and $x \in J_\alpha^{-1}(W)$, $\lambda \in \mathbf{C} \to f \circ (\phi|_V)^{-1}(\phi(x) + \lambda a)$ is a bounded entire function and therefore constant. It follows that $D_a f|_V \equiv o$, all $a \in \pi_\alpha^{-1}(o)$, so by Lemma 1.2 and Theorem 2.1 (2) $\mathfrak{X} \subset L(\pi_\alpha,\Omega) \subset {}^\alpha J(H(\Omega_\alpha))$.

Let $\mathfrak{X}_\alpha = ({}^\alpha J)^{-1}(\mathfrak{X})$ and let $\mathcal{C} = \{V \subset \Omega | V$ is open and $\mathfrak{X}$ is uniformly bounded on $V\}$. $\mathcal{C}$ covers Ω so $J_\alpha(\mathcal{C}) \equiv \{J_\alpha(V) | V \in \mathcal{C}\}$ is an open cover of Ω_α. Clearly, $\mathfrak{X}_\alpha$ is uniformly bounded on each $U \in J_\alpha(\mathcal{C})$, so $\mathfrak{X}_\alpha$ is equi-bounded in $H(\Omega_\alpha)$.

COROLLARY 2.1 *Every equi-bounded subset of* $H(\Omega)$ *is* τ_I*-bounded.*

PROOF If $\mathfrak{X} \subset H(\Omega)$ is equi-bounded, then for some $a \in A_\Omega$, $\mathfrak{X} \subset {}^\alpha J(H(\Omega_\alpha))$ and is equi-bounded there. Hence (see [6]) it is τ_o-bounded in $H(\Omega_\alpha)$, so by definition of τ_I, $\mathfrak{X}$ is τ_I-bounded.

3. DOMAINS SPREAD OVER A $\mathfrak{DFM}$ SPACE

PROPOSITION 3.1 *Let* F *be a* l.c.s., *then the following two statements are equivalent:*

(*a*) F *is countable at infinity and hereditarily Lindelöf.*

(*b*) *Each open subset of* F *is countable at infinity.*

If F *is also separable, then* (*a*) *and* (*b*) *are equivalent to:*

(*c*) *Every domain spread over* F *is separable and countable at infinity.*

PROOF (b) $\Longrightarrow$ (a) is trivial. Suppose (a) is satisfied and $U \subset F$ is open. Since U is regular and Lindelöf, there is a countable open cover of U, $\mathcal{C} = \{W_i\}_i$ such that $\overline{W}_i \subset U$, all $W_i \in \mathcal{C}$.

Let $\{K_n\}_n$ be a fundamental sequence of compact sets of F, then the set of all finite unions of sets of the form $K_n \cap \overline{W}_i$, $n,i \in \mathbb{N}$, is a countable fundamental sequence of compact sets of U, so (b) is satisfied.

(c) $\Longrightarrow$ (b) is obvious. Suppose F is separable, (b) is satisfied and (Ω,ϕ,F) is a domain spread over F. Let $x_0 \in \Omega$ be fixed, and let $\mathcal{C}$ be the set of all charts $W \subset \Omega$ such that $\phi(W)$ is convex and $\overline{W}$ is contained in a chart. Now define inductively: $X_1 = \{W \in \mathcal{C} \mid x_0 \in W\}$, $X_{n+1} = \{W \in \mathcal{C} \mid W \cap X_n \neq \emptyset\}$. Since $\mathcal{C}$ is an open cover and Ω is pathwise connected, it follows that $\Omega = \bigcup_{n \in \mathbb{N}} X_n$.

$\phi|_{X_1}$ is injective, since if $x,y \in X_1$, then there exists W_1, $W_2 \in \mathcal{C}$ such that $x, x_0 \in W_1$ and $y,x_0 \in W_2$ thus $W_1 \cap W_2 \neq \emptyset$ and $\phi(W_1) \cap \phi(W_2)$ is connected (since it is convex) so $\phi|_{W_1 \cup W_2}$ is injective (see e.g.: [10] lemma 1.6) and so $\phi(x) = \phi(y) \Longrightarrow x = y$. Therefore X_1 is homeomorphic to a open subset of F so it is separable and Lindelöf. Assume inductively that X_n is separable. Let $\{e_i\}_i$ be a dense sequence in X_n and let $Y_i = \{W \in \mathcal{C} \mid e_i \in W\}$, $i \in \mathbb{N}$. Clearly $X_{n+1} = \bigcup_{i \in \mathbb{N}} Y_i$ and, arguing as for X_1, each Y_i is separable and Lindelöf. Hence X_{n+1} is separable and Lindelöf.

Therefore Ω is separable and $\mathcal{C}$ has a countable subcover $\{W_i\}_i$. Each compact subset of Ω is contained in a finite union of compact sets of the form $(\phi|_{\overline{W}_i})^{-1}\,\overline{(\phi(W_i) \cap K_n)}$, $n,i \in \mathbb{N}$ where $\{K_n\}$ is a fundamental sequence of compact sets for F. Hence Ω is also countable at infinity.

COROLLARY 3.1 *If* F *is a* $\mathcal{DFM}$ *(resp: a* $\mathcal{DFN}$*)-space, and* (Ω,ϕ,F) *is a domain spread over* F, *then* $(H(\Omega),\tau_0)$ *is a Fréchet*

Montel (resp: Nuclear) space, τ_o-bounded sets are equi-bounded, and $\tau_o = \tau_\delta$.

PROOF Since F satisfies (a) of proposition 3.1, $(H(\Omega),\tau_o)$ is metrizable and since F is a k-space, $(H(\Omega),\tau_o)$ is complete. Let $\mathcal{C}$ be the set of all charts in Ω. It is easily verified that $(H(\Omega),\tau_o)$ has the projective limit topology induced by the restriction mappings $\{\rho_W : f \in H(\Omega) \to f|_W \in (H(W), \tau_o)\}_{W \in \mathcal{C}}$. Now each $W \in \mathcal{C}$ is homeomorphic to an open subset of F so each $(H(W),\tau_o)$ is a Montel(resp. Nuclear) space (see [8], resp. [5]) and a projective limit of Montel (resp. Nuclear) spaces is semi-Montel (resp. Nuclear). A semi-Montel Fréchet space is Montel. Since equi-boundedness is a local property, we may complete the proof as in [8] proposition 6 (see also [1]).

DEFINITION 3.1 *A sequence $\{y_n\}_n$ in a l.c.s. E is called very strongly convergent to o if for all continuous semi-norms α on E, $\alpha(y_n) = o$ for n sufficiently large.*

LEMMA 3.1 *Let (Σ,ψ,E) be a domain spread over a l.c.s E, ξ a point in Σ, $\{n_i\}_i$ a sequence in $\mathbb{N}$, $\{x_i\}_i$ a sequence in E and $\{y_i\}_i$ a very strongly convergent sequence in E. Then:*

$$p : f \in H(\Sigma) \to \sup_{i \in \mathbb{N}} \left| \frac{\hat{d}^{n_i}}{n_i!} (D_{y_i} f \quad (\xi)\ (x_i) \right|$$

is a τ_ω-continuous semi-norm ported by $\{\xi\}$.

PROOF Let $V \subset \Sigma$ be any neighbourhood of ξ. By shrinking V, we may identify it with an open subset of E. Let α be a continuous semi-norm on E whose unit ball centered at ξ, $B^\alpha(\xi, 1)$, is contained in V. Cauchy's inequalities imply that for each $i \in \mathbb{N}$,

$$\left|\frac{\hat{d}^{n_i}}{n_i!}(D_{y_i} f)(\xi)(x_i)\right| \leq 2^{n_i}(\alpha(x_i))^{n_i} \| D_{y_i} f \|_{B^\alpha(\xi,1/2)}$$

and $\| D_{y_i} f \|_{B^\alpha(\xi,1/2)} \leq 2 \| f \|_{B^\alpha(\xi,1)} \alpha(y_i)$. Since for some $N \in \mathbb{N}$, $i \geq N \Longrightarrow \alpha(y_i) = 0$, we have:

$p(f) \leq C \| f \|_{B^\alpha(\xi,1)} \leq C \| f \|_V$, where

$C = \max_{0 \leq i \leq N} \{2^{n_i+1}(\alpha(y_i))(\alpha(x_i))^{n_i}\}$, so p is $\{\xi\}$-ported.

THEOREM 3.1 *Let* $E = \text{surj} \lim_{n \in \mathbb{N}} (E_n, \pi_n, \pi_{mn})_{m \geq n}$ *be a compact countable surjective limit of* $\mathcal{DFM}$ *(resp.* $\mathcal{DFN}$*) spaces, and let* (Ω, ϕ, E) *be a domain spread over* E*. Then on* $H(\Omega)$ $\tau_{\omega b} = \tau_\delta = \tau_I$ *is a strict (LF)-Montel (resp. Nuclear) space and the* τ_I*-bounded sets are precisely the equi-bounded sets.*

PROOF An inductive limit of a sequence of Montel (resp. Nuclear) spaces is Montel (resp. Nuclear), hence it immediately follows from theorem 2.2 and corollary 3.1 that τ_I is a strict (LF)-Montel (resp. Nuclear) space. Since by corollary 3.1, $\tau_0 = \tau_\delta$ on the defining subspaces $\{H(\Omega_n)\}_{n \in \mathbb{N}_\Omega}$, and by lemma 1.1, the maps ${}^nJ : (H(\Omega_n), \tau_\delta) \to (H(\Omega), \tau_\delta)$, $n \in \mathbb{N}_\Omega$, are continuous, it follows that $\tau_\delta \leq \tau_I$ on $H(\Omega)$. It is also known (see [6]) that $\tau_{\omega b} \leq \tau_\delta$, and since every (LF)-space is ultrabornological and as (corollary 2.1) every equi-bounded subset of $H(\Omega)$ is τ_I-bounded, to complete the proof it suffices to show that every τ_ω-bounded subset of $H(\Omega)$ is equi-bounded.

Suppose $\mathcal{X} \subset H(\Omega)$ is τ_ω-bounded, then we claim that $\mathcal{X} \subset {}^nJ(H(\Omega_n))$ for some $n \in \mathbb{N}_\Omega$. If not, then we can find a sequence $\{f_i\}_i$ in $\mathcal{X}$ and an increasing sequence of positive integers $\{m_i\}_i$ such that

$f_i \in {}^{m_{i+1}}J(H(\Omega_{m_{i+1}})) \setminus {}^{m_i}J(H(\Omega_{m_i}))$, $i = 1, 2, 3, \ldots$.

Let $\xi \in \Omega$ be fixed. Since ${}^{m_i}J(H(\Omega_{m_i})) \supset L(\pi_{m_i}, \Omega)$, and $f_i \notin {}^{m_i}J(H(\Omega_{m_i}))$, it follows from lemma 1.2 that for some $a_i \in \pi_{m_i}^{-1}(o)$, $D_{a_i} f_i \not\equiv o$ in any neighbourhood of ξ. By taking a Taylor series expansion at ξ (in some chart about ξ), this implies that $\hat{d}^{n_i}(D_{a_i} f_i)(\xi)(x_i) \neq o$ for some $n_i \in \mathbb{N}$ and $x_i \in E$. $\hat{d}^n(D_a f)(\xi)(x)$ is linear in the variable a and n-linear in x, so we may assume that $|\hat{d}^{n_i}(D_{a_i} f_i)(\xi)(x_i)| > i$. We now have that: $\sup_{f \in \mathfrak{X}} (\sup_j |\hat{d}^{n_j}(D_{a_j} f)(\xi)(x_j)|) > i$ for all $i \geq \mathbb{N}$, but this is impossible if $\mathfrak{X}$ is τ_ω-bounded because of Lemma 3.1.
Therefore $\mathfrak{X} \subset {}^nJ(H(\Omega_n))$ for some $n \in \mathbb{N}_\Omega$.
$\tau_o \leq \tau_\omega$ so $\mathfrak{X}$ is also τ_o-bounded hence by corollary 1.1, $\mathfrak{X} \subset {}^nJ(H(\Omega_n), \tau_o)$ is bounded. Every bounded set in $(H(\Omega_n), \tau_o)$ is equi-bounded (corollary 3.1), so $({}^nJ)^{-1}(\mathfrak{X})$ is equi-bounded on Ω_n. It is immediate now that $\mathfrak{X}$ is equi-bounded so the proof is complete.

4. OTHER TOPOLOGIES ON $H(\Omega)$

In this section we study the τ_o and τ_ω topologies in relation to the τ_I topology on $H(\Omega)$ under the hypothesis of theorem 3.1. In [3] we showed that if E was a non-trivial countable surjective limit of $\mathcal{DFM}$ spaces then $\tau_{ob} \neq \tau_\omega \neq \tau_{\omega b}$ on $H(E)$ and, when restricted to the subspace of 2-homogeneous continuous polynomials $\mathcal{P}(^2E)$ neither τ_o nor τ_{ob} are barrelled. A small modification of the proof (see [2]) shows that the same is true on $H(\Omega)$ when Ω is a domain spread over such a l.c.s. E. We will use this information in proving:

THEOREM 4.1 *Let* (Ω,ϕ,E) *be a domain spread over a non-trivial compact countable surjective limit of* $\mathcal{DFM}$ *(resp.* $\mathcal{DFN}$*) spaces,* $E = \text{surj}\lim_{n\in\mathbb{N}} E_n$. *Then on* $H(\Omega)$:

(1) $\tau_{\omega b} = \tau_\delta = \tau_I$ *is a strict* (LF)-*Montel (resp. Nuclear) space.*

(2) The bounded τ_ω-*bounded sets are precisely the equi-bounded sets and* τ_o, τ_{ob}, τ_ω *and* $\tau_{\omega b}$ *all coincide on the equi-bounded sets.*

(3) $\tau_{ob} \neq \tau_\omega \neq \tau_{\omega b}$, *and* τ_ω *is semi-montel, hence quasi-complete (and if* Ω *is a balanced subset of* E, τ_ω *is complete).*

(4) τ_{ob} *is not barrelled, hence, not semi-complete.*

(5) τ_o *is not barrelled nor semi-complete.*

(6) There are no ultrabornological topologies weaker than $\tau_{\omega b}$.

PROOF Statement (1) is just theorem 3.1 where we also showed that a subset is τ_ω bounded iff it is equi-bounded. Since each equi-bounded set is contained in a defining subspace ${}^nJ(H(\Omega_n))$, and $\tau_{\omega b} = \tau_I$ is strict, $\tau_{\omega b}|{}^nJH(\Omega_n) = {}^nJ(H(\Omega_n),\tau_o) = \tau_o|{}^nJH(\Omega_n)$ (see Corollary 1.1) so statement (2) is verified. The τ_ω topology on τ_ω-closed and bounded subsets coincides by (2)with $\tau_{\omega b}$ which is compact on such subsets since $\tau_{\omega b}$ is Montel. Hence τ_ω is semi-montel. If Ω is a balanced subset of E, then τ_ω is complete by corollary 2.2 of [6] which uses a Taylor series argument. The remainder of statement (3) follows from the remarks beginning this section.

Let $\xi \in \Omega$ be fixed, then for $a \in E$, the map $f \in H(\Omega) \to \hat{d}^2 f(\xi)(a) \in \mathbb{C}$ is τ_o continuous because for sufficiently small $s > 0$,

$|d^2f(\xi)(a)| \leq 2! \ ^{-2}\|f\|_B$, where B is the appropriate homeomorphic image of the compact set $\{\phi(\xi) + \lambda a| \ |\lambda| = s\}$. Now if τ_o (resp. τ_{ob}) were barrelled then $f \to \|\hat{d}^2f(\xi))\circ\phi\|_K \equiv$ $\sup_{x \in K}|\hat{d}^2f(\xi)(\phi(x))|$ would be continuous for each $K \subset \Omega$ compact. Hence $\{f\circ\phi \ |f \in \mathcal{P}(^2E)\}$ would be a direct subspace of $H(\Omega)$ via the mapping $f \in H(\Omega) \to (\hat{d}^2f(\xi))\circ\phi$, so $\{f\circ\phi|f \in \mathcal{P}(^2E)\}$ would also be barrelled for the τ_o(resp. τ_{ob}) topology. But as previously remarked, this is not the case, so τ_o and τ_{ob} are not barrelled. As a semi-complete bornological space is barrelled, the proof of (4) is complete.

The bornological topology associated to a semi-complete space is semi-complete so the proof of (5) is complete.

Since a continuous map from an (LF)-space onto an ultrabornological space is necessarily open (6) is trivial.

REMARKS Examples 2.1 and 2.2 both satisfy the hypothesis of Theorem 4.1. In particular, we have an example of a space with two distinct non-trivial Nuclear topologies: If Ω is a domain spread over a space of distributions $\mathcal{D}'$, then $(H(\Omega), \tau_{\omega b})$ is a strict (LF)-Nuclear space, and, by the result of Boland and Waelbroeck [5], $(H(\Omega), \tau_o)$ is also Nuclear, but not barrelled or semi-complete.

FURTHER RESULTS If G is a normed l.c.s., we could have considered $H(\Omega,G)$, the space of holomorphic mappings from a domain spread Ω with values in G, in place of $H(\Omega)$. In that case all our results would remain valid (with virtually the same proofs), except for corollary 3.1, and theorems 3.1 and 4.1, where we would have to require the completeness of G and drop the words Montel and Nuclear from the conclusions (unless G were finite dimensional).

In theorem 4.1 we required E to be a countable surjective limit. If we allow the more general case of a compact symmetric i-surjective limit of $\mathcal{DFM}$ spaces (see [7] for definitions), then τ_I is no longer an (LF)-space, but we still have that $\tau_{\omega b} = \tau_\delta = \tau_I$ and statements (2), (3), (4) and (5) are still satisfied (see also [8]).

REFERENCES

1. J. Barroso, M. Matos, and L. Nachbin, *On bounded sets of holomorphic mappings*, Lecture notes in Math., Vol. 364, Springer-Verlag (1974), 123-133.
2. P. Berner, *Holomorphy on surjective limits of locally convex spaces*, Thesis, University of Rochester (1974).
3. P. Berner, *Sur la topologie de Nachbin de certains espace de fonctions holomorphes*, C.R. Acad. Sc. Paris, t. 280 (1975), 431-433.
4. P. Berner, *A global factorization property for holomorphic functions of a domain spread over a surjective limit*, Séminaire P.Lelong,1974/75.Notes in Math. 524, Springer-Verlag(1976).
5. P. Boland and L. Waelbroeck, *On the nuclearity of* H(U), Colloque D'analyse Fonctionell, Bordeaux, Mai 1975.
6. S. Dineen, *Holomorphic functions of locally convex spaces: I. locally convex topologies of* H(U), Ann. Inst. Fourier, Grenoble, t. 23, fasc 3 (1973), 155-185.
7. S. Dineen, *Surjective limits of locally convex spaces and their application to infinite dimensional holomorphy*,

Bull. Soc. math. France T.103 (1975).

8. S. Dineen, *Holomorphic functions on strong duals of Fréchet-Montel spaces*, These proceedings.

9. L. Nachbin, *Sur les espaces vectoriels topologiques d'applications continues*, C.R. Acad. Sci. Paris, t.271 (1970), 596-598.

10. M. Schottenloher, *Riemann domains; Basic results and open problems*, Lecture notes in Math., Vol. 364, Springer--Verlag (1974), 196-212.

School of Mathematics,
Trinity College,
Dublin 2, IRELAND.

Current address: Department of Mathematic,
Le Moyne College
Le Moyne Heights
Syracuse, New York 13214

Infinite Dimensional Holomorphy and Applications, Matos (ed.)

NUCLEARITY AND THE SCHWARTZ PROPERTY IN THE THEORY OF HOLOMORPHIC FUNCTIONS ON METRIZABLE LOCALLY CONVEX SPACES

By *KLAUS-DIETER BIERSTEDT AND REINHOLD MEISE*

PREFACE: In writing this paper, the authors were influenced by the two recent articles [10] and [22]. First, the work of Boland and Waelbroeck [10] on nuclearity of $(H(U),\tau_0)$ for open subsets U of quasi - complete dual - nuclear locally convex spaces E indicated that certain strong topological vector space properties of (E or, rather) E' - far from the Banach space case, however - might lead to interesting strong results for the spaces of holomorphic functions on subsets of E. (This idea was also supported by the theorem contained in [6] that $(H(U),\tau_0)$ is even s - nuclear if E is the strong dual of an s - nuclear (F)-space.) The main results of the present paper confirm the strength of the general principle.

The second source of information for our paper was Mujica's thesis [22] on H(K) and $(H(U),\tau_\omega)$ for compact subsets K and open subsets U of metrizable locally convex spaces E. Al-though we make use of many definitions, ideas, constructions, and results from [22] here, our general impression is that Mujica was chiefly interested in the Banach space case and hence his main conditions are so restrictive as to exclude even Fréchet - Schwartz spaces E. (How this can be remedied is shown in Prop. 4 below).

A combination of the ideas we gathered from a study of the two papers [10] and [22] lead us first to investigations on the space H(K) of holomorphic germs on a compact subset K of a metrizable Schwartz space E. It turned out in this case - by a rather elementary application of the Cauchy inequalities, and of the Arzelà-Ascoli theorem by the way - that H(K) with its usual topology is in fact a Silva space. (This first proof of Theorem 7 (a) [see 8.(a) here] was sketched in a announcement of some results of this paper in C.R. Acad. Sci. Paris Série A, t. 283, 325-327 (1976).) As we then proceeded to show the (DFN) - property of H(K) for nuclear metrizable E (Theorem 7 (b)), a (slightly) different proof of 7 (a) - analogous to the proof of part (b) of Theorem 7 and based on the "extension" lemma 6 that is also needed at some other places in this paper - was developed at the same time. It should be remarked that an essential part of the Boland-Waelbroeck theorem is made use of in the proof of 7 (b), too.

Theorem 7 is our fundamental result, and several consequences arise from it: For instance, the space $(H(U),\tau_\omega)$ of all holomorphic functions on an open subset U of a metrizable Schwartz [resp. nuclear] space E with Nachbin's ported topology τ_ω is a complete Schwartz [resp. s - nuclear] space (Prop. 16.). If H(U) is topologized by $\underset{\leftarrow}{\text{proj}}_{K \subset\subset U} H(K)$ for open $U \subset E$ - this topology is denoted by τ_π; as Mujica had remarked, τ_π and τ_ω coincide for open subsets "with the Runge property" - then the sheaf $\mathcal{H}$ of holomorphic functions on a metrizable Schwartz [resp. nuclear] space E is a locally convex (topological) sheaf of complete Schwartz [resp. s-nuclear] spaces (Theorem 19 (a)).

For some other corollaries, results of the authors from [9] and [7] are used. We obtain e. g. the representations of the ε-products $H(K_1)\,\varepsilon\, H(K_2) = H(K_1 \times K_2)$ resp. $(H(U_1),\tau_\pi)\varepsilon(H(U_2),\tau_\pi) = (H(U_1 \times U_2),\tau_\pi)$ for compact sets $K_j \subset E_j$ resp. open subsets $U_j \subset E_j$ of metrizable Schwartz spaces E_j, $j = 1,2$ (Cor. 22 (c) and Prop. 23.). Moreover, we would also like to mention the converse of Theorem 7 (Prop. 9), some remarks on the approximation property of H(K) and $(H(U),\tau_\omega)$ (Cor. 11, Prop. 12, Prop. 20) and on vector-valued functions (Prop. 21, Prop. 25).

The last part of the article deals with possible nuclearity types of H(U): Among the power series spaces $\Lambda_\infty(\alpha)$ (of infinite type), Proposition 26 gives a rather restrictive (necessary) condition for the sequences α with the property that $(H(U),\tau_o)$ [resp. $(H(U),\tau_\omega)$] is $\Lambda_\infty(\alpha)$-nuclear for all balanced open subsets U of an infinite dimensional Fréchet or Silva [resp. metrizable Schwartz] space E. On the other hand, Prop. 28. proves that $(H(U),\tau_\pi)$ is indeed $\Lambda_\infty(\alpha)$-nuclear for each open $U \subset \mathbb{C}^{\mathbb{N}}$, if α satisfies the condition of 26. (Examples of such sequences α are exhibited in Remark 27 (b)).

ACKNOWLEDGEMENT: We thank Richard Aron and Martin Schottenloher for some helpful correspondence (and several remarks) during the preparation of this paper and Dietmar Vogt for a remark which led to the present form of Prop. 9.

PRELIMINARIES: For our notation from the theory of locally convex (l.c.) topological vector spaces (which is quite standard), we refer to Horvāth [18] , Köthe [19], and Floret - Wloka [15]. The reader may also consult Grothendieck [16], Pietsch [27], and Martineau [20] concerning nuclearity and s-nuclearity. We mention [16], Schwartz [31], and [9] for topological tensor products and the ε-product. Concerning holomorphic functions and mappings on infinite dimensional space, the books of Nachbin [23] and Noverraz [24] and their notation are used. We remind that, for complex l.c. spaces E and F and $U \subset E$ open, a mapping $f : U \to F$ is called holomorphic if and only if it is G-analytic (i.e. Gâteaux-analytic) and continuous. We denote by $H(U,F)$ the space of all holomorphic mappings $f : U \to F$ and define $H(U) := H(U,\mathbb{C})$. Several natural topologies can be introduced on $H(U)$, e. g. the topology τ_0 (=co) of uniform convergence on compact subsets of U or Nachbin's ported topology τ_ω which is defined by all semi-norms p on $H(U)$ that are ported by a compact subset K of U. (p is ported by K if for each V open with $K \subset V \subset U$, there exists $C(V) > 0$ such that $p(f) \leq C(V) \sup_{x \in V} |f(x)|$ for all $f \in H(U)$.) For metrizable spaces E, for instance, it is known (cf. [12]) that τ_0 and τ_ω have the same bounded sets, but τ_ω need not be bornological. If E is a (DFS)-space (i.e. a Silva space in the terminology of [4]), then τ_0 and τ_ω coincide on $H(U)$. For more information on these topologies see Barroso-Matos-Nachbin [4]. For any locally convex space E, $\mathcal{P}(^m E)$ denotes the space of all continuous m-homogeneous complex-valued polynomials on E. If E is a (semi-)normed space, $\mathcal{P}(^m E)$ will be endowed with its natural (complete) norm topology. Other definitions will be given later on.

From now on, let always E be a metrizable (Hausdorff) l.c. space over $\mathbb{C}$. The topology of E is given by an increasing system $(p_n)_{n\in\mathbb{N}}$ of semi-norms. B_δ^n denotes the set $\{x \in E;\ p_n(x) < \delta\}$. Let $E_{(n)}$ be the space E, endowed with the semi-norm p_n only, and let E_n be the quotient space $E/p_n^{-1}(0)$ with the norm $\|\cdot\|_n$ induced by p_n; its completion is denoted by $(\hat{E}_n, \|\cdot\|_n)$. $\pi_n : E \to E_n$, $\hat{\pi}_n : E \to \hat{E}_n$, $\pi_{nm} : E_m \to E_n$ and $\hat{\pi}_{nm} : \hat{E}_m \to \hat{E}_n$ for $m > n$ are the canonical mappings. Obviously $\pi_n(B_\delta^n) = \{y \in E_n;\ \|y\|_n < \delta\} =: \tilde{B}_\delta^n$ holds true. Let $\hat{B}_\delta^n = \{y \in \hat{E}_n;\ \|y\|_n < \delta\}$. For a fixed compact set K in E we define $U_{n,\delta} = K + B_\delta^n$ and $\tilde{U}_{n,\delta} = \pi_n(U_{n,\delta}) = \pi_n(K) + \tilde{B}_\delta^n$.

1. DEFINITIONS: *K will always denote a non-empty compact subset of E, Y a complex Banach space.*

(a) *For $U \subset E$ open, $H^\infty(U,Y)$ is the (sup-normed) Banach space of all bounded holomorphic mappings from U into Y and $H^\infty(U) = H^\infty(U,\mathbb{C})$.*

(b) *Let $(r_n)_{n\in\mathbb{N}}$ be a strictly decreasing null sequence of positive numbers r_n. We define $U_n = U_{n,r_n}$, $\tilde{U}_n = \tilde{U}_{n,r_n}$, $H(U,Y) = \operatorname{ind}_{n\to} H^\infty(U_n,Y)$, and $H(K) = H(K,\mathbb{C})$. (As one can easily see, this definition does not depend on the sequence $(r_n)_{n\in\mathbb{N}}$ [nor on the semi-norms $p_1 < p_2 < \ldots$].)*

Def. 1 is taken from Mujica [22], where it is proved (in Prop. 2.5) that $H(K)$ is always Hausdorff. In the case we are mainly interested in a proof of this will also be contained in Theorem 7 below.

2. REMARKS:

(a) For each $n \in \mathbb{N}$ and $\delta > 0$, the mapping $A_n : H^\infty(\tilde{U}_{n,\delta}) \to H^\infty(U_{n,\delta})$, given by $A_n(f) = f \circ \pi_n$, is a (surjective) isometric isomorphism, see [22], Lemma 2.2. This is of course still true for Banach space valued mappings.

(b) We also have $H(K) = \operatorname{ind}_{n\to} (H(U_n), \tau_\omega)$ by [22], Def. 2.4 and Prop. 2.3.

In [22], Thm. 3.1, Mujica proves regularity of the inductive limit $H(K) = \operatorname{ind}_{n\to} H^\infty(U_n)$. In his Def. 1.5 (b) he defines the term "Cauchy regularity" for inductive limits and shows in Theorem 3.2 that $H(K)$ is even Cauchy regular, if E satisfies the condition (B) of [22], Def. 3.1 (See Remark 5 (a) below.) On the other hand, we introduced the notion of "strongly boundedly retractive inductive limit" in [8], § 1,1. The two notions coincide in many important cases:

3. LEMMMA: *Let* $\{F_\alpha, i_{\alpha\beta}\}$ *be an injective inductive system of Banach spaces* F_α. *Then the inductive limit is strongly boundedly retractive if and only if it is Cauchy regular.*

PROOF: From the definition it is immediate that each strongly boundedly retractive inductive limit of Banach spaces is quasicomplete*) and hence Cauchy regular. The converse can be seen

---

*) Note that two l.c. topologies which coincide on a convex balanced subset even induce the same uniform structure on this set.

as follows: Let B be the closed unit ball of F_α. Then $i_\alpha(B_\alpha)$ is bounded in F, hence there exists $\beta \geq \alpha$ such that B_α is bounded in F_β and such that each F - Cauchy net in B_α is already an F_β - Cauchy net. Therefore the completeness of F_β implies that the topology on B_α induced by F coincides with the one induced by F_β. □

Mujica's condition (B) is satisfied (trivially) for normed spaces. Among distinguished Fréchet spaces E with a continuous norm, however, only normable spaces satisfy (B). Therefore it is important to take into consideration once more that for a set M bounded in some $H^\infty(U_n)$, $\{\hat{d}^m g(\xi);\ g \in M, \xi \in K\}$ is bounded in $\mathcal{P}(^m E_{(n)})$ for $m = 0,1,\ldots$ by the Cauchy inequalities. (Here we follow the terminology of [22].) Hence (after analyzing the proofs of Lemmas 3.1, 3.4, and 3.5) it turns out that Mujica's main results (like Theorems 3.2 and 3.3 and their Corollaries in [22]) remain true under the weaker condition (BM) mentioned in our first proposition.

4. PROPOSITION: *Let* B *be a metrizable l.c. space satisfying the following condition:*

(BM) $\mathcal{P}(^m E) = \operatorname{ind}_{n\to} \mathcal{P}(^m E_n)$

[or $\operatorname{ind}_{n\to} \mathcal{P}(^m E_{(n)})$, *as* $\mathcal{P}(^m E_n) = \mathcal{P}(^m E_{(n)})$ *in a canonical way] is strongly boundedly retractive uniformly in* $m \in \mathbb{N}$ *(obvious notation)**

---

*) We thank J. Mujica and P. Aviles for pointing out that uniformness in m is necessary here. We would also like to note that Prop. 4 was proved independently in Patricio Aviles Thesis de Magister.

Then $H(K) = \underset{n\to}{\text{ind}}\, H^\infty(U_n)$ *is a strongly bounded retractive inductive limit of Banach spaces and hence a complete ultrabornological* (DF) - *space.*

5. REMARKS:

(a) Mujica's condition (B) requires that the inductive limits $\underset{n\to}{\text{ind}}\, \mathcal{P}(^mE_{(n)})$ are strict. (For $m = 1$, for instance, this means that $\underset{n\to}{\text{ind}}\,(E_n)'_b = \underset{n\to}{\text{ind}}\,(\hat{E}_n)'_b$ is strict). Hence (BM) is weaker than (B). And for (FS)-spaces E, (B) is satisfied if and only if all the spaces E_n are finite dimensional, i.e. if and only if $E = \mathbb{C}^{\mathbb{N}}$ or $E = \mathbb{C}^n$ for some $n \in \mathbb{N}$.

(b) Condition (BM) is always satisfied, however, if E is a metrizable Schwartz space: For then the inductive limits $\underset{n\to}{\text{ind}}\, \mathcal{P}(^mE_n)$ are even compact (i. e. Silva), as an argument similar to the one used in the proof of the theorem of Schauder (see Floret — Wloka [15], 19, 2.1) immediately shows. Yet we will prove much more for such spaces E in 7 below.

The ideas used in the proof of the next lemma are already well - known (see e. g. Schottenloher [29]). We shall need this lemma several times in the proofs of our main results, however.

6. LEMMA: *Let* F *be a normed space,* Y *a Banach space and* $K \subset F$ *compact. The completion of* F *will be denoted by* $\hat{F}$, *and we let again* $B_\rho = \{x \in F, \|x\| < \rho\}$ *and* $\hat{B}_\rho = \{x \in \hat{F}; \|x\| < \rho\}$ *for* $\rho > 0$. *If* r, s *satisfy* $0 < r < s$, *we define* $U = K + B_s$ *and* $V = K + \hat{B}_r$.

In this terminology there exists an injective continuous linear ("extension") mapping $F : H^\infty(U,Y) \to H^\infty(V,Y)$ *with* $F(f)|U \cap V = f|U \cap V$ *for each* $f \in H^\infty(U,Y)$.

PROOF: For $f \in H^\infty(U,Y)$, $x \in K$ and $h \in \overline{B_r} \subset F$, we have the convergent Taylor expansion $f(x + h) = \sum_{n=0}^{\infty} l_n^f(x, h)$, where

$l_n^f(x) = l_n^f(x,\cdot)$ is a (uniquely determined) n-homogeneous continuous polynomial on F with values in Y and

$$\| l_n^f(x)\| = \sup_{\|h\|\leq 1} \| l_n^f(x,h)\| \leq \frac{1}{r^n} \|f\|_{H^\infty(U,Y)}$$

by the Cauchy inequalities $(n = 0,1,\ldots)$.

Now extend $l_n^f(x)$ to some continuous n-homogeneous polynomial $\tilde{l}_n^f$ on $\hat{F}$; the estimates above are preserved. Hence the series $\sum_{n=0}^{\infty} \|\tilde{l}_n^f(x,\hat{h})\| \leq \|f\|_{H^\infty(U,Y)} \cdot \sum_{n=0}^{\infty} (\frac{t}{r})^n$ converges for all $\hat{h} \in \hat{B}_t$, $t < r$, and the function $\tilde{f}_x$ defined on $x + \hat{B}_r$ by $\tilde{f}_x(x + \hat{h}) = \sum_{n=0}^{\infty} \tilde{l}_n(x,\hat{h})$ is holomorphic with $\tilde{f}_x|(x+B_r) = f|(x+B_r)$. If $x,y \in K$ satisfy $(x + \hat{B}_r) \cap (y + \hat{B}_r) \neq \emptyset$, it is easy to see (by density of $(x + B_r) \cap (y + B_r)$ and by continuity of $\tilde{f}_x$ and $\tilde{f}_y$) that the restrictions of $\tilde{f}_x$ and $\tilde{f}_y$ to the intersection coincide. As holomorphy is a local property, there exists some $\tilde{f} \in H(V)$ with $\tilde{f}|V \cap U = f|V \cap U$, and

$$\sup_{x\in V} \|\tilde{f}(x)\| \leq \sup_{x\in U} \|f(x)\|$$

holds by density again. Then it is obvious that the mapping $F : f \to \tilde{f}$ is well-defined, continuous, linear, and injective ☐

The next is our main theorem on $H(K)$.

7. THEOREM: *Let* E *be a metrizable l.c. space and* $K \subset E$ *compact.*

(a) *If* E *is a Schwartz space, the spectrum* $\{H^\infty(U_n); \rho_{nm}\}$ *is compact and hence* $H(K)$ *a* (DFS)*-space.*

(b) *If* E *is even nuclear, then the spectrum is nuclear, too, and hence* $H(K)$ *is a* (DFN)*-space, i.e. even s - nuclear.*

PROOF: In the notation introduced in 1.(b) we may assume $r_1 \le 1$. It is enough to show that the system of semi-norms $p_1 \le p_2 \le \dots$ can be chosen in such a way as to get all the canonical mappings $\rho_{n,n+1} : H^\infty(U_n) \to H^\infty(U_{n+1})$ compact [resp. $\rho_{n,n+2}$ absolutely summing] ($n \in \mathbb{N}$). For then $H(K)$ is a (DFS)-[resp. (DFN)-] space (see Floret - Wloka [15]), because no component of U_n has a void intersection with U_{n+1} [resp. U_{n+2}]. We fix $n \in \mathbb{N}$ and $s \in \mathbb{R}$ with $r_{n+1} < s < r_n$ and define $V = \hat{\pi}_n(K) + \widehat{B_s^n} \subset \hat{E}_n$.

(a) In this case we may assume without loss of generality that $\pi_{n,n+1} : E_{n+1} \to E_n$ is precompact and hence $\hat{\pi}_{n,n+1}$ compact. Then

$$Q = \overline{\hat{\pi}_{n,n+1}(\tilde{U}_{n+1})}^{\hat{E}_n}$$

is a compact subset of V because of $p_n \le p_{n+1}$. Using 2.(a) and lemma 6., we notice, that $\rho_{n,n+1}$ can be represented $\rho_{n,n+1} = A_{n+1} \circ B \circ F \circ A_n^{-1}$ as follows:

$$H^\infty(U_n) \xrightarrow{A_n^{-1}} H^\infty(U_n) \xrightarrow{F} H^\infty(V) \xrightarrow{B} H^\infty(\tilde{U}_{n+1}) \xrightarrow{A_{n+1}} H^\infty(U_{n+1}).$$

Here A_n, A_{n+1} are like in 2.(a), F is defined as in 6., and B is given by $B(f) = f \circ \sigma, \sigma = \hat{\pi}_{n,n+1}|\tilde{U}_{n+1}$. B can be written in a natural way as $B = B_o \circ R$, where $R : H^\infty(V) \to C(Q)$ and $B_o : C(Q) \to CB(U_{n+1})$ [CB = continuous and bounded] are defined by $R(f) = f|Q$ and $B_o(f) = f \circ \sigma$. As all mappings are linear and continuous, it is enough to show compactness of R only. This, in turn, follows form a simple application of the Cauchy integral formula and of the theorem of Arzelà-Ascoli: Let dist$(Q, \complement V)$ in $\hat{E}_n$ be 3δ, and fix $x, y \in Q$ with $\|x - y\|_n < \delta$. Then for $f \in H^\infty(V)$ the following estimate holds:

$$|f(x) - f(y)| = \left|\frac{1}{2\pi i}\cdot\int_{|z|=2\delta} f\left(x + z\frac{y-x}{\|y-x\|_n}\right)\cdot\left(\frac{1}{z} - \frac{1}{z-\|y-x\|_n}\right)\cdot dz\right|$$

$$\leq 2\delta \| f \|_{H^\infty(V)} \cdot \| y-x \|_n \cdot \sup_{|z|=2\delta} \frac{1}{|z||z-\|y-x\|_n|} = \frac{1}{\delta} \|f\|_{H^\infty(V)} \|y-x\|_n .$$

Hence R maps the unit ball of $H^\infty(V)$ onto a uniformly bounded and equicontinuous family of continuous functions on the compact set Q, and so R is compact by Arzelà - Ascoli.

(b) The proof of (b) proceeds similarly. We assume without loss of generality that $(p_k)_{k \in \mathbb{N}}$ is chosen in such a way as to make all the E_k Hilbert spaces and to let all the canonical mappings $\hat{\pi}_{k,k+1} : \hat{E}_{k+1} \to \hat{E}_k$ be of class l^p with $p < \frac{1}{6}$ (cf.Pietsch [27], 8.6.1 Thm.). As mappings of class l^p are precompact ([27], Satz 8.2.6), the set $Q_1 = \pi_{n+1,n+2}(\tilde{U}_{n+2})$ E_{n+1} is precompact. Hence for $r \in \mathbb{R}$ with $r_{n+2} < r < r_{n+1}$ there exist $m \in \mathbb{N}$ and $k_j \in \pi_{n+1}(K)$ with $Q_1 \subset \bigcup_{j=1}^{m} (k_j + \tilde{B}_r^{n+1})$, and so we have:

$$\hat{\pi}_{n,n+2}(\tilde{U}_{n+2}) \subset \overline{\hat{\pi}_{n,n+1}(Q_1)}^{\hat{E}_n} \subset \bigcup_{j=1}^{m} \left[\hat{\pi}_{n,n+1}(k_j) + \overline{\hat{\pi}_{n,n+1}(\tilde{B}_r^{n+1})}^{\hat{E}_n}\right]$$

$$\subset \bigcup_{j=1}^{m} \left[\hat{\pi}_{n,n+1}(k_j) + \widehat{B}_s^n\right] \quad V.$$

After these preparations we are going to show now, how a suitable factorization allows us to use a result of Boland and Waelbroeck to prove that $\rho_{n,n+2}$ is absolutely summing. The factorization of $\rho_{n,n+2}$ we need is given as follows:

$$H^\infty(U_n) \xrightarrow{A_n^{-1}} H^\infty(\tilde{U}_n) \xrightarrow{F} H^\infty(V) \xrightarrow{J} (H(V),p_{\tilde{L}}) \xrightarrow{R} (H(V),p_L) \longrightarrow$$

$$\xrightarrow{B} H^\infty(\tilde{U}_{n+2}) \xrightarrow{A_{n+2}} H^\infty(U_{n+2}) .$$

(Here, for a compact subset S of V, we denote by $(H(V),p_S)$ the space $H(V)$, endowed with the semi - norm $p_S(f) = \sup_{x \in S}|f(x)|$

J denotes the continuous inclusion, R the identical mapping, F and B are as in part (a) of the proof with e.g. $B(f) = f \circ \sigma$, $\sigma = \hat{\pi}_{n,n+2} | \tilde{U}_{n+2}$, and L is the compact set

$$\bigcup_{j=1}^{m} \left[\hat{\pi}_{n,n+1}(k_j) + \overline{\hat{\pi}_{n,n+1}(\widetilde{B_r^{n+1}})}^{\hat{E}_n} \right] \supset \hat{\pi}_{n,n+2}(\tilde{U}_{n+2}) .$$

As for $\tilde{L} \supset L$ all mappings in the above factorization are continuous, it will suffice to choose $\tilde{L}$ in such a way that R becomes absolutely summing.

To do this, we remark first that by well-known theorems on the representation of compact operators in Hilbert spaces (cf. [27], 8.3) the condition "$\hat{\pi}_{n,n+1}$ of class l^p" implies that

$$\hat{\pi}_{n,n+1}(\widehat{B_1^{n+1}}) \subset \{x \in \hat{E}_n;\ x = \sum_{j=1}^{\infty} \Theta_j \cdot x_j,\ \sum_{j=1}^{\infty} |\Theta_j| \leq 1\}$$

for an appropriate sequence $(x_j)_{j \in \mathbb{N}}$ in $\hat{E}_n$ with the additional property that $\lim_{j \to \infty} (1 + j)^4 x_j = 0$ in $\hat{E}_n$. (For some orthonormal systems $(e_j)_{j \in \mathbb{N}}$ and $(f_j)_{j \in \mathbb{N}}$ in $\hat{E}_{n+1}$ resp. $\hat{E}_n$ and a sequence $(\lambda_j)_{j \in \mathbb{N}}$ with $\lambda_j \searrow 0$ and $\sum_{j=1}^{\infty} \lambda_j^p < \infty$, we get $\hat{\pi}_{n,n+1}(y) = \sum_{j=1}^{\infty} \lambda_j (y|e_j) f_j$. Now take $C = \sum_{j=1}^{\infty} \frac{1}{(1+j)^2}$, $x_j = C(1 + j)^2 \lambda_j \cdot f_j$ and notice that $\sup_{j \in \mathbb{N}} (1 + j)^{1/p} \cdot \lambda_j < \infty$.) Because of $r < r_{n+1} < 1$, $\lim_{j \to \infty} (1 + j)^4 x_j = 0$, and the above considerations, the proof of the theorem of Boland and Waelbroeck [10] implies that for $Q_2 = \overline{\hat{\pi}_{n,n+1}(\widetilde{B_r^{n+1}})}^{\hat{E}_n}$ there exist a compact set Q_3 with $Q_2 \subset Q_3 \subset \widehat{B_s^n}$ and a positive Radon measure μ on Q_3 such that for any $f \in H(\widehat{B_s^n})$ the inequality

$$\sup_{x \in Q_2} |f(x)| \leq \int_{Q_3} |f| \, d\mu$$

holds true. Translating the set Q_3 and the measure μ by $\hat{\pi}_{n,n+1}(k_j)$, we first obtain μ_j $(j = 1,\ldots,m)$ and finally a positive measure ν on the compact set $\tilde{L} = \bigcup_{j=1}^{m} \left[\hat{\pi}_{n,n+1}(k_j) + Q_3\right] \subset V$ such that for all $f \in H(V)$:

$$p_L(f) = \sup_{x\in L} |f(x)| \leq \sum_{j=1}^{m} \sup\{|f(x)|;\ x \in \overline{\hat{\pi}_{n,n+1}(k_j + \widetilde{B_r^{n+1}})}^{\hat{E}_n}\}$$

$$\leq \sum_{j=1}^{m} \int_{\hat{\pi}_{n,n+1}(k_j)+Q_3} |f|\,d\mu_j = \int_L |f|\,d\nu .$$

This proves that R and hence $\rho_{n,n+2}$ is absolutely summing.□

8. REMARKS:

(a) The application of lemma 6 in the proof of part (a) of the Theorem can be replaced by the observation that a uniformly equicontinuous family of functions on a precompact set P extends to an equicontinuous family of functions on the completion $\hat{P}$.

(b) The proof of the theorem indicates that for a compact set K in E (metrizable), by lemma 6 ,$H_{\hat{E}}(K)$ = $[H(K)$ considered as a space of functions in $\hat{E}]$ is topologically isomorphic (by restriction) to $H_E(K)$.

We turn to the converse of the theorem.

9. PROPOSITION: *Let* E *be a metrizable l.c. space. If for some non-void compact set* K *in* E, H(K) *is a* (a) (DFS)-*resp.* (b) (DFN)-*space, then* E *is a* (a) *Schwartz resp.* (b) *nuclear space.*

PROOF: H(K) and H(K + e) are topologically isomorphic by translation

($e \in E$ arbitrary), hence we may assume $0 \in K$. Then with $V_n = \Gamma_1^r$, we obtain $H^\infty(U_n) \cap E' = E'_{V_n^o}$: $B^n_{r_n} \subset U_n$ implies $H^\infty(U_n) \cap E' \subset E'_{V_n^o}$.

On the other hand, from $K \subset \lambda V_n$ for suitable $\lambda > 0$ we deduce $U_n \subset (\lambda + r_n)V_n$; so for $l \in E'_{V_n^o}$ the inequality

$\sup_{x \in U_n} |l(x)| \leq (\lambda + r_n) \sup_{x \in V_n} |l(x)|$ holds and $l \in H^\infty(U_n) \cap E'$. It is also clear that $H^\infty(U_n)$ induces on its (closed) subspace $E'_{V_n^o}$ the canonical (complete) norm topology of this space.

If $H(K)$ is a (DFS) - space, the inductive system

$$\{H^\infty(U_n), \rho_{nm}\}_{n \in \mathbb{N}}$$

is compact, and hence the system $\{E'_{V_n^o}, \rho_{nm}|E'_{V_n^o}\}_{n \in \mathbb{N}}$ is compact. The remark $\hat{E}'_n = E'_n = E'_{V_n^o}$ together with the theorem of Schauder and $\rho_{nm}|E'_{V_n^o} = {}^t\hat{\pi}_{nm}$ then implies that for every $n \in \mathbb{N}$, there exists an $m \geq n$ such that $\hat{\pi}_{nm}$ is compact and hence π_{nm} precompact. So E is a Schwartz space. If $H(K)$ is a (DFN) - space, a similar argument shows that E is a nuclear space. □

The main idea in the proof of the following proposition is due to Aron and Schottenloher [2], Thm.2.2, proof of (c) $\Longrightarrow$ (d):

10. PROPOSITION: *Let* E *be a l.c. metrizable Schwartz space. Then for any non-void compact set* K *in* E, $H(K)$ *contains a continuously projected topological subspace canonically isomorphic to* $(E', \beta(E',E))$.

PROOF: As in the proof of 9., we assume $0 \in K$. The mapping

$P : H(K) \to H(K)$ is defined by $P(f) = \hat{d}^1 f(0)$ $(= f'(0))$. Then obviously $P^2 = P$ and $P(H(K)) = E'$. For every $n \in \mathbb{N}$, $P|H^\infty(U_n) : H^\infty(U_n) \to H^\infty(U_n)$ is continuous, because, by the Cauchy inequalities, we have $\sup_{\|f\|\leq 1} \sup_{h \in B_1^n} |P(f)(h)| \leq \frac{1}{r^n}$, and by the proof of 9. this implies $\sup_{\|f\|\leq 1} \|P(f)\|_{H^\infty(U_n)} < \infty$. Hence $P : H(K) \to E'$ is continuous, if $E' \subset H(K)$ is given the induced topology. By Theorem 7 (a), the proof of Prop. 9, and Floret [14], 14, we know that $H(K)$ induces the topology of $\operatorname{ind}_{n\to} E'_{V_n^o}$ on E'. If we identify $\hat{E}'$ with E', it follows again from the arguments in the proof of 9 and the general theory that $(\hat{E}', \beta(\hat{E}', \hat{E})) = \operatorname{ind}_{n\to} E'_{V_n^o}$. Therefore the proof is finished, if we show $(\hat{E}', \beta(\hat{E}', \hat{E})) = (E', \beta(E', E))$. But this is a consequence of the following facts: The metrizable Schwartz space E is separable (see e.g. Pfister [26]) and hence by Köthe [19] § 29, 6 (1) every bounded subset in $\hat{E}$ is contained in the completion of a bounded subset of E. ◻

REMARK: The first part of the proof of 10 shows that for any $\emptyset \neq K \subset E$ metrizable, the subspace of $H(K)$ canonically isomorphic to E' is always complemented. (The assumption of "E Schwartz" was only needed to assure that $H(K)$ induces the topology $\beta(E', E)$ on this subspace.)

There is an immediate consequence of 10. for the approximation property (a.p.) of $H(K)$.

11\. COROLLARY: *In a space* E *as in 10, the a.p. of* $H(K)$ *for some non-void compact* $K \subset E$ *implies the a.p. of* $(E', \beta(E', E))$.

REMARKS:

(a) Under the assumptions of 11., $\hat{E} = (E'', \beta(E'', E'))$ and (see the proof of 10). $(\hat{E}', \beta(\hat{E}', \hat{E})) = (E', \beta(E', E).)$ Hence the remark after 1, Satz 6 in [8] shows that the a.p. of $(E', \beta(E', E))$ is then equivalent to the a.p. of $\hat{E}$. (And the a.p. of $\hat{E}$ implies of course the a.p. of E.)

(b) In particular, if E is an (FS)-space without a.p., then by 11. and (a), $H(K)$ does not have the a.p. for any $K \subset E$. (The existence of (FS)-spaces without a.p. was deduced from Enflo's counterexample by Hogbe - Nlend.)

(c) By 7 (b), $H(K)$ has always the a.p., if $E \supset K$ is nuclear. And from 4 and [8], 1, Satz 2 it is obvious that under condition (BM) on $E \supset K$, $H(K)$ has the a.p., if all the spaces $H^{\infty}(U_n)$ (or, by 2 (b), all spaces $(H(U), \tau_{\omega})$, $U \supset K$) can be chosen to have the a.p. - Another positive result will be proved next.

12. PROPOSITION: *Let E be a l.c. metrizable Schwartz space whose topology can be given by an increasing system $(p_n)_{n \in \mathbb{N}}$ of seminorms with the property that all spaces $\hat{E}_n$ have the a.p. Then $H(K)$ has the a.p. for each balanced compact set $K \subset E$.*

PROOF: We use the equivalence for the a.p. mentioned in [8]. In our case it implies that $H(K)$ has the a.p., if for each Banach space Y, $H(K) \otimes Y$ is dense in $H(K) \varepsilon Y$. We have $(H(K) \varepsilon Y)_{bor} = \operatorname{ind}_{n \to} (H^{\infty}(U_n) \varepsilon Y)$ by [9], 3, 13., and

$H^\infty(U_n)\varepsilon Y = H^{\infty,p}(U_n,Y) = \{f \in H^\infty(U_n,Y);\ f(U_n)$ is precompact in $Y\}$ follows from [5], 31.

[Now Remark 2 (a) remains true for Banach space valued bounded holomorphic mappings. Hence the construction used in the proof of 7.(a) is easily seen to show

$$\operatorname{ind}_{n\to} H^{\infty,p}(U_n,Y) = \operatorname{ind}_{n\to} H^{\infty,p}(\tilde{U}_n,Y) = \operatorname{ind}_{n\to} H^\infty(\tilde{U}_n,Y) = \operatorname{ind}_{n\to} H^\infty(U_n,Y),$$

because here $\pi_{n,n+1}(\tilde{U}_{n+1})$ is precompact in $\tilde{U}_n$ and lemma 6. applies. So $(H(K)\varepsilon Y)_{bor} = \operatorname{ind}_{n\to} H^\infty(U_n,Y)$.]

After these preparations, it obviously suffices to show density of $H^\infty(U_{n+1}) \otimes Y$ in $\rho_{n,n+1}(H^\infty(U_n,Y))$ to get the a.p. of $H(K)$. We fix a function $f \in H^\infty(U_n,Y)$ and take V as defined in the proof of 7. $Q = \overline{\hat{\pi}_{n,n+1}(\tilde{U}_{n+1})}^{\hat{E}_n}$ is then a compact subset of the balanced set V. By assumption, $\hat{E}_n$ has the a.p. , so by Aron - Schottenloher [2], Theorem 2.2, for any given $\varepsilon > 0$ one can find

$$g = \sum_{j=1}^{m} g_j \otimes y_j \in H(V) \otimes Y \text{ with } \sup_{x\in Q} \|F(\bar{A}_n^{-1}(f))(x) - g(x)\| < \varepsilon. \quad \text{As}$$

$$\tilde{g} = g \circ \hat{\pi}_{n,n+1} \mid \tilde{U}_{n+1} \in H^\infty(\tilde{U}_{n+1}) \otimes Y \quad \text{and hence}$$

$$A_{n+1}(\tilde{g}) \in H^\infty(U_{n+1}) \otimes Y,\ \sup_{x\in U_{n+1}} \|\rho_{n,n+1}(f)(x) - A_{n+1}(\tilde{g})(x)\| < \varepsilon$$

proves the desired density result. □

REMARK: 12 remains true for every compact set K such that each $\hat{\pi}_n(K) \subset \hat{E}_n$ has a neighbourhood basis of open, finitely Runge sets. For then one can conclude as above by taking V smaller and replacing, if necessary, $n+1$ by a suitable $m > n$. (Theorem 2.2 of [2] is still valid in this case.)

As Chae [11] and Mujica [22], Ch. 5 and 6 have done before

us, we will now apply the information on H(K) contained in the results above to prove some consequences for Nachbin's ported topology τ_ω on H(U). We are mainly interested in open subsets U of a metrizable Schwartz space, however, and the strong properties of H(K) derived in 7 allow better results than in the Banach space case.

It is obvious that an injective inductive system of normed spaces is boundedly retractive ([8], 1,1.) if and only if it is even strongly boundedly retractive. So, by 3 , the first part of the following lemma is nothing but a rewording of Mujica's lemmas 5.1. and 5.2. of [22] in (much) more general terms and can be proved by (almost literally) repeating his arguments.For the second part of the lemma, the nuclear case is established e. g. by combining Pietsch [27], 3.2.5, 3.2.4, 3.2.13, and 3.3.5 (in this order).

13. LEMMA: *Let* $\{X_n; j_{nm}\}$ *be a countable injective inductive system of Banach spaces which is boundedly retractive. For a linear subspace* Y *of* $X = \text{ind}_{n\to} X_n$ *we define* $Y_n = Y \cap X_n$ *with the induced norm topology and topologize* Y *as* $\text{ind}_{n\to} Y_n$. *Then the canonical inductive system* $\{\overline{Y_n}^{X_n} ; j_{nm} | \overline{Y_n}^{X_n}\}$ *is again (strongly) boundedly retractive and the completion* $\hat{Y}$ *of* Y *equals* $\text{ind}_{n\to} \overline{Y_n}^{X_n}$ *(topologically). If* $\{X_n ; j_{nm}\}$ *is compact [resp. nuclear],* $\hat{Y} = \text{ind}_{n\to} \overline{Y_n}^{X_n}$ *is also a* (DFS)- *[resp.* (DFN)-*] space.*

REMARK: Even in the case of (DFS)-spaces $X = \text{ind}_{n\to} X_n$ it is an open problem, whether $\hat{Y} = \text{ind}_{n\to} \overline{Y_n}^{X_n}$ as in 13. must be a topological subspace of X . There are several other equivalent formulations of this question which we do not state explicitly here.

The following definitions needed in the rest of this paper are taken from [22].

14. DEFINITIONS: *Let* E *be a metrizable l.c. space and* $U \subset E$ *open.*

(a) *The system* $\mathcal{K}_U = \{K \subset U;\ K \text{ compact}\}$ *is directed upward by inclusion.*

(b) *For* $K \in \mathcal{K}_U$ *we take* $X = H(K) = \underset{n \to}{\operatorname{ind}} H^{\infty}(U_n)$ *and define* $Y = H^K(U)$ *as the image* $\rho(H(U))$ *of* $H(U)$ *under the canonical restriction mapping* $\rho: H(U) \to H(K)$. Y *is topologized by* $Y = \underset{n \to}{\operatorname{ind}} Y_n$, *where* $Y_n = Y \cap H^{\infty}(U_n)$. *Then* $\underset{n \to}{\operatorname{ind}} \overline{Y}^{H^{\infty}(U_n)}$ *will be denoted by* $\tilde{H}^K(U)$.

(c) $K \in \mathcal{K}_U$ *is called* U - *Runge, if* $\rho(H(U))$ *is sequentially dense in* $H(K)$.

(d) *We say that* U *has the Runge property, if* U - *Runge sets are cofinal in* $\mathcal{K}_U$.

(e) *A subset* M *of* E *is said to be* ξ*-balanced (for* $\xi \in E$*), if* $M = \xi + M_o$ *for a balanced set* M_o .

15. REMARKS:

(a) By 13 and 4 we have $\tilde{H}^K(U) = \widehat{H^K(U)}$, if E satisfies condition (BM). If E is even a Schwartz [resp. nuclear] space, then $\tilde{H}^K(U)$ is a (DFS)-[resp. (DFN)-, hence s-nuclear] space by 7 and 13.

(b) $(H(U), \tau_\omega) = \underset{\leftarrow K \in \mathcal{K}_U}{\operatorname{proj}} H^K(U) = \underset{\leftarrow K \in \mathcal{K}_U}{\operatorname{proj}} \tilde{H}^K(U)$ holds for any open subset U of a metrizable l.c. space E, see Mujica [22], Lemmas 5.5, 5.6 (So $(H(U),\tau_\omega)$ is complete, if E satisfies condition (BM)). From these remarks and well-known permanence properties of Schwartz resp. s-nuclear spaces (cf. Martineau [20]) we obtain immediately:

16. PROPOSITION: *Let* E *be a metrizable l.c. Schwartz space and* U *open in* E. *Then* $(H(U),\tau_\omega)$ *is a complete Schwartz space. It is even* s-*nuclear, if* E *is nuclear.*

For nuclearity and s-nuclearity of $(H(U),\tau_0)$ see Boland-Waelbroeck [10], and [6], 1.12.

17. LEMMA: *If* E *is a metrizable l.c. space and* $U \subset E$ *open, we have algebraically* $H(U) = \operatorname{proj}_{\leftarrow K \in \mathcal{K}_U} H(K)$, *and* τ_ω *is stronger than the topology* τ_π *of* $\operatorname{proj}_{\leftarrow K \in \mathcal{K}_U} H(K)$.

PROOF: This is straightforward: $\{H(K);\ \pi_{KL}\}$ with the canonical (continuous) restrictions $\pi_{KL} : H(L) \to H(K)$ for $L \supset K$ is a projective system. The linear mapping $A : H(U) \to \operatorname{proj}_{\leftarrow K \in \mathcal{K}_U} H(K)$ defined by $A(f) = (f_K)_{K \in \mathcal{K}_U}$, where f_K denotes the germ on K induced by $f \in H(U)$, is well-defined, injective, and τ_ω-continuous by Remark 2.(b). Surjectivity of A is also clear (cf. Remark 15(b)). □

18. REMARKS:

(a) The topology τ_π had been considered in the case of Banach spaces by Hirschowitz [17] and Chae [11].

The problem of whether $\tau_\pi = \tau_\omega$ holds for arbitrary open subsets U of, say, a Schwartz space E is related by 15 (b) to the question mentioned in the remark after 13 (see [22], last lines of Ch. 5).

(b) For open sets U with the Runge property in a metrizable l.c. space satisfying condition (BM), we can find a system $\mathcal{K}$ cofinal in $\mathcal{K}_U$ such that $\tilde{H}^K(U) = H(K)$ for each $K \in \mathcal{K}$ (algebraically, and so even topologically by a general open mapping theorem, cf. [22], Lemma 6.1). Hence by 15.(b) we get $(H(U),\tau_\omega) = \operatorname{proj}_{\leftarrow K \in \mathcal{K}_U} H(K)$ also topologically in this case ([22], Thm. 6.1).

(c) Each ξ-balanced open set has the Runge property (cf. [22], Ch. 6, where other examples of open sets with the Runge property and some equivalent assertions are given, too). So there always exists a basis $\mathcal{U}$ of open sets with the property that $\tau_\omega = \tau_\pi$ on H(U) for each $U \in \mathcal{U}$. - M. Schottenloher [30] remarked recently that each pseudoconvex open set in an arbitrary product $\mathbb{C}^I$ has the Runge property.

It turns out below that the topology τ_π on H(U) has many pleasant properties, at least for (general) open subsets U of Schwartz spaces. So some of our next results are formulated in terms of this topology rather than τ_ω.

19. THEOREM: *Let* E *be a l.c. metrizable Schwartz space.*

(a) *The sheaf* $\mathcal{H}$ *of holomorphic functions on* E *, endowed* *with the topology* τ_π *on all spaces* $H(U)$ *for* $U \subset E$ *open, is a locally convex (topological) sheaf of complete Schwartz spaces. (For the definition of a l.c. sheaf and related remarks see e.g. [6], 1.1). If* E *is nuclear,* $\mathcal{H}$ *is even an s-nuclear sheaf.*

(b) τ_π *is the uniquely determined l.c. sheaf topology on* $\mathcal{H}$ *that coincides with the ported topology* τ_ω *on the spaces* $H(U)$ *for* ξ*-balanced open sets* $U \subset E$.

PROOF: (a) For open subsets $U,V \subset E$ with $U \supset V$, the canonical restriction $\rho_{UV} : (H(U),\tau_\pi) \to (H(V),\tau_\pi)$ is continuous, because $\pi_K \circ \rho_{UV} : H(U) \to H(K)$ is continuous for each $K \in \mathcal{K}_V$ by definition of the topology τ_π on $H(U)$.

Let $(U_i)_{i \in I}$ be a system of open subsets U_i of E with $U := \bigcup_{i \in I} U_i$. We notice already that τ_π is stronger on $H(U)$ than the projective topology τ with respect to the mappings $\rho_{U,U_i} : H(U) \to (H(U_i),\tau_\pi)$, and we have to show that the converse is also true. Let p be a continuous seminorm on $(H(U),\tau_\pi)$. By definition of τ_π there exists a compact $K \subset U$ and a continuous seminorm q on $H(K)$ with $p \leq q \circ \pi_K$. As K is compact and E metrizable, we can write $K = \bigcup_{j=1}^{m} K_j$ with compact subsets $K_j \subset U_{i_j}$ $(j = 1,\dots,m$; use the existence of a Lebesgue number for the covering $\bigcup_{i \in I} U_i)$. The natural (injective) linear mapping $A = \prod_{j=1}^{m} A_j : H(K) \to \prod_{j=1}^{m} H(K_j)$ is continuous. By regularity of the inductive limits in the definition of $H(K_j)$, $j=1,\dots,m$, it is easy to see that $A^{-1}(B)$ is bounded in $H(K)$ for each bounded set $B \subset \prod_{j=1}^{m} H(K_j)$. Furthermore $\prod_{j=1}^{m} H(K_j)$ is a (DF)-space and 7 (a) implies the Montel property for $H(K)$. Hence we can apply Baernstein's lemma from [3] : A is open, and so there exist continuous seminorms q_j on $H(K_j)$, $j = 1,\dots,m$, such that $q(g) \leq \max_{j=1}^{m} q_j(A_j(g))$ for all $g \in H(K)$. By $A_j \circ \pi_K = \pi_{K_j}$ we obtain $p(f) \leq q(\pi_K(f)) \leq \max_{j=1}^{m} q_j(\pi_{K_j} \circ \rho_{U,U_{ij}}(f))$ for all $f \in H(U)$.

But $\max_{j=1}^{m} q_j \circ \pi_{K_j}$ is a continuous seminorm on $\prod_{i\in I}(H(U_i),\tau_\pi)$, and hence τ_π is equal to the projective topology τ defined above. It follows that $\mathcal{H}$ is a l.c. sheaf with respect to τ_π. As in Prop. 16 the locally convex properties (Schwartz, s-nuclear) can be derived from the definition of τ_π, from 7, and from well-known permanence properties.

(b) It has already been remarked in 18 (c) that $\tau_\pi = \tau_\omega$ on $H(U)$ for all ξ- balanced open sets $U \subset E$ and that these sets are a basis for the open subsets of E. Hence (b) follows from [6], 1.2.b). □

$(H(U),\tau_\pi)$ or $(H(U),\tau_\omega)$ has the a. p. for each open subset U of a nuclear metrizable l.c. space E by 19 and 16. We give another result corresponding to what we proved in 12.

20. PROPOSITION: *Let* E *be as in 12. and assume that* U *is a* ξ-*balanced open subset of* E. *Then* $(H(U),\tau_\pi) = (H(U),\tau_\omega)$ *has the a. p.*

PROOF: If U is an open subset of E with the Runge property, then (by the very definition) the projective system $\{H(K);\ \pi_{KL}\}_{K\in\mathcal{K}_U}$ is equivalent to the system $\{H(K);\ \pi_{KL}\}_{K\in\mathcal{K}'}$ where $\mathcal{K}' = \{K \in \mathcal{K}_U;\ K \text{ is U-Runge}\}$. And the last system is reduced by definition of U-Runge.

If U is even ξ-balanced for some $\xi \in E$, one can replace $\mathcal{K}$ by the system $\tilde{\mathcal{K}}$ of ξ-balanced compact subsets of U. But for $K \in \tilde{\mathcal{K}}$ it immediately follows from 12 that H(K) has the a. p. Hence 20 follows from the remark (see e.g. [8], Introduction) that the projective limit of a reduced system of quasi-complete spaces with a.p. again has the a.p. □

By use of methods involving topological tensor products and the ε-product (see [21] and [9]), the topological vector space properties of $H(K) = \operatorname{ind}_{n\to} H^{\infty}(U_n)$ derived in the first part of this article (4 and 7) allow to treat vector-valued holomorphic germs as well, at least in certain cases.

For the following proposition we remark that compactly regular (injective) inductive systems were defined in [8], 1,1. And for open subsets U in a metrizable l.c. space E and any complete l.c. space F we denote by $H^{\infty}(U,F)$ again (cf. 1.(a)) the space of all bounded holomorphic mappings from U into F with the topology of uniform convergence on U . - The proof of Prop. 21 follows from 4 and 7 above together with [9], 4, Satz 6, 4, Satz 3, and 3, Remark before Korollar 11 as well as, say, [5], 31. (Notice that one can show the equivalence of the inductive systems $\{H^{\infty}(U_n,F_n); \rho_{nm}\}$ and $\{H^{\infty,p}(U_n,F_n); \rho_{nm}\}$ below in the case of Schwartz spaces E by aid of 7 (a) just as in the proof of 12.)

21. PROPOSITION: *Let* E *be a metrizable l.c. space and* $K \subset E$ *compact. Assume that the complete (Hausdorff) l.c. space* F *is the inductive limit of a countable injective inductive system of Banach spaces* F_n.

(a) *If* E *satisfies condition* (BM) *and if* $F = \text{ind}_{n\to} F_n$ *is compactly regular, then* $(H(K)\,\varepsilon\,F)_{bor} = \text{ind}_{n\to} H^{\infty,p}(U_n,F_n)$, *where* $(\ldots)_{bor}$ *denotes the associated bornological space and* $H^{\infty,p}(U_n,F_n) = \{f \in H^{\infty}(U_n,F_n);\ f(U_n)$ *is precompact*$\}$ *with the induced sup-norm topology.*

(b) *If* E *is a Schwartz space, no further assumption on* F *is needed to assure the equality*

$$(H(K)\,\varepsilon\,F)_{bor} = \text{ind}_{n\to} H^{\infty}(U_n,F_n).$$

(c) *If* E *is a Schwartz space and if* F *is even a* (DFS)-*space, we get the following improvement:*

$$H(K)\,\varepsilon\,F = \text{ind}_{n\to} H^{\infty}(U_n,F_n) = \text{ind}_{n\to} H^{\infty}(U_n,F).$$

(d) *For nuclear* E *one obtains:*

$$H(K)\,\hat{\otimes}_{\pi}\,F = H(K)\,\check{\otimes}_{\varepsilon}\,F = H(K)\varepsilon F = \text{ind}_{n\to} H^{\infty}(U_n,F_n) = \text{ind}_{n\to} H^{\infty}(U_n,F).$$

REMARK: For Banach spaces E and F, Aron and Schottenloher [2], Sect. 6 denote $\text{ind}_{n\to} H^{\infty,p}(U_n,F)$ by $H_K(K,F)$ and topologize this space by taking the induced topology τ of $\text{ind}_{n\to} H^{\infty}(U_n,F)$. If K is even balanced, they are able to show $H(K)\,\varepsilon\,F = (H_K(K,F),\tau)$ (even) topologically. (In this case the topology of $H(K)$ can be characterized explicitly by representation of its semi-norms, cf. Aron [1], Sect. 4) - It is still an open question whether τ is bornological.

22. COROLLARY: *Let* E_j *be metrizable l.c. spaces with* $K_j \subset E_j$ *compact* $(j = 1,2)$.

(a) *If both* E_1 *and* E_2 *satisfy condition* (BM), *we get* $(H(K_1)\,\varepsilon\,H(K_2))_{bor} = \text{ind}_{n\to} H^{\infty,p}(U_n^1 \times U_n^2)$, *where the notation* U_n^1 *and* U_n^2 *is obvious and where*

$H^{\infty,p}(U_n^1 \times U_n^2) = \{f \in H^\infty(U_n^1 \times U_n^2);$

$\{f(x_1, \cdot);\ x_1 \in U_n^1\}$ *is precompact in* $H^\infty(U_n^2)$ *and*

$\{f(\cdot, x_2);\ x_2 \in U_n^2\}$ *is precompact in* $H^\infty(U_n^1)\}$

with the induced sup-norm.

(b) *If* E_1 *is a Schwartz space, we get for any* E_2 *which satisfies* (BM):

$(H(K_1) \varepsilon H(K_2))_{bor} = H(K_1 \times K_2).$

(c) *If both* E_1 *and* E_2 *are Schwartz spaces, there is an equality* $H(K_1) \varepsilon H(K_2) = H(K_1 \times K_2)$.

PROOF: (a) is an easy consequence of (the proof of) 21 (a) and [5], 43.

(b) By 21. (b) we have $(H(K_1) \varepsilon H(K_2))_{bor} = \operatorname{ind}_{n\to} H^\infty(U_n^1, H^\infty(U_n^2))$.

But there is a canonical norm isomorphism

$$H^\infty(U_n^1, H^\infty(U_n^2)) \cong H^\infty(U_n^1 \times U_n^2):$$

In fact, this isomorphism is given by J defined as

$$J(f)(x_1,x_2) = f(x_1)(x_2).$$

Separate G-analyticity of $J(f)$ is obvious, so $J(f)$ is G-analytic by the classical Hartogs theorem and hence (by boundedness) $J(f) \in H^\infty(U_n^1 \times U_n^2)$. For the converse let $g \in H^\infty(U_n^1 \times U_n^2)$ and put $I(g)(x_1) : x_2 \to g(x_1,x_2)$. Then one can show $I(g) \in H^\infty(U_n^1, H^\infty(U_n^2))$ by analyzing e.g. 3, Lemma 2 of [21]. Obviously $J \circ I = id$, $I \circ J = id$, hence the isomorphism above is proved and (b) follows immediately.

(c) This is a consequence of (b) and of the authors' result from [9] (already used in the proof of 21 (c)) that the

ε-product of two (DFS)-spaces is again (DFS) and hence bornological. □

23. PROPOSITION: *Let* E_j *be l. c. metrizable Schwartz spaces with* $U_j \subset E_j$ *open* $(j = 1,2)$. *Then the equality*

$$(H(U_1),\tau_\pi) \,\varepsilon\, (H(U_2),\tau_\pi) = (H(U_1 \times U_2),\tau_\pi) \quad \textit{holds.}$$

PROOF: The system $\mathcal{K}^P_{U_1 \times U_2} = \{K_1 \times K_2;\ K_j \quad U_j \text{ compact}, l=1,2\}$ is cofinal in $\mathcal{K}_{U_1 \times U_2}$, hence $(H(U_1 \times U_2),\tau_\pi) = \underset{\leftarrow}{\text{proj}}_{K \in \mathcal{K}^P_{U_1 \times U_2}} H(K_1 \times K_2)$ holds already. But by a simple reasoning using the fundamental properties of projective limits and [7], 4.4 as well as 22 (c), we get (canonically):

$$\underset{\leftarrow\, K\in\mathcal{K}^P_{U_1 \times U_2}}{\text{proj}} H(K_1 \times K_2) = \underset{\longleftarrow\, K_1\in\mathcal{K}_{U_1}}{\text{proj}}\ \underset{\longleftarrow\, K_2\in\mathcal{K}_{U_2}}{\text{proj}} H(K_1 \times K_2) =$$

$$= \underset{\leftarrow}{\text{proj}}\ \underset{\longleftarrow}{\text{proj}}\ H(K_1) \,\varepsilon\, H(K_2) = \underset{\longleftarrow}{\text{proj}}\ H(K_1) \,\varepsilon\, (H(U_2),\tau_\pi) =$$

$$= (H(U_1),\tau_\pi) \,\varepsilon\, (H(U_2),\tau_\pi). \ \square$$

24. COROLLARY: *For* E_1, E_2 *as in 23. and* ξ_j*-balanced open sets* $U_j \subset E_j$ $(j = 1,2)$, *it follows that* $(H(U_1),\tau_\omega)\varepsilon(H(U_2),\tau_\omega) = (H(U_1 \times U_2),\tau_\omega)$. This is obvious from 23 , because the product of ξ_j- balanced sets is (ξ_1, ξ_2)-balanced, and hence $\tau_\pi = \tau_\omega$ on $H(U_1 \times U_2)$, too.

There are more representation theorems for vector - valued holomorphic germs and functions of which we only mention the following one (as an example):

25. PROPOSITION: *Let* E *be a metrizable nuclear l. c. space and* $U \subset E$ *open. Assume that the complete l. c. space* F *is*

projective limit of the system $\{F_\alpha ; \pi_{\alpha\beta}\}$ *of Banach spaces* F_α. *Thin the following representation holds:*

$$(H(U),\tau_\pi) \hat{\otimes}_\pi F = (H(U),\tau_\pi) \check{\otimes}_\varepsilon F = (H(U),\tau_\pi) \varepsilon F =$$

$$\underset{\leftarrow \alpha}{\mathrm{proj}} (H(U),\tau_\pi) \varepsilon F_\alpha = \underset{\leftarrow \alpha}{\mathrm{proj}} \ \underset{\leftarrow}{\mathrm{proj}}_{K \in \mathcal{K}_U} H(K) \varepsilon F_\alpha =$$

$$\underset{\leftarrow}{\mathrm{proj}}_{K \in \mathcal{K}_U} H(K) \varepsilon F = \underset{\leftarrow}{\mathrm{proj}}_{K \in \mathcal{K}_U} \underset{\leftarrow \alpha}{\mathrm{proj}} H(K) \varepsilon F_\alpha =$$

$$\underset{\leftarrow}{\mathrm{proj}} \ \underset{\leftarrow}{\mathrm{proj}} H(K,F_\alpha).$$

(Again, one can replace τ_π *by* τ_ω *here, if the open set* U *has the Runge property).*

PROOF: Obvious from a repeated application of [7], 4.4 and from 21 (d). □

The rest of the article is dedicated to the question of "how good" the nuclearity of H(U) under the topologies τ_o and τ_ω can be. As s-nuclearity of $(H(U),\tau_o)$ [resp. of $(H(U),\tau_\omega)$] was already proved in some cases (see [6], 1.12 [resp. Prop.16 of this paper]), it is natural to ask more generally, what types of λ-nuclearity (cf. Dubinsky-Ramanujan [13]) can occur for H(U), if U is an open subset of an infinite dimensional l. c. space E. To start with, the following Proposition 26 confirms the obvious conjecture that the type of nuclearity cannot be better than for open subsets U of $\mathbb{C}^N$, where N may of course be any natural number. The main tool in the proof of 26 is the result I, 4. 12 of Petzsche [25].

So let $\alpha = (\alpha_n)_{n \in \mathbb{N}}$ be an "exponent sequence" of positive numbers, i.e. $\alpha_n \leq \alpha_{n+1}$ for all $n \in \mathbb{N}$ and $\lim_{n\to\infty} \alpha_n = \infty$. Let $\Lambda_\infty(\alpha)$ be the associated power series space of infinite type of which we assume nuclearity. We do not define $\Lambda_\infty(\alpha)$-(and $\Lambda_{\mathbb{N}}(\alpha)$-) nuclearity here, but refer to the Memoir [13] of Dubinsky and

Ramanujan and to the article [28] of Ramanujan and Terzioglu.

26. PROPOSITION: *Let $\mathcal{E}$ be a class of l. c. spaces with the following three properties:*

(1) *$E \in \mathcal{E}$ implies $F \in \mathcal{E}$ for each continuously projected subspace F of E.*

(2) *For any $E \in \mathcal{E}$ and any open subset U of E, a l.c. topology τ is defined on $H(U)$ such that τ coincides with τ_o, if E is finite dimensional.*

(3) *If E_1, E_2 and $E_1 \times E_2$ belong to $\mathcal{E}$ and if U_j is open in $E_j (j = 1,2)$, $(H(U_1),\tau) \varepsilon (H(U_2,\tau) = (H(U_1 \times U_2),\tau)$ holds.*

Then if $E \in \mathcal{E}$ is infinite dimensional and if $(H(U),\tau)$ is $\Lambda_\infty(\alpha)$- (or only $\Lambda_{\mathbb{N}}(\alpha)$-) nuclear for each balanced open subset U of E, we have:

$$(*) \qquad \lim_{n\to\infty} \frac{\alpha_n}{\sqrt[N]{n}} = 0 \quad \textit{for each } n \in \mathbb{N}.$$

PROOF: A well-known corollary to the Hahn-Banach theorem implies that E can be represented (topologically) $E = E_N \times E_N^\infty$ for arbitrary $n \in \mathbb{N}$, where E_N with $\dim E_N = N$ and E_N^∞ are subspaces of E. E_N and E_N^∞ belong to $\mathcal{E}$ by (1). Let $j : E_N \to \mathbb{C}^N$ be a topological linear isomorphism and define $U_N = j^{-1}(D^N)$, where D is the open unit disk of $\mathbb{C}$. If $U_N^\infty \neq \emptyset$ is some balanced open subset of E_N^∞, $U = U_N \times U_N^\infty$ is balanced and open and hence, by (3), $(H(U),\tau) = (H(U_N),\tau) \varepsilon (H(U_N^\infty),\tau)$. Moreover, $(H(U),\tau)$ is $\Lambda_\infty(\alpha)$-(or at least $\Lambda_{\mathbb{N}}(\alpha)$-) nuclear by assumption, $(H(U_N),\tau) = (H(U_N),\tau_o)$ by (2), and this space is also $\Lambda_\infty(\alpha)$-[resp. $\Lambda_{\mathbb{N}}(\alpha)$-] nuclear as a topological linear subspace of $(H(U),\tau)$.

On the other hand, $(H(U_N),\tau_o)$ is topologically isomorphic to $(H(D^N),\tau_o) = (H(D),\tau_o) \hat{\otimes}_\pi \ldots \hat{\otimes}_\pi (H(D),\tau_o) = N$ - fold projective tensor product of $\Lambda_1(n)$. Now Petzsche [25], I,4.12 proves that this N - fold tensor product is topologically isomorphic to $\Lambda_1(\beta)$, where $\beta_n = [\sqrt[N]{n}]$ (= largest integer $\leq \sqrt[N]{n}$). Finally, it follows now from an application of Ramanujan-Terzioglu [28], Prop. 2.12, that $\lim_{n\to\infty} \frac{\alpha_n}{\beta_n} = 0$ and hence $\lim_{n\to\infty} \frac{\alpha_n}{\sqrt[N]{n}} = 0$, where $N \in \mathbb{N}$ was arbitrary. □

27. REMARKS:

(a) As the proof shows, we actually need much less than (1) and (3) of 26. Yet assumptions (1) - (3) in Prop. 26 are satisfied for $\tau = \tau_o$, if $\mathcal{E}$ = Fréchet spaces or $\mathcal{E}$ = Silva spaces (see e.g. [5], 43). By (23 and) 24 , it is also allowed to take $\tau = \tau_\omega$ (or τ_π), if $\mathcal{E}$ = metrizable Schwartz spaces.

(b) Of course, condition (*) restricts the possible $\Lambda_\infty(\alpha)$-nuclearity types for H(U) considerably. But (*) is still satisfied by the following exponent sequences $\alpha^{(p)}$, $1 < p < \infty$, which lead to different nuclearity types stronger than s-nuclearity: $\alpha_n^{(p)} = [\ln(1 + n)]^p$. ($\lim_{n\to\infty} \frac{\alpha_n^{(p)}}{\alpha_n^{(q)}} = 0$ holds for $1 < p < q < \infty$, and so $\Lambda_\infty(\alpha^{(q)})$, but not $\Lambda_\infty(\alpha^{(p)})$, is $\Lambda_\infty(\alpha^{(p)})$-nuclear by [28], Prop. 2.12 and Cor. 2.13. [In particular, $\Lambda_\infty(\alpha^{(p)})$ and $\Lambda_\infty(\alpha^{(q)})$ cannot be topologically isomorphic].)

After the examples in 27 (b), the next question is whether the $\Lambda_\infty(\alpha)$-nuclearity types not excluded by condition (*) of 26 are in fact realized, i. e.: Do there exist infinite dimensional l.c. spaces E such that for each (say balanced) open subset U of E the space $(H(U),\tau)$ is $\Lambda_\infty(\alpha)$-nuclear, where the topology τ agrees with conditions (2) and (3) of Prop. 26 and, for instance, $\alpha = \alpha^{(p)}$, $1 < p < \infty$, like in 27 (b) ? We answer this question in the affirmative for $\tau = \tau_\pi$ by taking $E = \mathbb{C}^{\mathbb{N}}$ in our last proposition.

An exponent sequence α is called stable, if it satisfies $\sup_{n\in\mathbb{N}} \frac{\alpha_{2n}}{\alpha_n} < \infty$. (The sequences $\alpha^{(p)}$ are all stable).

28. PROPOSITION: *Let the stable exponent sequence α satisfy condition (*) of 26. Then the sheaf $\mathcal{H}$ of holomorphic functions on $\mathbb{C}^{\mathbb{N}}$, endowed with the topology τ_π (on all $H(U)$, $U \subset \mathbb{C}^{\mathbb{N}}$ open), is $\Lambda_\infty(\alpha)$-nuclear.*

PROOF: As $(\mathcal{H},\tau_\pi)$ is a locally convex sheaf by Theorem 19, it is enough (cf. [6], 1.2.c), [13], Thm. 2.10, and/or [25], I, 5.4 - here we need that α is stable) to show $\Lambda_\infty(\alpha)$-nuclearity of $(H(U),\tau_\pi)$ only for open sets U in a suitable neighbourhood basis $\mathcal{U}$ of zero. By the permanence properties of $\Lambda_\infty(\alpha)$-nuclear spaces that we have already used, it then suffices to prove $\Lambda_\infty(\alpha)$-nuclearity for each $H(K)$, where K is taken from an appropriate cofinal subsystem $\tilde{\mathcal{K}}_U$ of $\mathcal{K}_U$, $U \in \mathcal{U}$. Hence, by choosing $\mathcal{U}$ and the systems $\tilde{\mathcal{K}}_U$, $U \in \mathcal{U}$, in an obvious way, we need only show $\Lambda_\infty(\alpha)$-nuclearity of $H(K)$, if $K = \prod_{j\in\mathbb{N}} K_j$, where $K_j = \{z \in \mathbb{C};\ |z| \leq s_j\}$ with $s_j \geq 0$. In this case, $H(K) = \operatorname{ind}_{n\to} H^\infty(U_n)$ with $U_n = \prod_{j\in\mathbb{N}} U_j^n$, $U_j^n = \mathbb{C}$ for $j > n$ and

$U_j^n = \{z \in \mathbb{C}; |z| < s_j + r_n\}$ for $j \leq n$, and hence $\tilde{U}_n = \prod_{j=1}^{n} U_j^n \subset \mathbb{C}^n$.

We represent the canonical mapping $\rho_{n,n+1} : H^\infty(U_n) \to H^\infty(U_{n+1})$, $n \in \mathbb{N}$, as follows to get its $\Lambda_\infty(\alpha)$-nuclearity:

$$H^\infty(U_n) \xrightarrow{A_n^{-1}} H^\infty(\tilde{U}_n) \xrightarrow{id} (H(\tilde{U}_n), \tau_o) \xrightarrow{R} H^\infty(\tilde{U}_{n,\sigma_n}) \xrightarrow{B}$$

$$\longrightarrow H^\infty(\tilde{U}_{n+1}) \xrightarrow{A_{n+1}} H^\infty(U_{n+1}).$$

Here σ_n satisfies $r_{n+1} < \sigma_n < r_n$, R is the canonical restriction, and $B(f) = f \circ \pi_{n,n+1}|\tilde{U}_{n+1}$, where $\pi_{n,n+1} : \mathbb{C}^{n+1} \to \mathbb{C}^n$ is defined by $\pi_{n,n+1}(z_1,\ldots,z_{n+1}) = (z_1,\ldots,z_n)$. (Any other notation is obvious.) Similarly as in the proof of 26 , it follows from [25], I, 4. 12 that: $(H(\tilde{U}_n),\tau_o) = (H(U_1^n),\tau_o)\hat{\otimes}_\pi \ldots \hat{\otimes}_\pi (H(U_n^n),\tau_o) =$

$$= \Lambda_{s_1+r_n}(n) \,\hat{\otimes}_\pi \ldots \hat{\otimes}_\pi \Lambda_{s_n+r_n}(n) \cong \Lambda_1(n) \,\hat{\otimes}_\pi \ldots \hat{\otimes}_\pi \Lambda_1(n) \cong \Lambda_1(\beta) ,$$

where $\beta_k = [\sqrt[n]{k}]$, and hence this space is certainly $\Lambda_\infty(\alpha)$-nuclear for any α satisfying (*) (by [28], Prop. 2.12). As R maps $(H(\tilde{U}_n), \tau_o)$ into a Banach space, R is $\Lambda_\infty(\alpha)$-nuclear. The factorization of $\rho_{n,n+1}$ given above implies the $\Lambda_\infty(\alpha)$-nuclearity of this mapping, as claimed.

As the transposed mapping ${}^t\rho_{n,n+1}$ is again $\Lambda_\infty(\alpha)$-nuclear for each $n \in \mathbb{N}$, we obtain from the representation $H(K)_b' = \text{proj}_{\leftarrow n} H^\infty(U_n)_b'$ (by the very definition) that $H(K)_b'$ is $\Lambda_\infty(\alpha)$-nuclear by Ramanujan - Terzioglu [28], Cor. 3.7. □

PROBLEM: *Let* E *be an infinite dimensional* $\Lambda_{\mathbb{N}}(\alpha)$ *- nuclear Fréchet space for a stable exponent sequence* α *satisfying* *condition* (*) *in* 26. (*Remark that* E'_b *is already* $\Lambda_\infty(\alpha)$-*nuclear by* [28], *Cor.* 3.7). *Is then* $(H(U),\tau_\pi)$ *a* $\Lambda_\infty(\alpha)$-*nuclear space for each open subset* U *of* E ?

FINAL REMARKS:

(a) In a private communication, Richard Aron mentioned that, some time ago, he had already given a proof (unpublished) for regularity and the Montel property of H(K) for each compact K in a l.c. metrizable Schwartz space E .

(b) Generalizing a result of Barroso (1970)[for $U = \mathbb{C}^{\mathbb{N}}$], Martin Schottenloher (in a private communication) has recently proved $\tau_o = \tau_\omega$ (and hence $\tau_o = \tau_\pi = \tau_\omega$) on H(U) for each open subset U of $\mathbb{C}^{\mathbb{N}}$. So Prop. 28 remains true with τ_π replaced by τ_ω or τ_o .

(c) In connection with b), the authors would like to ask: For which (FS)-[resp. (FN)-] spaces E does the equality $\tau_o = \tau_\omega$ always hold on H(U), $U \subset E$ balanced and open ? (We know of no examples with $\tau_o \neq \tau_\omega$ in this case).

REFERENCES

[1] R. ARON: Tensor products of holomorphic functions, Indag. Math. 35, 192 - 202 (1973).

[2] R. ARON, M. SCHOTTENLOHER: Compact holomorphic mappings on Banach spaces and the approximation property, J. Functional Analysis 21, 7-30 (1976).

[3] A. BAERNSTEIN II: Representation of holomorphic functions by boundary integrals, Transact. Amer. Math. Soc. 160, 27 - 37 (1971).

[4] J. A. BARROSO, M. C. MATOS, L. NACHBIN: On bounded sets of holomorphic mappings, Proceedings on infinite dimensional holomorphy, University of Kentucky 1973, Springer Lecture Notes Math. 364, p. 123-134 (1974).

[5] K. - D. BIERSTEDT: Tensor products of weighted spaces, Function spaces and dense approximation, Proc. Conference Bonn 1974, Bonner Math. Schriften 81, p.26-58 (1975).

[6] K. - D. BIERSTEDT, B. GRAMSCH, R. MEISE: Approximations - eigenschaft, Lifting und Kohomologie bei lokalkonvexen Produktgarben, manuscript math. 19 (1976).

[7] K. - D. BIERSTEDT, R. MEISE: Lokalkonvexe Unterräume in topologischen Vektorräumen und das ε-Produkt, manuscripta math. 8, 143-172 (1973).

[8] K. - D. BIERSTEDT, R. MEISE: Bemerkungen über die Approximationseigenschaft lokalkonvexer Funktionenräume, Math. Ann. 209, 99-107 (1974).

[9] K. - D. BIERSTEDT, R. MEISE: Induktive Limites gewechtiter Räume stetiger und holomorpher Funktionen, J. reine angew. Math. 282, 186-220 (1976).

[10] P. J. BOLAND, L. WAELBROECK: Holomorphic functions on nuclear spaces, to appear (preprint, Dublin 1975).

[11] S. B. CHAE: Holomorphic germs on Banach spaces, Ann. Inst. Fourier 21, 3, 107 - 141 (1973).

[12] S. DINEEN: Holomorphic functions on locally convex topological vector spaces I, Ann. Inst. Fourier 23, 1, 19 - 54 (1973).

[13] E. DUBINSKY, M. S. RAMANUJAN: On λ-nuclearity, Memoirs Amer. Math. Soc. 128 (1972).

[14] K. FLORET: Lokalkonvexe Sequenzen mit kompakten Abbildungen, J. reine angew. Math. 247, 155 - 195 (1971).

[15] K. FLORET, J. WLOKA: Einführung in die Theorie der lokal konvexen Räume, Springer Lecture Notes Math. 56 (1968).

[16] A. GROTHENDIECK: Produits tensoriels topologiques et espaces nucléaires, Memoirs Amer. Math. Soc. 16, reprint 1966.

[17] A. HIRSCHOWITZ: Bornologie des espaces de fonctions analytiques en dimension infinie, Séminaire P. Lelong 1970, Springer Lecture Notes Math. 275, p.21-53(1971).

[18] J. HORVÁTH: Topological vector spaces and distributions I, Addison-Wesley 1966.

[19] G. KÖTHE: Topological vector spaces I, Springer Grundlehren der Math. 159, 1969.

[20] A. MARTINEAU: Sur une propriété universelle de l'espace des distributions de M. Schwartz, C. R. Acad. Sci. Paris A 259, 3162 - 3164 (1964).

[21] R. MEISE: Räume holomorpher Vektorfunktionen mit Wachstumsbedingungen und topologische Tensorprodukte, Math. Ann. 199, 293 - 312 (1972).

[22] J. MUJICA: Spaces of germs of holomorphic functions, to appear in Advances in Math. (cf. also Bull. Amer. Math. Soc. 81, 904 - 906(1975)).

[23] L. NACHBIN: Topology on spaces of holomorphic mappings, Springer Ergebnisse der Math. 47, 1969.

[24] P. NOVERRAZ: Pseudo - convexité, convexité polynomiale et domains d'holomorphie en dimension infinie, North Holland 1973.

[25] H. J. PETZSCHE: Darstellung der Ultradistributionen vom Beurlingschen und Roumieuschen Typ durch Randwerte holomorpher Funktionen, Dissertation Düsseldorf 1976.

[26] H. PFISTER: Bemerkungen zum Satz über die Separabilität der Fréchet-Montel-Räume, Arch. der Math. 27,86-92 (1976).

[27] A. PIETSCH: Nukleare lokalkonvexe Räume,Akademie-Verlag, 2. Aufl, 1969.

[28] M. S. RAMANUJAN, T. TERZIOGLU: Power series spaces $\Lambda_k(\alpha)$ of finite type and related nuclearity,Studia Math. 53, 1-13 (1975).

[29] M. SCHOTTENLOHER: Holomorphe Vervollständigung metrisierbarer lokalkonvexer Räume, Bayer. Akad. d. Wiss., Math. - naturw. Klasse, S. - ber. 1973, 57-66(1974).

[30] M. SCHOTTENLOHER: Polynomial approximation on compact sets, Coference Campinas (São Paulo, Brazil) 1975, these proceedings.

[31] L. SCHWARTZ: Théorie des distributions à valeurs vectorielles I, Ann. Inst. Fouries 7, 1-142 (1957).

(K. - D. Bierstedt)
Fachbereich 17, Mathematik,
Gesamthochschule D 2
Warburger Str. 100 Postfach 1621
D - 4790 Paderborn
Bundesrepublik Deutschland

(R. Meise)
Mathematisches Institut
der Universität
Universitätsstr. 1
D - 4000 Düsseldorf
Bundesrepublik Deutschland

Infinite Dimensional Holomorphy and Applications, Matos (ed.)

DUALITY AND SPACES OF HOLOMORPHIC FUNCTIONS

By *PHILIP J. BOLAND*

I. INTRODUCTION

A classical problem of continual interest in holomorphic function theory is the following: when may one represent the dual of a space of holomorphic functions as a space of holomorphic functions ? This problem was elegantly solved in the one complex variable case by Grothendieck, Silva - Dias, and Koethe in 1952 - 53. Their result states that if U is an open subset of the complex plane and H(U) is the space of holomorphic functions on U with the compact open topology, then the dual of H(U) may be characterized as the space of holomorphic germs on the complement of U in the Riemann sphere which vanish at infinity. The several complex variable case is much harder, but Martineau (1966) has proven some results in this direction for U an open convex subset of C^n.

In this paper we will consider some spaces of holomorphic functions on open absolutely convex subsets of dual of Frechet nuclear spaces ($\mathcal{D}$FN spaces), and classify their duals as spaces of holomorphic functions.

II. PRELIMINARIES

For notation and terminology we refer primarily to [B1], [B2], [Mu], [N], and [P]. The proofs of the propositions in this section may be found in [B1], and [B2].

E will denote any $\mathcal{D}FN$ space (the strong dual of a Frechet-nuclear space) and E' its strong dual. Examples of $\mathcal{D}FN$ spaces are $\sum_N C$, C^n, $H'(C)$, $\mathcal{S}'$, and $\mathcal{E}'$. If $A \subset E$, then $A^{\circ} = \{\phi : \phi \in E', |\phi|_A \leq 1\}$ and $A^{\bullet} = \{\phi : \phi \in E', |\phi|_A < 1\}$.

$P(^mE)$ and $P(^mE')$ are respectively the spaces of continuous m-homogeneous polynomials on E and E'. $P(E)$ is the space of continuous polynomials on E. If $p \in P(^mE)$, then there exist $(\phi_n) \subset E'$ such that $p(x) = \sum_{n=1}^{\infty} \phi_n^m(x)$ for all $x \in E$ and $\sum_{n=1}^{\infty} |\phi_n^m|_K < +\infty$ for all compact $K \subset E$.

We may define two equivalent topologies ε & π on $P(^mE)$. ε is the compact open topology defined by the family of semi-norms ε_K where K is a compact subset of E and $\varepsilon_K(p) = \sup_{x \in K} |p(x)|$. π is generated by all semi-norms of the type π_K where $\pi_K(p) = \inf\{ \sum_{n=1}^{\infty} |\phi_n^m|_K : p(x) = \sum_{n=1}^{\infty} \phi_n^m(x)$ for all $x \in E$, $\sum_{n=1}^{\infty} |\phi_n^m|_L < +\infty$ for all L compact in $E \}$.

PROPOSITION 1. *π and ε are equivalent topologies on $P(^mE)$. In fact given K compact in E, there exist $C > 0$ and K_1 compact in E such that $K \subset K_1$ and*

$$\varepsilon_K(p) \leq \pi_K(p) \leq C^m \varepsilon_{K_1}(p)$$

for all $p \in P(^mE)$ and all $m \geq 1$.

PROPOSITION 2. *$P'({}^mE)$, the dual of $P({}^mE)$, is isomorphic to $P({}^mE')$ via the mapping $\beta : T \in P'({}^mE) \to \beta T \in P({}^mE')$ where $\beta T(\phi)$ $T(\phi^m)$ for $\phi \in E'$.*

III. SPACES OF HOLOMORPHIC FUNCTIONS

Let U denote an absolutely convex open set in the $\mathcal{D}FN$ space E, and $H(U)$ the space of holomorphic functions on U.

DEFINITION 1. *The ε topology on $H(U)$ is that defined by all semi-norms ε_K (K compact in U) where $\varepsilon_K(f) = \sum_{n=0}^{\infty} \varepsilon_K\left(\frac{\hat{d}^n f(0)}{n!}\right)$ for $f \in H(U)$. In fact the ε topology is the compact open topology on $H(U)$, and $H(U),\varepsilon$ is a Fréchet nuclear space* ([B3]).

DEFINITION 2. $H_N(U) = \{f : f \in H(U)$, *and for all compact* $K \subset U$, $\pi_K(f) = \sum_{n=0}^{\infty} \pi_K\left(\frac{\hat{d}^n f(0)}{n!}\right) < +\infty\}$. *The π topology on $H_N(U)$ is that generated by all semi-norms π_K, where K is compact in U. $H_N(U),\pi$ is a Fréchet space, and $H_N(U),\pi \to H(U),\varepsilon$ is a continuous injection.*

REMARK 1. $H_N(U),\pi$ *is an attractive space because one can nicely characterize its dual. I would like to conjecture that in fact $H_N(U) = H(U)$ (and in this case it would follow that $\pi = \varepsilon$ on $H(U)$).*

REMARK 2. *The following comments are in order concerning* $H_N(U)$:

(a) *The space $H(E)$ of entire functions on E is a subspace of $H_N(U)$ (and hence if $U=E$, then $H_N(U)=H(U)$).*

(b) *If* K *is compact in* U, *there exists* $\delta > 0$ *such that* $\pi_{\delta K}(f)$ *is finite for all* $f \in H(U)$.

DEFINITION 3. *Let* U *be an absolutely convex open subset of* E. *Let* (K_n) *be an increasing sequence of absolutely convex compact subsets of* U *such that* $U = \bigcup_n K_n$ *and if* K *is compact in* U *then* $K \subseteq K_n$ *for some* n. *Now* U° *is compact in* E', *and hence we let* $H(U^{\circ})$ *be the space of holomorphic germs on* U°. *For each* n, $K_n^{\bullet}$ *is an absolutely convex open set in* E' *such that* $U^{\circ} \subseteq K_n^{\bullet}$ *and* $U^{\circ} = \bigcap_n K_n^{\bullet}$. $H^{\infty}(K_n^{\bullet})$ *will denote the Banach space of bounded holomorphic functions on* $K_n^{\bullet}$ *endowed with the supremum norm. We endow* $H(U^{\circ})$ *with the inductive limit topology* $\lim\limits_{\rightarrow} H^{\infty}(K_n^{\bullet})$. $H(U^{\circ})$ *is a* $\mathcal{D}F$ *space (see* [Muj]*).*

DEFINITION 4. *Let* U *and* (K_n) *be as given in Definition 3. For each* n, *choose* $a_n > 1$ *such that* $L_n = a_n K_n \subseteq U$. *Without loss of generality (by replacing* (K_n) *with a subsequence if necessary), we may assume* (L_n) *is an increasing sequence of absolutely convex compact subsets of* U *such that* $U^{\circ} = \bigcap_n L_n^{\bullet}$ *and* $H(U^{\circ}) = \lim\limits_{\rightarrow} H^{\infty}(L_n^{\bullet})$.

Now suppose $T \in H_N'(U)$. Then there exist $C > 0$ and K_n such that $|T(f)| \leq C\pi_{K_n}(f)$ for all $f \in H_N(U)$. For each m we let $T_m = T|_{P(^mE)}$. Then $T_m(p) \leq C\pi_{K_n}(p)$ for all $p \in P(^mE)$, and $T_m \in P'(^mE)$. Now let $F_m = \beta T_m \in P(^mE')$ (see Proposition 2).

Then $|F_m(\phi)| = |T_m(\phi^m)| \le C\pi_{K_n}(\phi^m)$ for all $\phi \in E'$. Finally define F on $L_n^\bullet$ by $F(\phi) = \sum_{m=0}^{\infty} F_m(\phi)$ for all $\phi \in L_n^\bullet$. Now $|F(\phi)| \le \sum_{m=0}^{\infty} |F_m(\phi)| \le C \sum_{m=0}^{\infty} (\frac{1}{a_n})^m = (Ca_n)/(a_n - 1)$. It is clear that $F \in H^\infty(L_n^\bullet)$ and $\|F\| \le (Ca_n)/(a_n - 1)$. We define $\hat{\beta} T = \tilde{F}$, where $\tilde{F}$ is the germ of F on U^o. $\hat{\beta}$ is well defined, and is a linear map from $H_N'(U)$ to $H(U^o)$.

THEOREM 1. *Let* U *be an absolutely convex open subset of the* $\mathcal{D}$FN E. *Then* $\hat{\beta}$ *defines a homeomorphism between* $H_N'(U)$ *(strong topology) and* $H(U^o)$. *Hence the dual of* $H_N(U)$ *may be identified with the space of germs on* U^o.

PROOF. (a) $\hat{\beta}$ is an isomorphism. Since by Proposition 2 β is an isomorphism between $P'(^mE)$ and $P(^mE')$ for each m, it follows that $\hat{\beta}$ is $1-1$. To show that $\hat{\beta}$ is an isomorphism, we need only show $\hat{\beta}$ is onto.

Let $\tilde{F} \in H(U^o)$. There exist n and $F \in H^\infty(K_n^\bullet)$ such that $\tilde{F}$ is the germ of F on U^o. Let $M = \|F\|$, and $F = \sum_{m=0}^{\infty} F_m$ be the Taylor series representation of F in $K_n^\bullet$. For each m let $T_m \in P'(^mE)$ be such that $\beta T_m = F_m$ It follows that $T_m(p) \le M\pi_{K_n}(p)$ for all $p \in P(^mE)$. We now define T on $H_N(U)$ by

$$T(f) = \sum_{m=0}^{\infty} T_m\left(\frac{\hat{d}^m f(0)}{m!}\right).$$

Clearly $|T(f)| \le M\pi_{K_n}(f)$ for all $f \in H_N(U)$, and hence $T \in H_N'(U)$. Since $\hat{\beta} T = \tilde{F}$, we have shown that $\hat{\beta}$ is an isomorphism between $H_N'(U)$ and $H(U^o)$.

(b) $\hat{\beta}$ is a homeomorphism. First we show that $\hat{\beta}$ is continuous. Let $T_\alpha \to 0$ in $H_N'(U)$, and we will show that $\hat{\beta} T_\alpha = \tilde{F}_\alpha \to 0$

in $H(U^{\circ})$. Suppose now that W is a neighborhood of 0 in $H(U^{\circ})$, and let $W_n = I_n^{-1}(W)$ for each n where $H^{\infty}(L_n^{\bullet}) \overset{I_n}{\to} H(U^{\circ})$. Therefore there exists a sequence (b_n) of positive constants such that $V_n = \{F : F \in H^{\infty}(L_n^{\bullet}),\ \|F\| \leq b_n\} \subseteq W_n$ for each n. It suffices to show there exists α' such that when $\alpha \geq \alpha'$, then $\tilde{F}_{\alpha} = \hat{\beta} T$ has a representation $F_{\alpha} \in H^{\infty}(L_n^{\bullet}) \cap V_n$ for some n.

Define $B = \{f : f \in H_N(U),\ \pi_{K_n}(f) < C_n = a_n / (b_n(a_n - 1))$ for each $n\}$. Then B is bounded in $H_N(U)$ and we may find α' such that when $\alpha \geq \alpha'$, $T_{\alpha} \in B^{\circ}$. Suppose now $\alpha \geq \alpha'$. Then there exist $C > 0$ and K_n such that $|T_{\alpha}(f)| \leq C\pi_{K_n}(f)$ for all $f \in H_N(U)$. But as $T_{\alpha} \in B^{\circ}$, $|T_{\alpha}(f)| \leq (1/C_n)\pi_{K_n}(f)$ for all $f \in H_N(U)$. From the construction in Definition 4 it follows that there exists $F_{\alpha} \in H^{\infty}(L_n^{\bullet})$ such that $\|F_{\alpha}\| \leq (a_n/C_n)/(a_n - 1) = b_n$ and $\tilde{F}_{\alpha} = \hat{\beta} T_{\alpha}$. Since $F_{\alpha} \in H^{\infty}(L_n^{\bullet}) \cap V_n$, this completes the proof that $\hat{\beta}$ is continuous.

We complete the proof by showing that $\hat{\beta}^{-1} : H(U^{\circ}) \to H_N'(U)$ is continuous. As $H(U^{\circ})$ is bornological, it suffices to show that if $\{\tilde{F}_{\alpha}\}$ is bounded in $H(U^{\circ})$, then $\{\hat{\beta}^{-1}\tilde{F}_{\alpha}\}$ is bounded in $H_N'(U)$.

Let $\{\tilde{F}_{\alpha}\}$ be bounded in $H(U^{\circ})$. Then there exist K_n and $\{F_{\alpha}\}$ bounded in $H^{\infty}(K_n^{\bullet})$ such that $\tilde{F}_{\alpha}$ is the germ of F_{α} on U° for each α (see [Muj]). Let $M = \sup_{\alpha} \|F_{\alpha}\|$. From the first part of this proof, it follows that there exist $T_{\alpha} \in H_N'(U)$ for each α such that $\hat{\beta} T_{\alpha} = \tilde{F}_{\alpha}$ and $|T_{\alpha}(f)| \leq M\pi_{K_n}(f)$ for all $f \in H_N(U)$. Hence we see that if $V = \{f : f \in H_N(U), \pi_{K_n}(f) \leq 1/M\}$, then V is a neighborhood of 0 in $H_N(U)$ such that $T_{\alpha} \in V^{\circ}$ for each α. Hence $\{T_{\alpha}\}$ is bounded in $H_N'(U)$.

BIBLIOGRAPHY

[BMN] J. A. BARROSO, M. C. MATOS AND L. NACHBIN - *On bounded sets of holomorphic mappings*, Proc. 1973 Internat. Conf. on Infinite Dimensional Holomorphy, Lecture Notes in Math., vol. 364, Springer-Verlag, Berlin and New York, 1974.

[B1] P. J. BOLAND - *Malgrange Theorem for entire functions on nuclear spaces*, Proc., 1973 Internat. Conf. on Infinite Dimensional Holomorphy, Lecture Notes in Math., vol. 364, Springer-Verlag, Berlin and New York, 1974.

[B2] P. J. BOLAND - *Holomorphic Functions on nuclear spaces*, TAMS, vol. 209, 1975.

[B3] P. J. BOLAND - *An example of a nuclear space in infinite dimensional holomorphy*, Arkiv för matematik, 15:1 (1977).

[G] A. GROTHENDIECK - *Sur certains espaces de fonctions holomorphes*, I and II, Crelles Journal, vol. 192, 1953.

[K] G. KOETHE - *Dualitat in der Funktiontheorie*, Crelles Journal, vol. 191, 1953.

[M] A. MARTINEAU - *Sur la topologie des espaces de fonctions holomorphes*, Math. Annalen, Band 163, Heft 1, 1966.

[Muj] J. MUJICA - *Spaces of germs of holomorphic functions*, to appear in Advances in Mathematics.

[N] L. NACHBIN - *Topology on spaces of holomorphic mappings*, Ergebnisse der Mathematik und ihrer Grenzgebiete, Ban 47, Springer-Verlag New York, 1969.

[P] A. PIETSCH - *Nuclear locally convex spaces*, Ergebnisse der Mathematik und ihrer Grenzgebiete, Band 66, Springer-Verlag, New York, 1972.

[S] C. L. DA SILVA DIAS - *Espaços Vectoriais Topologicos e sua applicações nos espaços funcionais analiticos*, Boletim da Sociedade de Matematica de São Paulo, vol. 5, 1952.

Department of Mathematics
University College,
Dublin 4, Ireland.

A HOLOMORPHIC CHARACTERIZATION OF BANACH SPACES WITH BASES

By *Soo Bong Chae*

ABSTRACT: Let E be a Banach space with a monotone normalized basis $\{b_n\}$. Every holomorphic automorphism on the open unit ball E_1 of E is of the form $\sum_{n=1}^{\infty} x_n b_n \to \sum_{n=1}^{\infty} \lambda_n \frac{x_{\pi(n)}-\alpha_n}{1-\bar{\alpha}_n x_{\pi(n)}} b_n$ where $\sum_{n=1}^{\infty} \alpha_n b_n \in E_1$; $|\lambda_n| = 1$ $(n \in N)$; π a permutation of N if and only if E is isometrically isomorphic to c_o.

AMS (1970) Subject Classifications. Primary 32A30, 46A45, 46B15, 46G99.

Key Words and Phrases. Holomorphic maps on Banach Spaces, Basis, Möbius transformations, automorphism, isometry.

1. INTRODUCTION: On the open unit ball E_1 of a complex Banach space E with a normalized basis $\{b_n\}$, we define the *Möbius transformation*

$$\phi_\alpha : E_1 \to E \quad \text{by}$$

$$\phi_\alpha\left(\sum_{n=1}^{\infty} x_n b_n\right) = \sum_{n=1}^{\infty} \frac{x_n - \alpha_n}{1-\bar{\alpha}_n x_n} b_n$$

where $\alpha = \sum_{n=1}^{\infty} \alpha_n b_n \in E_1$. Then ϕ_α is an injective holomorphic (Fréchet differentiable) map on E_1. Standart results about holomorphic functions on Banach spaces may be found in [2]. If $E = C$, it is well known that the Möbius transformation characterizes the injective analytic maps (i.e., the conformal maps) of the open unit disk onto itself.

In this paper we show that for the Möbius transformations to characterize the holomorphic automorphisms of E_1 onto itself it is necessary and sufficient that the Banach space E is isometrically isomorphic to the Banach space c_o of sequences converging to 0.

2. AUTOMORPHISMS: Let E be a complex Banach space and let $U \subset E$ be a nonempty open set. A mapping $f : U \to U$ is said to be *holomorphic automorphism* if f is a bijective holomorphic map with the holomorphic inverse. Aut(U) will denote the space of the automorphisms on U and Iso(E) will denote thet set of the linear isometries of E onto itself. Unlike the finite dimensional case, a bijective map may not have the holomorphic inverse.

LEMMA 1: *Let* E_1 *be the open unit ball of a Banach space* E *such that for every* $\alpha \in E_1$ *there exists* $f_\alpha \in \mathrm{Aut}(E_1)$ *with* $f_\alpha(0) = \alpha$. *Then for every* $g \in \mathrm{Aut}(E_1)$ *there exists* $S \in \mathrm{Iso}(E)$ *such that*

$$g = f_\alpha \circ S, \; g(0) = \alpha .$$

PROOF: We have $f_\alpha^{-1}(\alpha) = 0$ and $f_\alpha^{-1} \circ g(0) = 0$. By Schwarz's lemma [1] there exists $S \in \mathrm{Iso}(E)$ such that $S = f_\alpha^{-1} \circ g$ on E_1.

Let E be a Banach space with an unconditional basis $\{b_n\}$. The norm $\|x\|$ is called *symmetric* if for any permutation ϕ on N and for any sequence $\{\lambda_n\}$ in C with $|\lambda_n| = 1$, the following equality holds:

$$\left\| \sum_{n=1}^{\infty} x_n b_n \right\| = \left\| \sum_{n=1}^{\infty} \lambda_n x_{\pi(n)} b_n \right\| .$$

We state the following lemma from [3], p. 265.

LEMMA 2: *Let* E *be a Banach space with a basis* $\{b_n\}$ *and the symmetric norm* $\|x\|$. *If* E *is not a Hilbert space, then each linear isometry* ϕ *is of the from:*

$$\phi\left(\sum_{n-1}^{\infty} x_n b_n\right) = \sum_{n=1}^{\infty} \lambda_n x_{\pi(n)} b_n$$

where π *is a permutation of* N *and* $|\lambda_n| = 1$.

THEOREM 1: *Let* f *be a holomorphic mapping of the open unit ball* B *of* c_o *into itself. Then* $f \in \mathrm{Aut}(B)$ *if and only if there exist a permutation* π *of* N *and* $\alpha = (\alpha_n) \in B$ *such that if* $f = (f_n)$ *and* $x = (x_n) \in B$ *then*

$$f_n(x) = \lambda_n \frac{x_{\pi(n)} - \alpha_n}{1 - \bar{\alpha}_n x_{\pi(n)}}$$

for some constant λ_n *with* $|\lambda_n| = 1$.

PROOF: Let $\alpha = (\alpha_n) \in B$. Then the Möbius transformation

$$\phi_\alpha(x) = \sum_{n=1}^{\infty} \frac{x_n - \alpha_n}{1-\bar{\alpha}_n x_n} e_n$$

is an automorphism and $\phi_\alpha(0) = -\alpha$ ($\{e_n\}$ denotes the standard basis for c_o). It is an easy matter to check that

$$\phi_\alpha \circ \phi_{-\alpha} = \phi_{-\alpha} \circ \phi_\alpha = id_B .$$

Let $f \in Aut(B)$. Then there exists, $S \in Isc(c_o)$ such that $f = \phi_\alpha \circ S$ by Lemma 1, and hence we obtain the desired representation of f as a consequence of Lemma 2.

3. A CHARACTERIZATION OF c_o: Let E be a Banach space with a basis $\{b_n\}$. $\{b_n\}$ is said to be *monotone* if

$$\left\| \sum_{n=1}^{k} x_n b_n \right\| \leq \left\| \sum_{n=1}^{k+1} x_n b_n \right\|$$

for all k [4].

THEOREM 2: *Let* E *be a Banach space with a monotone normalized basis* $\{b_n\}$. *If every automorphism* f *on the open unit ball* E_1 *of* E *is of the form*

$$f\left(\sum_{n=1}^{\infty} x_n b_n\right) = \sum_{n=1}^{\infty} \lambda_n \frac{x_{\pi(n)} - \alpha_n}{1-\bar{\alpha}_n x_{\pi(n)}} b_n$$

(where π, α_n, λ_n *are as in Theorem 1), then* E *is isometrically isomorphic to* c_o.

We need the following lemmas.

LEMMA 3: *Let* $n \geq 2$ *be a fixed integer.*

(a) *For* $0 \leq \lambda \leq \frac{2}{n+2}$, $\alpha = \frac{n}{n^2+2n-2}$, *there exists* $x \in R$ *such that* $|x| \leq \frac{1}{n}$ *and* $\lambda = \left|\frac{x-\alpha}{1-\alpha x}\right|$.

(b) *Let* $m \in N$. *For each* λ, $0 \leq \lambda \leq \frac{m+2}{m+n+2}$, $\alpha = \frac{m}{m+n}$, *there exists* $x \in R$ *such that* $|x| \leq \frac{m+1}{m+n+1}$ *and* $\lambda = \left|\frac{x-\alpha}{1-\alpha x}\right|$.

PROOF: We use the fact the Möbius transformation on the open unit disk of the complex plane maps circles to circles and line segments to line segments. In particular, this transformation maps a real line segment to another real line segment. We prove only (b) since (a) can be shown in exactly the same way as (b).

For $\alpha = \frac{m}{m+n}$, let

$$\phi_\alpha(z) = \frac{z-\alpha}{1-\alpha z}, \quad z \in C, \ |z| < 1.$$

We denote by $S(r)$ the circle $\{z \in C : |z| = r\}$.

Then $\operatorname{card} \phi_\alpha\left(S\left(\frac{m+1}{m+n+1}\right)\right) \cap S\left(\frac{m+2}{m+n+2}\right) = 2$.

In fact,

$$-1 < a = \phi_\alpha\left(-\frac{m+1}{m+n+1}\right) < -\frac{m+2}{m+n+2} < 0 < b = \phi_\alpha\left(\frac{m+1}{m+n+1}\right) < \frac{m+2}{m+n+2}$$

and the interval $[a,b]$ is the diagonal of the circle $\phi_\alpha\left(S\left(\frac{m+1}{m+n+1}\right)\right)$. Therefore, we can find $x \in R$ such that

$$|x| \leq \frac{m+1}{m+n+1} \quad \text{and} \quad -\lambda = \frac{x-\alpha}{1-\alpha x}.$$

LEMMA 4: *Under the hypothesis of Theorem 2, we have*

$\| b_1 + b_2 + \ldots + b_n \| = 1$ *for each* $n \in N$.

PROOF: It is sufficient to show that for each λ, $0 < \lambda < 1$, $\lambda \|b_1 + b_2 + \ldots + b_n\| < 1$. Let $0 < \lambda < 1$. Then $0 < \lambda \leq \frac{m+2}{m+n+2}$ for some m.

We use induction on m.

If $\lambda \leq \frac{2}{n+2}$, then take $\alpha = \frac{n}{n^2 + 2n - 2}$ and $x \in R$ such that $|x| < \frac{1}{n}$ and $\phi_\alpha(x) = -\lambda$. Since $0 < \alpha < \frac{1}{n}$ and $|x| < \frac{1}{n}$, we have $\| \alpha(b_1 + \ldots + b_n) \| < 1$; $\| x(b_1 + \ldots + b_n)\| < 1$. By the hypothesis of Theorem 2, if $0 < \lambda \leq \frac{2}{n+2}$ then

$$\lambda \| b_1 + \ldots + b_n \| = \| \phi_\alpha(x)(b_1 + \ldots + b_n)\| < 1.$$

Let $m = 1$, i.e., $0 < \lambda \leq \frac{3}{n+3}$. Take $\alpha = \frac{1}{n+1}$. By Lemma 3, there exists $x \in R$ such that $|x| \leq \frac{2}{n+2}$ and $\lambda = \left| \frac{x - \alpha}{1 - \alpha x} \right|$. Since $\|\alpha(b_1 + \ldots + b_n)\| < 1$ and $\|x(b_1 + \ldots + b_2)\| < 1$ (by the preceding case), we must have

$$\lambda \|b_1 + \ldots + b_n \| < 1 .$$

Inductively we obtain for each $0 < \lambda \leq \frac{m+2}{m+n+2}$, $\lambda \|b_1 + \ldots + b_n\| < 1$. Now take $\lambda \to 1$ and we have $\| b_1 + \ldots + b_n\| \leq 1$. Since $\{b_n\}$ is monotone and normalized, we conclude that $\| b_1 + \ldots + b_n \| = 1$.

PROOF OF THEOREM: Since every automorphism fixing 0 is an isometry by Schwarz's lemma [1], it is immediately apparent that

$$\| \sum_{n=1}^{\infty} x_n b_n \| = \| \sum_{n=1}^{\infty} \lambda_n x_{\pi(n)} b_n \|$$

for any permutation π on N and $|\lambda_n| = 1$. From this fact we have

$$\|\sum_{n=1}^{\infty} x_n b_n\| = \|\sum_{n=1}^{\infty} |x_n| b_n\| .$$

Therefore, the basis $\{b_n\}$ is unconditional (see [4], p. 500).

Since $\|b_1 + \ldots + b_n\| = 1$ for all $n \in N$, $\{b_n\}$ is equivalent to the unit vector basis $\{e_n\}$ of c_o, i.e., there exist $M, N > 0$ such that

$$M \sup |x_n| \leq \|\sum_{n=1}^{\infty} x_n b_n\| \leq N \sup |x_n|$$

(see [4], p. 504). Therefore, $\sum_{n=1}^{\infty} x_n b_n \in E$ if and only if $\sum_{n=1}^{\infty} x_n e_n \in c_o$ by the definition of equivalent bases (see [4], p. 68).

Let $T : E \to c_o$ be defined by $T(\sum_{n=1}^{\infty} x_n b_n) = \sum_{n=1}^{\infty} x_n e_n$. Then T is an isomorphism. We want to show that T is an isometry.

Let $|\beta_n| \leq 1$. We can find numbers λ_n and μ_n on the unit circle of C satisfying $\beta_n = 1/2(\lambda_n + \mu_n)$. Then

$$(*) \quad \|\sum_{n=1}^{\infty} \beta_n x_n b_n\| = \|\sum_{n=1}^{\infty} 1/2(\lambda_n + \mu_n) x_n b_n\|$$

$$\leq 1/2(\|\sum_{n=1}^{\infty} \lambda_n x_n b_n\| + \|\sum_{n=1}^{\infty} \mu_n x_n b_n\|) = \|\sum_{n=1}^{\infty} x_n b_n\|.$$

This inequality shows that if $|y_n| \leq |x_n|$,

$$\|\sum_{n=1}^{k} y_n b_n\| \leq \|\sum_{n=1}^{k} x_n b_n\| \text{ (for all } k \in N).$$

From this and Lemma 4 we obtain

$$\|\sum_{n=1}^{k} x_n b_n\| = \|\sum_{n=1}^{k} |x_n| b_n\| \leq \sup_{1\leq n\leq k} |x_n| ,$$

Therefore,

$$\left\| \sum_{n=1}^{\infty} x_n b_n \right\| \le \sup_n |x_n| \quad .$$

From (*) we also obtain

$$\left\| \sum_{n=1}^{\infty} x_n b_n \right\| \ge |x_k| \quad \text{for all} \quad k \in N,$$

or

$$\left\| \sum_{n=1}^{\infty} x_n b_n \right\| \ge \sup_n |x_n|$$

This proves that T is an isometry.

REFERENCES

[1] L. A. HARRIS, Schwarz's lemma in normed linear spaces, Proc. Nat. Acad. Sc. 62(1969), pp. 1014 - 1017.

[2] L. NACHBIN, *Topology on spaces of holomorphic mappings*, Springer - Verlag, Berlin, 1969.

[3] S. ROLEWICZ, *Matric linear spaces*, PWN-Polish Scientific Publishers, Warszawa, 1972.

[4] I. SINGER, *Bases in Banach spaces* I, Springer - Verlag, Berlin, 1970.

New College

University of South Florida

Sarasota, Fla. 33580

Infinite Dimensional Holomorphy and Applications, Matos (ed.)

HOLOMORPHIC FUNCTIONS ON STRONG DUALS OF FRÉCHET – MONTEL SPACES

By *SEÁN DINEEN*

1. INTRODUCTION

Many interesting results have been obtained about holomorphic functions on strong duals of Fréchet-Schwartz ($\mathcal{DFS}$) spaces and in [4] and [6] a number of these results are extended to holomorphic functions on strong duals of Fréchet-Montel ($\mathcal{DFM}$) spaces. However the methods of proof differ greatly, on $\mathcal{DFS}$ spaces frequent use is made of the fact that such spaces are countable inductive limits by compact linear mappings in the category of topological spaces and continuous mappings while on $\mathcal{DFM}$ spaces great reliance is placed on the fact that $\mathcal{DFM}$ spaces are hereditary Lindelöf k-spaces. It is easily seen that the property of $\mathcal{DFS}$ spaces, quoted above, characterises $\mathcal{DFS}$ in the collection of $\mathcal{DFM}$ spaces (which is strictly larger) and consequently the non-linear properties of $\mathcal{DFM}$ spaces are an essential tool in the study of holomorphic functions on such spaces.

In this paper we investigate various questions concerning

holomorphic functions defined on open subsets of $\mathcal{DFM}$ spaces and show that all the usual topologies coincide on such spaces and the pseudo-convex open subsets of a $\mathcal{DFM}$ space are domains of existance of a plurisubharmonic function. E will, unless otherwise stated, denote a $\mathcal{DFM}$ space over the complex field and our notation, will, generally, be the standard notation of [4], [9] and [10] .

2. $\mathcal{DFM}$ SPACES

A locally convex infrabarrelled space in which every bounded set is relatively compact is called a Montel space. A metrizable Montel space is a Fréchet space and its strong dual is also a Montel space and is called a $\mathcal{DFM}$ space. Thus E, a $\mathcal{DFM}$ space, has a fundamental sequence of compact sets, $(B_n)_{n=1}^{\infty}$, which we may, and will, suppose are convex, balanced and increasing. For each n let E_{B_n} denote the vector space spanned by B_n and endowed with the norm generated by the Minkowski functional of B_n. E_{B_n} is a Banach space.

E is isomorphic to the inductive limit $\varinjlim_{n} E_{B_n}$ in the category of locally convex spaces and continuous linear mappings. Since Fréchet-Montel spaces are separable ([9] p370) and reflexive it follows that the compact subsets of E are complete separable metrizable spaces. A topological space is a Souslin space ([12]) if it is the continuous image of a complete separable metrizable space. Countable inductive limits of Souslin spaces are also Souslin spaces. Since E is the continuous image of $\varinjlim_{n} B_n$, the inductive limit of the spaces B_n endowed with the topology induced by E in the category of topological spaces and continuous mappings, it follows that E is a Souslin space and consequently a hereditary Lindelöf space. Using once more the above inductive limit we see that if U is an open subset of E then there exists a countable dense subset of the boundary of U, δU, of which each point is the limit of a sequence in U ([4]).

A topological space is a k-space if continuity on compact sets implies continuity.

PROPOSITION 1 *E is a k-space.*

PROOF Let X denote a topological space and let $f : U \longrightarrow X$ denote a mapping which is continuous on compact sets. Let $x_0 \in U$ be arbitrary and let V denote an open neighbourhood of $f(x_0)$. It suffices to show that $f^{-1}(V)$ is a neighbourhood of x_0. Since B_1 is compact we may choose $\lambda_1 > 0$ such that $f(x_0 + \lambda_1 B_1) \subset V$. Suppose $\lambda_2, \ldots, \lambda_n$, positive scalars, have been chosen so that

$$f(x_0 + \sum_{i=1}^{n} \lambda_i B_i) \subset V.$$

If for each integer m there exists $(x_{i,m})_{i=1}^{n+1}$ such that $x_{i,m} \in B_i$ and $f(x_0 + \sum_{i=1}^{n} \lambda_i x_{i,m} + \frac{1}{m} x_{n+1,m}) \notin V$, then, since B_i is a compact metrizable space, we may use a diagonal process to find a subsequence of $(\sum_{i=1}^{n} \lambda_i x_{i,m} + \frac{1}{m} x_{m,n+1})_{m=1}^{\infty}$, $(y_n)_{n=1}^{\infty}$, which converges to $y \in \sum_{i=1}^{n} \lambda_i B_i$. Hence $f(y_n) \to f(y)$ as $n \to \infty$ but $f(y_n) \notin V$ for any n and V is a neighbourhood of $f(y)$. This is impossible and hence we can find $\lambda_{n+1} > 0$ such that

$$f(x_0 + \sum_{i=1}^{n+1} \lambda_i B_i) \subset V.$$

By induction we may now find $(\lambda_i)_{i=1}^{\infty}$ such that $f(\sum_{i=1}^{\infty} \lambda_i B_i)^{(*)} \subset V$.

(*) This means all finite sums.

Since $\sum_{i=1}^{\infty} \lambda_i B_i$ is convex, balanced and absorbs all the bounded subsets of E and E is a bornological space f is continuous at x_0. Hence E is a k-space

COROLLARY 2 $E = \varinjlim_n B_n$ *in the category of topological spaces and continuous mappings.*

COROLLARY 3 *Sequentially continuous mappings from open subsets of* E *are continuous.*

PROOF Follows from proposition 1 and the fact that the compact subsets of E are metrizable.

COROLLARY 4 *Sets of the form* ($\sum_{n=1}^{\infty} \lambda_n B_n$, $\lambda_n > 0$, *all* n) *form a neighbourhood base at* 0 *in* E.

PROOF The proceeding proof applied to the identity mapping on E shows that every neighbourhood of 0 in E contains a neighbourhood of the form $\sum_{n=1}^{\infty} {}_n B_n$ for some sequences of positive scalars.

3. HOLOMORPHIC FUNCTIONS $\mathcal{DFM}$ SPACES

(H(U;F) (resp. H(U)) will denote the set of all F (resp. C) valued holomorphic functions U.

PROPOSITION 5 ([4]) (H(U), T_0) *is a Frêchet space.*

PROOF Since E is a k-space (H(U), T_0) is complete. Since U is Lindelöf we have $U = \bigcup_n U_n$ where each U_n is a translate of a convex balanced open subset of E. Let $U_n = x_n + V_n$. The sequence $x_n + \{B_j \cap \lambda_\kappa V_n\}_{j,\kappa}$ forms a fundamental sequence of compact subsets of U_n when $(\lambda_n)_{n=1}^{\infty}$ denotes an increasing sequence of positive numbers and $\lim_{n\to\infty} \lambda_n = 1$.
Hence U contains a fundamental family of compact sets and

$(H(U), T_0)$ is metrizable. This completes the proof.

$L_s(^nE)$ is the vector space of continuous symmetric n-linear functionals on E endowed with the topology of uniform convergence on the compact subsets of E. $L_s(^nE)$, and $P(^nE)$ are isomorphic as locally convex topological vector space.

PROPOSITION 6 *The bounded subsets of* $(H(U), T_0)$ *are locally bounded.*

PROOF We may assume, without loss of generality, that U is a convex balanced open subset of E. Let B denote a bounded subset of $(H(U), T_0)$. We complete the proof by finding a neighbourhood of 0, V, such that $\sup_{f\in B} \| f \|_V < \infty$. Choose $\lambda_1 > 0$ such that $\lambda_1 B_1 \subset U$ and

$$\sup_{f\in B} \| f \|_{\lambda_1 B_1} = M < \infty$$

Let $\varepsilon > 0$ be arbitrary and suppose $\lambda_2, \dots \lambda_\kappa$ have been chosen so that $\sum_{i=1}^{\kappa} \lambda_i B_i \subset U$ and

$$\sup_{f\in B} \| f \|_{\sum_{i=1}^{\kappa} \lambda_i B_i} \leq M + \varepsilon$$

Let $L_1 = \sum_{i=1}^{\kappa} \lambda_i B_i$ and let $\varepsilon' > 0$ be arbitrary. Choose $\delta_1 > 0$ so that $L_2 = \sum_{i=1}^{\kappa} \lambda_i B_i + \delta_1 B_{\kappa+1} \subset U$. Since L_2 is a compact subset of we can find $\delta_2 > 1$ such that $\delta_2 L_2 \subset U$. Hence

$$p(f) = \sum_{n=0}^{\infty} \left\| \frac{\hat{d}^n f(0)}{n!} \right\|_{\delta_2 L_2} = \sum_{n=0}^{\infty} \delta_2^n \left\| \frac{\hat{d}^n f(0)}{n!} \right\|_{L_2}$$

is a T_0-continuous semi-norm on H(U) (to check this use Cauchy's inequalities and the fact that U is balanced). Since B is bounded we can find a positive integer N such that

$$\sup_{f\in B} \sum_{N}^{\infty} \left\| \frac{\hat{d}^n f(o)}{n!} \right\|_{L_2} \leq \frac{\varepsilon'}{4}.$$

For $f \in H(U)$ let $\frac{d^n f(0)}{n!}$ denote the continuous symmetric n-linear form which is canonically associated with the n-homogeneous polynomial $\frac{\hat{d}^n f(0)}{n!}$.

For $\lambda > 0$, we have, by expanding each polynomial,

$$\sup_{f\varepsilon B} \left|\left| \sum_{n=0}^{n-1} \frac{\hat{d}^n f(0)}{n!} \right|\right|_{L_1 + \lambda B_{\kappa+1}}$$

$$\leq \sup_{f\varepsilon B} \left|\left| \sum_{n=0}^{n-1} \frac{\hat{d}^n f(0)}{n!} \right|\right|_{L_1}$$

$$+ |\lambda| \sum_{n=1}^{n-1} \sup_{\substack{f\varepsilon B \\ x\varepsilon L_1 \\ y\varepsilon B_{\kappa+1}}} \left| \sum_{i=1}^{n} \binom{n}{i} \frac{d^n f(0)}{n!} (x)^{n-i} (\lambda y)^{i-1} (y) \right|$$

Since $L_1 \subset L_2$ we have

$$\sup_{f\varepsilon B} \left|\left| \sum_{n=0}^{n-1} \frac{\hat{d}^n f(0)}{n!} \right|\right|_{L_1} \leq M + \varepsilon + \frac{\varepsilon'}{4}.$$

Since B is bounded

$$\sum_{n=1}^{n-1} \sup_{\substack{f\varepsilon B \\ x\varepsilon L_1 \\ y\varepsilon B_{\kappa+1}}} \left| \sum_{i=1}^{n} \binom{n}{i} \frac{d^n f(0)}{n!} (x)^{n-i} (\lambda y)^{i-1} (y) \right| < \infty$$

Hence we can choose $\lambda_{\kappa+1}$ so that $\lambda_{\kappa+1} < \delta_1$ and

$$\sup_{f\varepsilon B} \left|\left| \sum_{n=0}^{n-1} \frac{\hat{d}^n f(0)}{n!} \right|\right|_{L_1 + \lambda_{k+1} B_{k+1}} \leq M + \varepsilon + \frac{\varepsilon'}{2}$$

It now follows that

$$\sup_{f\varepsilon B} || f ||_{L_1 + \lambda_{\kappa+1} B_{\kappa+1}} \leq M + \varepsilon + \varepsilon'$$

Since ε and ε' were arbitrary we may follow an inductive process to find a sequence of positive integers, $(\lambda_n)_{n=1}^{\infty}$, such that

$$\sup_{f\varepsilon B} || f ||_{\sum_{n=1}^{\infty} \lambda_n B_n} \leq 2M$$

Hence B is locally bounded at 0 and this completes the proof.

We now obtain a result which was proved for $\mathcal{DFS}$ spaces in [1]. We refer to [5] for the definitions of the different topologies on H(U).

PROPOSITION 7 *On* H(U), $T_0 = T_\omega = T_\delta$.

PROOF We always have $T_0 \leq T_\omega \leq T_\delta$. By proposition 6 every T_0-bounded subset of H(U) is equibounded and hence T_δ-bounded. Since T_0 is a bornological topology it follows that $T_0 = T_\delta$.

A locally convex is said to be semi-Montel if its bounded subsets are relatively compact.

PROPOSITION 8 (H(U), T_0) *is a Fréchet-Montel space.*

PROOF Semi-Montel spaces are closed under arbitrary products and closed subspaces, hence, we may assume that U is a convex balanced subset of E. Since (H(U), T_0) is metrizable it suffices to show that any bounded sequence in H(U), $(f_n)_{n=1}^{\infty}$, has a convergent subsequence. By taking subsequences and using a diagonal process if necessary, we may assume, since U is separable, that $(f_n)_{n=1}^{\infty}$ converges pointwise on a dense subset of U.

Since $(f_n)_{n=1}^{\infty}$ is equibounded we can, given $x_0 \in U$, find a convex balanced open set, V, such that

$$\sup_{n\geq 1} \sum_{m=0}^{\infty} 2^m \left\|\frac{\hat{d}^m f_n(x_0)}{m!}\right\|_V \leq M < \infty$$

Hence, given $\varepsilon > 0$, we can find $\delta > 0$ such that

$$\sup_{\substack{n\geq 1 \\ x\in\delta V}} |f_n(x_0) - f_n(x_0 + x)| \leq \varepsilon.$$

This shows that the sequence $(f_n)_{n=1}^{\infty}$ is equicontinuous. By a simple argument it follows that $(f_n)_{n=1}^{\infty}$ converges at all points of U to a function which we call f_0. By the classical Montel theorem f_0 is G-holomorphic and since the sequence $(f_n)_{n=1}^{\infty}$ is locally bounded the function f_0 is also locally bounded and hence continuous.

By Ascoli's theorem the sequence $(f_n)_{n=0}^{\infty}$ is a compact subset of (H(U), T_0). Hence $(f_n)_{n=1}^{\infty}$ contains a convergent subsequence and this completes the proof.

REMARK The above method shows that equibounded sets of holomorphic function on arbitrary locally convex spaces are equicontinuous.

We now show that weak and strong holomorphic functions coincide on open subsets of $\mathcal{DFM}$ spaces.

LEMMA 9 *Let* E *and* F *denote arbitrary locally convex spaces. If for each open subset* U *of* E *the bounded subsets of* $(H(U), T_0)$ *are equibounded then* $H(U:F) = H(U:(F, \sigma(F', F))$.

PROOF We may suppose that F is a normed linear space. Let B denote the unit ball of F'. Suppose $f \in H(U; (F, \sigma(F',F))$. If K is a compact subset of U and $\phi \in F'$ then $(\phi \circ f)(K)$ is a bounded subset on C. Hence f(K) is a weakly bounded subset of F and by Mackey's theorem this implies f(K) is a strongly bounded subset of F.

Thus $(\phi \circ f)_{\phi \in B}$ is a bounded subset of $(H(U), T_0)$. By our hypothesis this implies that $(\phi \circ f)_{\phi \in B}$ is an equibounded subset of H(U) and hence we can find, for each $x_0 \in U$, a neighbourhood of x_0, V, such that

$$\sup_{x \in V} ||f(x)|| = \sup_{\substack{x \in V \\ \phi \in B}} |\phi \circ f(x)| \leq M,$$ i.e. f is locally bounded and hence continuous. Since $H(U;F) \subset H(U;(F, \sigma(F', F))$ for any pair of locally convex spaces E and F we have completed the proof.

COROLLARY 10 *Let* F *denote an arbitrary locally convex space then* $H(U;F) = H(U;(F, \sigma(F', F))$ *if* U *is an open subset of a* $\mathcal{DFM}$ *space.*

COROLLARY 11 *Let* F *denote an arbitrary locally convex space and let* U *denote an open subset of a* $\mathcal{DF}\!\mathcal{M}$ *space. Then* $f: U \longrightarrow F$ *is holomorphic if and only if* f *is bounded on the compact subsets of* U *and is G-holomorphic.*

PROOF We may assume that U is convex and balanced. If $f:U \longrightarrow F$ is G-holomorphic and bounded on the compact subsets of U it suffices by proposition 1 and corollary 10 to show $\phi \circ f$ is continuous on each compact subset of U for each ϕ in F'. Let ϕ denote a fixed element of F' and let B denote a compact subset of U. By Cauchy's inequalities there exists a $\lambda > 1$ such that

$$\sum_{n=0}^{\infty} \lambda^n \left|\left|\frac{\hat{d}^n(\phi \circ f)}{n!}(0)\right|\right|_B \leq M \leq \infty$$

Hence it suffices to show $\frac{\hat{d}^n(\phi \circ f)}{n!}(0)$ is continuous for each n. Let $P_n = \frac{\hat{d}^n(\phi \circ f)}{n!}(0)$ and let $\tilde{P}_n$ denote the associated symmetric n-linear form. As in proposition 6 it suffices to prove the following; if K and L are convex balanced compact subsets of E and $||P_n||_K \leq M$ then for each $\varepsilon > 0$ we can find $\lambda > 0$ such that $||P_n||_{K+\lambda L} \leq M + \varepsilon$.

Since $||P_n||_{K+\lambda L} \leq ||P_n||_K + |\lambda| \sum_{i=1}^{n} \binom{n}{i} |\lambda|^{i-1} \sup_{\substack{x \in L \\ y \in K}} |\tilde{P}_n(x)^{n-i}(y)^i|$

and $\sup_{\substack{x \in L \\ y \in K}} |\tilde{P}_n(x)^{n-i}(y)^i| < \infty$ for all i, $0 \leq i \leq n$, this follows immediately.

COROLLARY 12 *A locally convex valued polynomial defined on* E *is continuous if and only if its restriction to the Banach spaces* E_{B_n} *is continuous for each* n.

PROOF A polynomial on a Banach space is continuous if and only if it is bounded on bounded sets and each bounded subset of E is contained and norm bounded in some B_n. Restating corollary 11 we have the following result.

COROLLARY 13 $E = \varinjlim_n E_{B_n}$ *in the category of locally convex spaces and continuous polynomial mappings.*

COROLLARY 14 *Separately continuous polynomials defined on a product of $\mathcal{DF}\mathcal{M}$ spaces are continuous.*

PROOF If $E = \varinjlim_n E_{B_n}$ and $F = \varinjlim_n F_{C_n}$ are $\mathcal{DF}\mathcal{M}$ spaces then $E \times F$ is also a $\mathcal{DF}\mathcal{M}$ space and $E \times F = \varinjlim_n E_{B_n} \times F_{C_n}$ (all inductive limits being taken in the category of locally convex spaces and continuous linear mappings). If P is a separately continuous polynomial on $E \times F$ then P restricted to $E_{B_n} \times F_{C_n}$ is separately continuous for each n and hence is continuous (by Hartogs' theorem on separate analyticity for Banach spaces (see [10])). Hence P is continuous by corollary 11.

COROLLARY 15 *If F is a sequentially complete locally convex space and (U,V) is a $\mathbb{C}$-extension pair*(*) *of domains spread over E then (U,V) is an F-extension pair.*

PROOF Apply corollary 10.

COROLLARY 16 *If* $(\phi_n)_{n=1}^{\infty}$ *is a sequence in E' then* $\sum_{n=1}^{\infty} \phi_n^n \in H(E)$ *if and only if* $\phi_n \to 0$ *as* $n \to \infty$ *in* $(E', \sigma(E,E'))$.

PROOF If $\sum_{n=1}^{\infty} \phi_n^n \in H(E)$ then $\sum_{n=1}^{\infty} |\phi_n(x)|^n < \infty$ for all all $x \in E$. Hence $\phi_n \to 0$ as $n \to \infty$ in $(E', \sigma(E,E'))$.

Conversely suppose $\phi_n \to 0$ as $n \to \infty$ pointwise on E. By [9] p. 370 $\phi_n \to 0$ as $n \to \infty$ uniformly on the bounded subsets of E

(*) (U,V) is a F-extension pair if each $f \in H(U;F)$ there exists a unique $\tilde{f} \in H(V;F)$ such that $\tilde{f}|_U = f$.

Hence $||\sum_{n=1}^{\infty}\phi_n^n||_B \leq \sum_{n=1}^{\infty}(||\phi_n||_B)^n$ is bounded if B is a bounded subset of E. By corollary 11 $\sum_{n=1}^{\infty}\phi_n^n \in H(E)$. This completes the proof.

We now show that the bornological topology associated with T_ω, $T_{\omega,b}$, is equal to the T_δ topology on certain surjective limits of $\mathcal{DFM}$ spaces. Examples of spaces which satisfy our criteria are $\mathcal{D}'(\Omega)$, Ω an open subset of $\mathbb{R}^n$, and $\prod_{(\alpha)} E_\alpha$, E_α a $\mathcal{DFM}$ space for each α and (α) may have any cardinality. We refer to [4] for background material to this result (especially sections 7 and 8). E will denote an arbitrary locally convex space. $H(E_0)$ is the vector space of germs of complex valued holomorphic functions at 0 in E. We endow $H(E_0)$ with the inductive limit topology $\varinjlim_V H_b(V)$ where V ranges over all open subsets of E which contain 0 and $H_b(V) = \{f, f \in H(V), ||f||_V < \infty\}$ is endowed with its norm topology. Since $H_b(V)$ is a Banach space $H(E_0)$ is barrelled and bornological (in fact ultrabornological) and the canonical injection $(H(V), T_\omega) \longrightarrow H(E_0)$ is continuous for each open neighbourhood of 0, V.

PROPOSITION 16 *Let $\theta = (E_\alpha, \pi_\alpha)_{\alpha \varepsilon A}$ denote a compact, open, symmetric, i (resp. j) surjective representation of* E *and let* U *denote a convex balanced open subset of* E. *If each holomorphic function on* U *has minimal θ-support and the T_0-bounded subsets (resp. sequences) in $H(\pi_\alpha(U))$ are equibounded for each α in* A *then the bounded subsets (resp. sequences) of $(H(U), T_\omega)$ are equibounded and $(H(U), T_\delta) = (H(U), T_{\omega,b})$.*

PROOF Let B denote a bounded subset (resp. sequence) in $(H(U), T_\omega)$. For each f in B let A(f) denote a minimal θ-support for f. If $A_1 = \bigcup_{f \varepsilon B} A(f)$ is an E-bounded subset of A then the required result follows immediately. Otherwise, using the fact that θ is an i (resp. j) surjective representation when B is a set (resp. a sequence), we can find a sequence of E-open subsets of A, $(W_n)_{n=1}^{\infty}$, with the following properties;

(1) $A_1 \cap W_n \neq \phi$ for each n,

(2) if $\alpha \varepsilon A$ then there exists a positive integer $n(\alpha)$ such that $\alpha \varepsilon W_n^c$ for all $n \geq n(\alpha)$.

Hence we may choose of elements in B, $(f_n)_{n=1}^{\infty}$, such that

(3) $f_n(x_n + y_n) \neq f_n(x_n)$ for all n and

(4) $S(y_n) \subset W_n$

By (2) the sequence $(y_n)_{n=1}^{\infty}$ converges very strongly to 0. Hence, by Liouville's theorem, we can suppose

(5) $f_n(x_n + y_n) - f(x_n) \geq n$ for all n.

Since B is bounded $\sup_{f \varepsilon B} \left| \frac{\hat{d}^n f(0)}{n!}(x) \right| < \infty$ for all n and thus by (5) we can find, using induction, two increasing sequences of positive integers, $(\ell_n)_{n=1}^{\infty}$ and $(\kappa_n)_{n=1}^{\infty}$, such that

$$(6) \quad \sum_{\kappa=\kappa_n+1}^{\kappa_{n+1}} \left| \frac{\hat{d}^\kappa f_{\ell_n}(0)}{\kappa!}(x_{\ell_n} + y_{\ell_n}) - \frac{\hat{d}^\kappa f_{\ell_n}(x_{\ell_n})}{\kappa!} \right| \geq n$$

Now consider the following seminorm on $H(E_0)$

$$p(f) = \sum_{n=1}^{\infty} \sum_{\kappa=\kappa_n+1}^{\kappa_{n+1}} \sup_{j \geq \ell_n} \left| \frac{\hat{d}^\kappa f(0)}{\kappa!}(x_j + y_j) - \frac{\hat{d}^\kappa f(0)}{\kappa!}(x_j) \right|$$

If $f \varepsilon H(E_0)$ then, since $(y_j)_{j=1}$ is a very strongly convergent sequence, there exists j_0 such that

$$\frac{\hat{d}^\kappa f(0)}{\kappa!}(x_j + y_j) - \frac{\hat{d}^\kappa f(0)}{\kappa!}(x_j) = 0 \text{ for all } \kappa \text{ and all } j \geq j_0.$$

Hence $p(f) = \sum_{n=1}^{j_0} \sum_{\kappa=\kappa_n+1}^{\kappa_{n+1}} \sup_{\substack{j \geq \ell_n \\ j \leq j_0}} \left| \frac{\hat{d}^\kappa f(0)}{\kappa!}(x_j + y_i) - \frac{\hat{d}^\kappa f(0)}{\kappa!}(x_j) \right|$

is finite for all f in $H(E_0)$.

Since $p(f) = \left|\frac{\hat{d}^n f(0)}{n!}(x)\right|$ is a continuous semi-norm on $H(E_0)$ for any x in E and $H(E_0)$ is barrelled it follows that p is a continuous semi-norm on $H(E_0)$. By (6) the image of B in $H(E_0)$ is not bounded and this contradiction implies that A_1 is an E-bounded subset of A. Hence there exists an α in A and $B=(\tilde{f})_{f\in B}$, a set of G-holomorphic functions on $\pi_\alpha(U)$ such that $f = \tilde{f} \circ \pi_\alpha$ for all $f \in B$. Since θ is a compact surjective limit it follows that B is a T_0-bounded and hence equicontinuous subset of $H(\pi_\alpha(U))$. Hence B is an equibounded subset of H(U) and this completes the proof.

COROLLARY 17 *Let $\theta = (E, \pi_\alpha)_{\alpha\in A}$ denote a compact, open symmetric representation of E by $\mathcal{DFM}$ spaces. If each holomorphic function defined on a convex balanced open subset of E has minimal θ-support then the following results hold for open subsets of E.*

(a) *If θ is a j-surjective representation then*

$$(H(U), T_{\omega,b}) = (H(U), T_\delta).$$

(b) *If θ is an i-surjective representation then the bounded subsets of $(H(U), T_\omega)$ are equibounded, $(H(U), T_\omega)$ is a semi-Montel space, $(H(U), T_{\omega,b})$ is a Montel space and T_0, T_ω and $T_{\omega,b}$ induce the same topology on the T_ω-bounded subsets of H(U).*

PROOF Since the restriction mapping $(H(U), T_\omega) \longrightarrow (H(V), T_\omega)$ $(U \supset V)$ is continuous we apply proposition 16 to complete the proof of (a) and the first part of (b). Let $H_\alpha(U)$ denote the set of all f in H(U) which factor locally through E_α. The bounded subsets of $(H_\alpha(U), T_0)$ are equibounded and hence $(H_\alpha(U), T_0)$ is a complete barrelled bornological locally convex space (use the methods of propositions 6 and 7). Hence, if θ is an i-surjective representation, the identity mapping;

$$(H_\alpha(U), T_0) \longrightarrow (H(U), T_{\omega,b})$$

is continuous. If B is a bounded subset of $(H(U), T_{\omega,b})$ then B is contained and bounded in $(H_\alpha(U), T_0)$ for some α in A. Since $(H_\alpha(U), T_0)$ is a Montel space (use proposition 7 and the fact that $(H_\alpha(U), T_0)$ is isomorphic to a closed subspace of $\prod_i (H(V_i), T_0)$ where V_i is an open subset of E_α for all i) $(H(U), T_\omega)$ is a semi-Montel space and $(H(U), T_{\omega,b})$ is a Montel space. We have also shown, since $T_0 \leq T_\omega \leq T_{\omega,b}$, that T_0, T_ω and $T_{\omega,b}$ induce the same topology on the T_ω-bounded subsets of H(U).

REMARK If θ satisfies the "countable stability" condition of [11] then θ is trivially a j-surjective representation and it is possible to prove a result similar to Corollary 17 without the θ-minimal support property on H(U).

EXAMPLE 18 $(H(\mathcal{D}'(\Omega), T_{\omega,b})$ is a Montel space.

It is possible (see [5]) that $T_o = T_\omega$ even when all the conditions of proposition 16 are satisfied.

An open subset of a locally convex space is said to be pseudo-convex if its finite dimensional sections are pseudo-convex.

PROPOSITION 19 *Pseudo-convex open subsets of $\mathcal{DFM}$ spaces are domains of existance of plurisubharmonic functions.*

PROOF Let U denote a pseudo-convex open subset of the space E. Let $(K_n)_{n=1}^{\infty}$ denote an increasing sequence of compact subsets of U such that $\bigcup_n K_n = U$. By theorem 2.3.6 of [10] we may suppose that each K_n is equal to its plurisubharmonic hull. Let $(x_{n,m})_{n,m=1}^{\infty}$ denote a double sequence of points in U such that $x_{n,m} \longrightarrow x_n \in \delta U$ (boundary of U) as $m \longrightarrow \infty$ for each n and the sequence $(x_n)_{n=1}^{\infty}$ is a dense subset of the boundary of U. Let $\phi : N \longrightarrow N$ denote a function from the integers into itself such that $\phi^{-1}(n)$ is infinite for all n. Now choose m_1 such that $x_{\phi(1),m_1} \notin K_1$. By theorem 2.3.6 of [10] there exists a plurisub harmonic function on U, f_1, such that

$$f_1(x_{\phi(1),m_1}) > \sup_{x\in K_1} f_1(x)$$

Since a $\exp(b\, f_1)$ is plurisubharmonic for all positive a and b we may suppose

$$f_1(x_{\phi(1),m_1}) > 2 > \tfrac{1}{2} > \sup_{x\in K_1} f_1(x) \geq 0$$

Suppose $(\ell_i)_{i=1}^{n}$ and $(m_i)_{i=1}^{n}$, two increasing sequences of integers, and $(f_i)_{i=1}^{n}$, a sequence of plurisubharmonic functions on U, have been chosen such that $\ell_1 = 1$,

$$f_i(x_{\phi(i),m_i}) \geq 2^i \geq \frac{1}{2^i} > \sup_{x\in K_{\ell_i}} f_i(x) \geq 0$$

for $i = 1, \ldots, n$ and $K_{\ell_{i+1}} \supset K_{\ell_i} \cup \{x_{\phi(i),m_i}\}$ for $i=1,\ldots,n-1$

Choose ℓ_{n+1} such that $\ell_{n+1} > \ell_n$ and $K_{\ell_{n+1}} \supset K_{\ell_n} \cup \{x_{\phi(n),m_n}\}$.

Next choose m_{n+1} such that $m_{n+1} > m_n$ and $x_{\phi(n+1),m_{n+1}} \notin K_{\ell_{n+1}}$ and f_{n+1} a plurisubharmonic function on U such that

$$f_{n+1}(x_{\phi(n+1),m_{n+1}}) > 2^{n+1} \geq \frac{1}{2^{n+1}} \geq \sup_{x\in K_{\ell_{n+1}}} f_{n+1}(x) \geq 0.$$

By induction we then define the sequence $(f_n)_{n=1}^{\infty}$. Let $f = \sum_{n=1}^{\infty} f_n$. By our construction this sum converges at all points of U and is unbounded on each neighbourhood of x_n, n arbitrary. Each f_n is a positive function and a finite sum of plurisubharmonic functions is plurisubharmonic, hence it suffices to show f is upper semi-continuous to complete the proof. Since U is a k-space it suffices to show that f is upper semi-continuous on each compact subset of U. Let K denote an arbitrary compact subset of U and let C denote some real number. By our construction we can find a positive integer N such that $||f_n||_K \leq \frac{1}{2^n}$ for all $n \geq N$. Let $V = \{x \in K, f(x) < C\}$. Let $x_0 \in V$. Choose $M \geq N$ such that $\frac{1}{2^M} < C - f(x_0)$. Since $\sum_{n=1}^{M+2} f_n$ is plurisubharmonic there exists a neighbourhood of x_0 in K, W, such that

$$\sup_{x\in W} \sum_{n=1}^{M+2} f_n(x) < c - \frac{1}{2^{M+1}}$$

$$\text{Hence } \sup_{x\in W} f(x) \leq \sup_{x\in W} \sum_{n=1}^{M+2} f_n(x) + \sum_{n=M+3}^{\infty} \frac{1}{2^n}$$

$$< (C - \frac{1}{2^{M+1}}) + \frac{1}{2^{M+2}} < C$$

Thus $f|_K$ is plurisubharmonic and U is the natural domain of existance of f.

PROPOSITION 19 *If E, a locally convex space, has an open surjective representation by $\mathcal{DFM}$ spaces then the pseudo-convex open subsets of E are domains of existance of plurisubharmonic functions.*

PROOF Let $\theta = (E_\alpha, \pi_\alpha)_{\alpha\in\Lambda}$ denote the open surjective representation of E by $\mathcal{DFM}$ spaces and suppose U is a pseudo-convex open subset of E. By [4] there exists an α in Λ such that

$U = \pi_\alpha^{-1}(\pi_\alpha(U))$ and $\pi_\alpha(U)$ is a pseudo-convex open subset of E_α. By proposition 18 there exists a plurisubharmonic function on $\pi_\alpha(U)$, f, which is unbounded on each neighbourhood of each boundary point of $\pi_\alpha(U)$. Let $\tilde{f} = f \circ \pi_\alpha$. $\tilde{f}$ is a plurisubharmonic function on U. If $\xi \in \delta U$ and V is a neighbourhood of ξ in E then $\pi_\alpha(\xi) \in \delta(\pi_\alpha(U))$ and $\pi_\alpha(V)$ is a neighbourhood of $\pi_\alpha(\xi)$ in $\pi_\alpha(E)$. Hence $||f||_{\pi_\alpha(V)} = \infty$ and $||\tilde{f}||_V = ||f \circ \pi_\alpha||_V = ||f||_{\pi_\alpha(U)} = \infty$. Thus U is the natural domain of existance of f. This completes the proof.

For the sake of completeness we include the following results. (a) is proved in [4] and (b) is proved for $\mathcal{DFM}$ spaces in [6].

PROPOSITION 19 (a) *A holomorphically convex open subset of a* $\mathcal{DFM}$ *space is the domain of existance of a holomorphic function.*

(b) *If the locally convex space* E *has an open surjective representation by* $\mathcal{DFM}$ *spaces each of which has a Schauder basis then the pseudo-convex open subsets of* E *are domains of existance of holomorphic functions.*

PROOF (b) Use the result in [6] for $\mathcal{DFM}$ spaces and exactly the same method as used in proposition 18.

We have been unable to prove or disprove the following conjecture.

CONJECTURE *Silva (or Mackey) continuous G-holomorphic functions on* $\mathcal{DFM}$ *spaces are continuous.*

If this conjecture were true then it would follow that $\mathcal{DFM}$ spaces are Zorn spaces (i.e. the set of points of continuity of G-holomorphic functions on open subsets of $\mathcal{DFM}$ spaces is open and closed).

This conjecture requires a deep study of convergent sequences which are not Mackey convergent. Indeed it is equivalent to showing that convergent sequences are bounding subsets for Silva holomorphic functions and a counterexample may not be found by using the usual techniques (this follows by corollary 15). Grothendieck's example of a $\mathcal{DFM}$ space E which is not a $\mathcal{DFS}$ space does not provide a counterexample.

This results from the following facts about E (which do not appear to be common to all $\mathcal{DFM}$ spaces which are not $\mathcal{DFS}$);

1) $E = \varinjlim_n E_n$ and each E_n is isometrically isomorphic to ℓ_∞.
2) If B denotes the unit ball in ℓ_∞ then $\overline{T_n(B \cap c_0)}\,{}_{n=1}^{\infty}$ is a fundamental sequence of bounded subsets of E.
3) Every element of $H(\lambda B)$, $\lambda > 1$, is bounded on $B \cap c_0$ ([8]).

The results on surjective limits parallel some of those in section 7 of [4] and loosely speaking we have shown that results for the T_0 topology can be extended to the T_ω topology without the extension requirement on the surjective limit.

The method of proposition 16 may also be combined with techniques in [2] to study holomorphic functions on domains spread over surjective limits of $\mathcal{DFM}$ spaces and this investigation has subsequently been carried out in [2].

BIBLIOGRAPHY

[1] J. BARROSO, M. MATOS and L. NACHBIN; On bounded sets of holomorphic mappings, Lecture Notes in Maths, Vol. 365, Springer-Verlag, (1973), 216-224.

[2] P. BERNER; A global factorization property for holomorphic functions of a domain spread over a surjective limit, Seminaire P.Lelong, 1974/75. Lecture Notes in Maths, 524 Springer-Verlag (1976)

[3] P. BERNER; Topologies on spaces of holomorphic functions of certain surjective limits (this proceedings).

[4] S. DINEEN; Surjective limits of locally convex spaces and their application to infinite dimensional holomorphy. Bull. Soc. Math. Fr. t103, 1975 (to appear).

[5] S. DINEEN; Holomorphic Functions on locally convex spaces I, Locally convex topologies on H(U), Ann Inst. Fourier, Grenoble, t23, 3, (1973), 155-185.

[6] S. DINEEN; PH. NOVERRAZ and M. SCHOTTENLOHER; Le probleme de Levi dens certains espace vectoriels topologiques localement convexes, Bull Soc. Math. Fr. t. 104 (1976).

[7] A. GROTHENDIECK; Sur les espaces (F) et (DF). Summa Bras. Math. 3, 57-123, (1954).

[8] B. JOSEFSON; Bounding Subsets of $\ell^{\infty}(A)$, Thesis, Uppsala, 1975.

[9] G. KOETHE; Topological vector spaces I, Springer-Verlag, Bd 159, 1969.

[10] PH. NOVERRAZ; Pseudo-Convexite, convexite polynomiale et domains d'holomorphie en dimension infinil, North-Holland, 1973.

[11] PH. NOVERRAZ; On a particular case of surjective limit (this proceedings).

[12] L. SCHWARTZ; Radon measures on arbitrary topological spaces and cylindrical measures, Oxford University Press, 1973.

DEPARTMENT OF MATHEMATICS
UNIVERSITY COLLEGE DUBLIN
DUBLIN 4, IRELAND.

Infinite Dimensional Holomorphy and Applications, Matos (ed.)

DIFFERENTIAL EQUATIONS OF INFINITE ORDER IN VECTOR-VALUED HOLOMORPHIC FOCK SPACES

By *THOMAS A. W. DWYER, III*

CONTENTS

INTRODUCTION

Various situations where power series in infinite dimensional domains naturally arise also involve infinite-dimensional ranges: e.g., the Volterra series representation of the outputs of non-linear systems as analytic functions of input signal [Al,2,3], [Bol.1,2,3,4,5], [Br.1,2,3], [W], and the variational equations related to the representation of solutions of well posed boundary value problems as functional power series, where the variable is the boundary value function [DL]. This last reference especially shows the desirability of extending the existence and approximation theorems on convolution equations and partial differential equations in infinite dimension of [G 1,2,3] [N 2,3,4,5], [Dil,2], [Mat 1,2], [Dw 1,2,3,5,6,7,8], [Bol 1,2,3,4,5] and [BD] to vector-valued functions.

Existence theorems do not hold for general convolution equations $\vec{T} * \vec{f} = \vec{g}$, where $\vec{f}$ and $\vec{g}$ are mappings from a (dual) vector space E' to a vector space F and $\vec{T}$ is an F-valued linear operator on functions from E' into F, even in finite dimension. The case when $\vec{T} = T \otimes A$, where T is a scalar-valued form acting on scalar-valued functions on E' and A is a linear operator on F, was shown in [Dw 9,10] to be more manageable: in the first reference the Malgrange-Gupta existence and approximation theorems were shown to hold for $T \otimes A *$ in the space $\mathcal{H}_{Nb}(E';F)$ of F-valued entire functions on E' of nuclear bounded type, when T is in the dual of $\mathcal{H}_{Nb}(E';F)$ and A is the identity operator on F (where E and F are Banach spaces). In the second reference those results were extended to surjective bounded linear operators A, and a basis was constructed for a dense subspace of the space of solutions of the homogeneous equation, associated with the zeros of A and those of the Fourier-Borel transform of T: the problem was approached by the representation of $T \otimes A*$ in the form $g'(d) \otimes A$, where g' is the Fourier-Borel transform of T and the "differential operator of infinite order" $g'(d)$ is defined as the sum of the homogeneous operators $g'_n(d)$ given by $g'_n(d)f(x') = \langle \hat{d}^n f(x'), \frac{1}{n!}\hat{d}^n g'(0)\rangle_n$, where $\langle\ ,\ \rangle_n$ is the bilinear form on the pair $P_N({}^nE') \times P({}^nE)$, determined by the isometry between the dual of $P_N({}^nE')$ (nuclear n-homogeneous polynomials on E') and $P({}^nE)$ (all continuous n-homogeneous polynomials on E). Here $\hat{d}^n f(x')$ is the n-th Fréchet derivative polynomial of $f: E' \to \mathbb{C}$ (complex field) at x', and $\hat{d}^n g'(0)$ is similarly defined on E. The adjoint of $g'(d) \otimes A$ relative to the Fourier-Borel isomorphism between the dual of $H_{Nb}(E';F)$ and $Exp(E;F')$ (F' - valued entire functions of exponential type on E) was shown to be the operator $f' \mapsto A' \circ g' \cdot \vec{f}'$ (where A' is the transpose of A), and the Malgrange-Gupta division theorem for g' on $Exp(E)$ was extended to $A' \circ g' \cdot Exp(E;F')$. The existence and approximation theorems then followed by standard duality arguments.

In the present article we consider the operators $g'(d) \otimes A$ in the spaces $F^p_{N,\rho}(E';F)$ of F-valued entire functions $\vec{f}$ on E' with the "p-summable" growth conditions $\Sigma_n \frac{\rho^n}{n!} \| \hat{d}^n \vec{f}(0) \|^p_N < \infty$ for weights $\rho > 0$ (N denoting the nuclear norm). The similarly defined spaces $F^p_{\theta,\rho}(E')$ of scalar-valued functions for an appropriate holomorphy type θ introduced in [Dw 5,7] serve to classify entire functions of finite order in infinite dimension, replacing the classification in terms of exponential growth estimates $|f(x')| \le C \exp \rho \| x' \|^{p'}$ used in [Mar 1,2], which have no analogue for holomorphy types other than the current type: cf [Tr 1] Ch. 11 for their equivalence when $p = 2$. We refer to [Bol 1,2] for a similar re-casting, in the form of weighted power series estimates, of the exponential estimates employed in [Ta]. When $p = 2$ and θ is the Hilbert - Schmidt type one gets the "Fischer-Fock" spaces introduced in [Dw 1,2,3].: the existence theorems in these references hold only for convolution operators

T_* for which the Fourier - Borel transform g' of T is a Hilbert-Schmidt polynomial, and follow the method of [Tr 1], Ch. 11. The results in the Hilbert-Schmidt case can also be extended to vector-valued solutions, but will not be considered here: cf [K 1,2,3,4,5], [Dw 4] and [Bon] for related topics.The "unbounded" case of $H_N(E')$ in [G 1,2,3] and [N 2,3,4] requires results in [Ar], and will also be omitted.

We now outline the results in the present article: following the scalar-valued case of [Dw 5,7], the Fourier-Borel dual of $F^p_{N,\infty}(E';F)$ (projective limit of the $F^p_{N,\rho}(E';F)$ for all weights ρ) is now shown to be $F^{p'}_0(E;F')$ (inductive limit of the $F^{p'}_{\rho'}(E;F)$, defined for the current holomorphy type by p'-summable estimates analogous to those for the nuclear type with $\rho^{p'}\rho'^{p}=1$), where $\frac{1}{p}+\frac{1}{p'}=1$ (corollary of proposition 1.1). The adjoint of $g'(d)\otimes A$ on $F^p_{N,\infty}(E';F)$, where $g'\in F^{p'}_0(E)$ and A is a continuous linear operator on F, is shown to be $A'\circ g'\cdot$ on $F^{p'}_0(E;F')$ with respect to the Fourier-Borel duality (Proposition 2.3). By use of the similar duality between the operators g(d) on $F^{p'}_0(E)$ and $g\,\cdot$ on $F^p_{N,\infty}(E')$ (where $g\in F^p_{N,\infty}(E')$), applied to the special cases $g=\exp\circ\langle u,\ \rangle$, $u\in E$ and $g=\langle v,\ \rangle^n$, $v\in E$, the space $F^{p'}_0(E;F')$ is shown to be translation-invariant (corollary to Proposition 2.2'), and the functions $\vec{f}=\exp\circ\langle u,\ \rangle\cdot\langle v,\ \rangle^n\cdot y$ are shown to be solutions of $g'(d)\otimes A\vec{f}=0$ if and only if either u is a zero of g' with order higher than n in the direction of $v\neq 0$ or y is in the kernel of A in F (Proposition 2.4). The division theorem for g' in $F^{p'}_{\theta'}(E)$ given in [Dw 6,8] is then extended to the operator $A'\circ g'$ in $F^{p'}_0(E;F')$ when A is surjective (Theorem 3.1, in which all propositions mentioned above are used, together with Boland's extension to Banach spaces [Bol 1, 2] of Taylor's estimate [Ta] on the maximum modulus of the quotient of two entire functions of finite order). The existence and approximation theorems are then extended to $g'(d) \otimes A$ on $\mathcal{F}^p_{N,\infty}(E';F)$ when A is surjective (Theorems 4.2 and 4.1: as in the case p=1, i.e., $\mathcal{H}_{Nb}(E';F)$, treated in [Dw 10], the approximation theorem provides a basis for a dense subspace of the kernel of $g'(d) \otimes A$). The preceding results are then extended to the case when F is a Fréchet space, leading to existence and approximation theorems for the operator $g_1'(d) \otimes g_2'(d)$ on $\mathcal{F}^{p_1}_{N,\infty}(E_1';\mathcal{F}^{p_2}_{N,\infty}(E_2'))$ for Banach spaces E_1 and E_2, where $g_i' \in \mathcal{F}_0^{p_i'}(E_i)$ and $\frac{1}{p_i} + \frac{1}{p_i'} = 1$, $i = 1,2$ (Theorems 5.1 and 5.2). Relations with operators $G'(d,d)$ on $\mathcal{F}^p_{N,\infty}(E_1' \times E_2')$, where $G' \in \mathcal{F}_0^{p'}(E_1 \times E_2)$; are then indicated, as well as applications to equations of the from

$$(\partial/\partial t)\ \vec{f}(t;x') = g'(d)\ \vec{f}(t;x') + \vec{g}(t;x'),$$

with $t \in \mathbb{C}$, $x' \in E'$ and $g' \in \mathcal{F}_0^{p'}(E)$ (remarks following Theorem 5.2). The space $\mathcal{F}^p_{N,\infty}(L^\infty(M);F)$ and $\mathcal{F}_0^{p'}(L^1(M);F')$, where M is an appropriate measure space, are then characterized by $L^1(M)$- (resp. $L^\infty(M)$-) growth conditions on the variational derivatives $\delta^n\vec{f}(x')/\delta x'(t_1)\ldots\delta x'(t_n)$ (resp. $\delta^n f'(x)/\delta x(t_1)\ldots\delta x(t_n)$) of functions $\vec{f} : L^\infty(M) \to F$ (resp. $\vec{f}' : L^1(M) \to F'$) (corollaries to Propositions 6.1 and 6.2, where the casting of the domain of $\vec{f}$ as a dual Banach space E' permits the use of the Dunford-Pettis theorem). Finally, the existence and approximation theorems are then applied to variational equations of infinite order of the form

$$\Sigma_{n=0}^{\infty} \frac{1}{n!} \int\cdots\int A\delta^n\vec{f}(x')/\delta x'(t_1)\ldots\delta x'(t_n)x_n'(t_1,\ldots,t_n)dm(t_1)\ldots dm(t_n) = \vec{g}(x')$$

(where m is the measure on M), for functions $\vec{f}: L^\infty(M) \to F$ and $\vec{g}: L^\infty(M) \to F$ with variational derivatives $\delta^n\vec{f}/\delta x'^n$ and $\delta^n\vec{g}/\delta x'^n$ in $L^1(M^n)$, and kernels x_n' in $L^\infty(M^n)$ (Proposition 6.3).

1. VECTOR-VALUED HOLOMORPHIC FOCK SPACES AND THEIR DUALS.

We use the notation of [Dw 9,10]: in particular, E, F are complex Banach spaces, E',F' their duals, $\mathcal{P}_N({}^nE';F)$ (resp. $\mathcal{P}({}^nE;F')$) the n-homogeneous nuclear polynomials $\vec{P}_n: E' \to F$ (resp. continuous polynomials $\vec{P}'_n : E \to F'$), $\|\vec{P}_n\|_{N,n} = \|\vec{P}_n\|_N$ the nuclear norm derived from the nuclear completion of the tensor product $E \otimes \ldots \otimes E \otimes F$ (resp. $\|\vec{P}'_n\|_n = \|\vec{P}'_n\|$ the current norm = sup. on the unit ball of E), $\mathcal{H}(E';F)$ is the space of entire mappings $\vec{f}: E' \to F$ with derivative polynomials $\hat{d}^n\vec{f}(x') \in \mathcal{P}({}^nE';F)$ (and similarly for $\vec{f}' : E \to F'$). We omit arrow superscripts and the explicit indication of the range spaces F when $F = \mathbb{C}$ = complex field. The canonical isometry between $\mathcal{P}_N({}^nE';F)'$ and $\mathcal{P}({}^nE;F')$ (not $\mathcal{P}({}^nE'';F')$) is represented by the bilinear form $\langle\, ,\, \rangle_{n,F}$ on $\mathcal{P}_N({}^nE';F) \times \mathcal{P}({}^nE;F')$ characterized by $\langle x^n \cdot y, \vec{P}'_n\rangle_{n,F} = \langle y, \vec{P}'_n(x)\rangle$, with $x^n := \langle x, \rangle^n$, so that $|\langle \vec{P}_n, \vec{P}'_n\rangle_{n,F}| \leq \|\vec{P}_n\|_N \|\vec{P}'_n\|$: cf [Dw 9], Prop. I. 1.

The holomorphic Fock space with degree $p > 1$, weight $\rho > 0$ and holomorphy type N(nuclear) from E' into F is the Banach space $\mathcal{F}^p_{N,\rho}(E';F)$ of functions $\vec{f}: E' \to F$ such that

$$|||\vec{f}|||_{N,\rho,p} := \{\Sigma_{n=0}^{\infty} \frac{1}{n!} \rho^n \|\hat{d}^n \vec{f}(0)\|_N^p\}^{1/p} < \infty ,$$

equipped with the norm $|||\ \ |||_{N,\rho,p}$ thus defined. The corresponding space $\mathcal{F}^{p'}_{\rho'}(E;F')$ with $\frac{1}{p} + \frac{1}{p'} = 1$ and $\rho^{1/p}\rho'^{1/p'}=1$ for the current holomorphy type is similarly defined and the corresponding norm is denoted by $|||\ \ |||_{\rho',p'}$. We write

$$\mathcal{F}^p_{N,\infty}(E';F) := \bigcap_{\rho>0} \mathcal{F}^p_{N,\rho}(E';F) ,$$

which is a Fréchet space with respect to the norms $(|||\ \ |||_{N,\rho,p})_{\rho>0}$. We also write $\mathcal{F}^{p'}_0(E;F') := \bigcup_{\rho'>0} \mathcal{F}^{p'}_{\rho'}(E;F')$, equipped with the locally convex inductive limit topology induced by the natural in jections $\mathcal{F}^{p'}_{\rho'}(E;F) \to \mathcal{F}^{p'}_{0}(E;F')$. The space $\mathcal{F}^{p'}_{N,0}(E)$ (defined as $\mathcal{F}^{p'}_{0}(E)$ but for the nuclear holomorphy type on E) is not representable as the dual of $\mathcal{F}^{p}_{\infty}(E')$ (defined as $\mathcal{F}^{p}_{N,\infty}(E')$ but for the current type), and the question of the regularity of its open sets is as yet unsettled, except when E is a Hilbert space or a Fréchet-Schwartz space. However, analogues of all the results on convolution operators given in this article are also valid on $\mathcal{F}^{p'}_{0,N}(E;F')$ although this case will be only briefly outlined (cf. [Dw 8], sec.1.6 through 1.9, 2.2, 2.5 and 2.6). A detailed study of the spaces $\mathcal{F}^{p}_{\theta}(E')$ and $\mathcal{F}^{p}_{\theta'}(E)$ for locally convex domains E and very general holomorphy types θ and dual types θ' is treated in [Dw 6,8].

We begin by extending the Fourier-Borel duality to the pairs $\mathcal{F}^{p}_{N,\rho}(E';F)$, $\mathcal{F}^{p'}_{\rho'}(E;F')$ and $\mathcal{F}^{p}_{N,\infty}(E';F)$, $\mathcal{F}^{p'}_{0}(E;F')$, where as in [Dw 9,10] the Fourier-Borel transform $\mathcal{B}T: E \to F^*$ (algebraic dual of F) of an analytic functional $T: \mathcal{F}^{p}_{N,\rho}(E';F) \to \mathbb{C}$ is defined by $\langle y, \mathcal{B}T(x)\rangle := T(e^{x}\cdot y)$ for $y \in F$ and $x \in E$, where $e^{x} := \exp \circ \langle x, \rangle$:

PROPOSITION 1.1. *The Fourier-Borel transformation* $\mathcal{B}$ *is an isometry from* $\mathcal{F}^{p}_{N,\rho}(E';F)'$ *onto* $\mathcal{F}^{p'}_{\rho'}(E;F')$.

PROOF: One first shows, parallel to the case $p=1$ ([Dw 9], Prop. II. 2) that $|\langle y, \mathcal{B}T(x)\rangle| \leq \|T\| \exp(\frac{\rho}{p}\|x\|^{p})\|y\|$ for all $x \in E$ and $y \in F$, i.e., $\|\mathcal{B}T(x)\| \leq \|T\| \exp(\frac{\rho}{p}\|x\|^{p}) < \infty$, so that $\mathcal{B}T(E) \subset F'$ whenever $T \in \mathcal{F}^{p}_{N,\rho}(E';F)'$. Letting then $\vec{P}_{n'}$ be the polynomial transform of the restriction of T to $\mathcal{P}_{N}(^{n}E';F)$ one checks that $\langle y, \Sigma^{m}_{n=0}\frac{1}{n!}P'_{n}(x)\rangle \to \langle y, \mathcal{B}T(x)\rangle$ as $m \to \infty$, getting weak*-convergence to $\mathcal{B}T(x)$. Moreover, the convergence of the series is strong because $\|\Sigma^{m}_{n=0}\frac{1}{m!}\vec{P}_{n'}(x)\| \leq \|T\| \exp\frac{\rho}{p}\|x\|^{p}$

for all m, hence $\mathcal{B}T(x) = \Sigma_{n=0}^{\infty} \frac{1}{n!} \vec{P}_n'(x)$ in F. To show that $\mathcal{B}T \in \mathcal{F}^{p'}(E;F')$ one employs the F-valued analogue of [Dw 7], Lemma 2.1.1 = [Dw 5], Lemma on p. A1441, which has an identical proof, and proceeds as in the proof of [Dw 7], Prop. 2.1.3 = [Dw 5], Prop. 2.1. Finally, to show that $\mathcal{B}$ is surjective and $|||\mathcal{B}T|||_{\rho',p'} = ||T||$ (hence $\mathcal{B}$ injective) one follows verbatim the scalar-valued proof of [Dw 5,7], loc. cit.

As in the scalar case of [Dw 7], Props. 2.1.3', 2.4.1' = [Dw5], Prop. 2.2, letting $\ll \vec{f}, \mathcal{B}T \gg_F := T(\vec{f})$ one has:

COROLLARY: $\mathcal{F}^p_{N,\rho}(E';F)$ *(resp.* $\mathcal{F}^p_{N,\infty}(E';F)$*) and* $\mathcal{F}^{p'}_{\rho'}(E;F;)$ *(resp.* $\mathcal{F}^{p'}_0(E;F')$ *are in separating duality with respect to the bilinear form* $\ll , \gg_F$ *defined by* $\ll \vec{f}, \vec{f}' \gg_F = \Sigma_{n=0}^{\infty} \frac{1}{n!} < \hat{d}^n\vec{f}(0), \hat{d}^n\vec{f}'(0)>_{n,F'}$ *and* $|\ll \vec{f}, \vec{f}' \gg_F| \leq |||\vec{f}|||_{N,\rho,p} |||\vec{f}'|||_{\rho',p'}$ *when* $\vec{f} \in \mathcal{F}^p_{N,\rho}(E';F)$ *and* $\vec{f}' \in \mathcal{F}^{p'}_{\rho'}(E;F')$.

2. VECTOR-VALUED CONVOLUTION OPERATORS AND THEIR ADJOINTS.

The continuous linear operators on $\mathcal{F}^p_{N,\infty}(E';F)$ that commute with translations $\tau_{u'}$, $u' \in E'$ (where $\tau_{u'}\vec{f}(x') := \vec{f}(x'-u')$) are the convolution operators $f \to \vec{T} * \vec{f}$ given by continuous linear mappings $\vec{T}: \mathcal{F}^p_{N,\infty}(E';F) \to F$ (where $\vec{T} * \vec{f}(x') := \vec{T}(\tau_{-x'}\vec{f})$): cf [Dw9], Ch. III and [Dw 10], Sec. 2. If $F = \mathbb{C}$ all non-zero convolution operators are surjective on $\mathcal{H}_{Nb}(E') = \mathcal{F}^1_{N,\infty}(E')$, [G1,2,3], [N 2,3,4], and on various weighted subspaces of $\mathcal{H}_{Nb}(E')$, [B1,2], as well as on $\mathcal{F}^p_{N,\infty}(E')$ and $\mathcal{F}^p_{N,0}(E')$ [Dw 6,8]. If dim $F > 1$ then not all convolution operators are surjective: indeed, if $E = \mathbb{C}$, $F = \mathbb{C}^2$, let $\vec{T}$ be defined on $\vec{f} = (f_1, f_2)$ with f_i scalar-valued, $i = 1,2$ by $\vec{T}(\vec{f}) := (f_1(0), 0)$: then $\vec{T} * \vec{f}(x') = (f_1(x'), 0)$, and the range of $\vec{T} *$ excludes all $\vec{f}$ with $f_2 \neq 0$. Following the case $p = 1$

treated in [Dw 9,10], we shall consider here the operators $\vec{T} *$ with $\vec{T} = T \otimes A$, where $T \in F^{p}_{N,\infty}(E')'$ and $A \in L(F;F)$ (continuous linear operators on F). Similar considerations hold for convolution operators on $F^{p'}_{N,0}(E;F')$, as indicated where appropriate.

We first represent convolution operators by differential operators of infinite order, starting from the homogeneous differential operators as in [Dw 9,10]: given $P'_n \in P(^nE)$, by $P'_n(d)$ is meant the linear operator on $P_N(^{m+n}E')$ given by

$$P'_n(d)Q_{m+n}(x') := \langle \hat{d}^n Q_{m+n}(x'), P'_n \rangle_n$$

in terms of the canonical bilinear for $\langle\, , \rangle_n$ on $P_N(^nE') \times P(^nE)$: cf [Dw 9]. Ch. III. Given $A \in L(F;F)$ we denote by $P'_n(d) \otimes A$ the canonical linear operator on $P_N(^{m+n}E') \otimes F$: cf [Dw 10]. Def. 2.1. We have:

PROPOSITION 2.1. *$P'_N(d) \otimes A \in L(P_N(^{m+n}E';F); P_N(^mE';F))$, and the following holds for all $\vec{Q}_{m+n} \in P_N(^{m+n}E';F)$ and $\vec{Q}'_m \in P(^mE;F')$:*

$$\text{(i)} \quad \frac{1}{n!} \| P'_n(d) \otimes A\, \vec{Q}_{m+n} \|_{N,m} \leq \binom{m+n}{m} \| \vec{Q}_{m+n} \|_{N,m+n} \| P'_n \|_n \| A \|$$

$$\text{(ii)} \quad \frac{1}{n!} \langle P'_n(d) \otimes A\, \vec{Q}_{m+n}, \vec{Q}'_m \rangle_{m,F} = \binom{m+n}{m} \langle \vec{Q}_{m+n}, A' \circ P'_n \cdot Q'_m \rangle_{m+n,F}$$

(where A' is the transpose of A).

PROOF: [Dw 9], Prop. III.1' and [Dw 10], Prop. 2.1.

We now fix $g' \in F^{p'}_0(E)$ and write $g'_n := \frac{1}{n!} \hat{d}^n g'(0)$ to define the differential operator of infinite order $g'(d) \otimes A$ by

$$g'(d) \otimes A\, \vec{f} := \Sigma_{n=0}^{\infty} g'_n(d) \otimes A\, \vec{f}$$

wherever the defining series converges.

PROPOSITION 2.2. *Given $\rho > 0$ and choosing $\sigma \geq \rho$ such that $g' \in F^{p'}_{\sigma'}(E)$ we have*

$$\||\Sigma_{n=0}^{m} g'_n(d) \otimes A\, \vec{f}|\|_{N,\rho,p} \leq 2^{1/p} \||g'|\|_{\sigma',p'} \|A\| \, \||\vec{f}|\|_{N,4\sigma,p}$$

for every m *and* $\vec{f} \in \mathcal{F}^p_{n,\infty}(E';F)$, *hence* $g'(d) \otimes A$ *is a continuous linear operator on* $\mathcal{F}^p_{N,}\ (E';F)$.

PROOF: By use of the estimate (i) in Proposition 2.1 applied to $P'_n = g'_n$ one can check as in the case $p = 1$ of [Dw 10], Prop. 2.1 (cf. also [Dw 8], Prop. 1.2.2), that

$$|||g'_n(d) \otimes A\ \vec{f}|||_{N,\sigma,p} \leq (n!\sigma^{-n})^{1/p} \| g'_n \|\ \| A \|\ \ |||\vec{f}|||_{N,2\sigma,p}\ .$$

It is enough then to apply this inequality term by term to the expansion of $g'(d) \otimes A\ \vec{f}$ (cf. [Dw 10], Prop. 2.2 and [Dw 8], Prop. 1.3.1 = [Dw 6] Prop. 1.1, setting in the latter

$$r = \log_2 \rho,\ \|\ \ \|^p_r = 2^r \|\ \ \|^p = \rho \|\ \ \|^p$$

on $E_r = E$, with dual norm $\|\ \ \|'_r{}^{p'} = \rho' \|\ \ \|'^{p'}$ (dual norm of E'), and similarly for $s = \log_2 \sigma$).

It follows from the expansion of $\langle\!\langle\ ,\ \rangle\!\rangle_F$ in terms of the bilinear forms $\langle\ ,\ \rangle_{n,F}$ in the corollary of Proposition 1.1 that the operators $g'(d) \otimes A$ are the convolution operators $\vec{T}\ *$ with $\vec{T} = T \otimes A$, where g' is the Fourier-Borel transform of T.

Given $v' \in E'$ and setting $g' = e^{-v'}$ as well as $A = 1_F$ (identity operator on F) we get $g'(d) \otimes A = \tau_{v'} \otimes 1_F$, hence we have:

COROLLARY: $\mathcal{F}^p_{N,\infty}(E';F)$ *is translation-invariant.*

PROPOSITION 2.3. *Given* $\vec{f} \in \mathcal{F}^p_{N,\infty}(E';F)$ *and* $\vec{h}' \in \mathcal{F}^{p'}_0(E;F')$ *we have* $\langle\!\langle g'(d) \otimes A\ \vec{f}, \vec{h}' \rangle\!\rangle_F = \langle\!\langle \vec{f}, A' \circ g' \cdot \vec{h}' \rangle\!\rangle_F$, *i.e., the* $\langle\!\langle\ ,\ \rangle\!\rangle_F$-*adjoint of* $g'(d) \otimes A$ *is multiplication by* g' *followed by composition with* A'.

PROOF: Follows from the term by term application of the identity (ii) in Proposition 2.1 to the expansion of $\langle\!\langle\ ,\ \rangle\!\rangle_F$ in the corollary of Proposition 1.1: cf [Dw 10], Prop. 2.4 and [Dw 8], Prop. 1.5.1.

We shall need the translation-invariance of $\mathcal{F}_0^p(E;F')$ (which does not follow from Proposition 2.2) and the fact that the $\ll\ ,\ \gg$-adjoint of $(\partial/\partial v)^n$ (directional derivative along $v \in E$) is multiplication by $v^n = \langle v\,,\ \rangle^n$ (which does not follow from Proposition 2.3).

We derive these results from the analogues of Propositions 2.1., 2.2 and 2.3 for the operators $P_n(d)$ on $P(^{m+n}E)$ and $g(d)$ on $\mathcal{F}_0^{p'}(E)$ defined below.

Given $P_n \in P_N(^nE')$, by $P_n(d)$ we mean the linear operator on $P(^{m+n}E)$ given by $P_n(d)Q'_{m+n}(x) := \langle P_n, \hat{d}^n Q'_{m+n}(x)\rangle_n$.

PROPOSITION 2.1'. $P_n(d) \in L(P(^{m+n}E); P(^mE))$ *and for each* $Q'_{m+n} \in P(^{m+n}E)$, $Q_m \in P_N(^mE')$ *we have*

(i) $\frac{1}{n!}\ \|P_n(d)Q'_{m+n}\|_m \leq \binom{m+n}{m}\ \|Q'_{m+n}\|_{m+n}\ \|P_n\|_{N,n}$

(ii) $\frac{1}{n!}\ \langle Q_m, P_n(d)Q'_{m+n}\rangle_m = \binom{m+n}{m}\ \langle P_n \cdot Q_m, Q'_{m+n}\rangle_{m+n}$

PROOF: The argument is different from that for Proposition 2.1 in using the Hahn-Banach theorem on the bidual of $P_N(^mE')$ followed by Alaoglu's theorem (density of $P_N(^mE')$ in its bidual), to find polynomials $Q_{m,\varepsilon} \in P_N(^mE')$ such that $\|Q_{m,\varepsilon}\|_N \leq \varepsilon$ and $\frac{1}{n!}\ \|P_n(d)\ Q'_{m+n}\|_m \leq \varepsilon + \binom{m+n}{m}\ |\langle P_n \cdot Q_{m,\varepsilon}, Q'_{m+m}\rangle_m|$ for each $\varepsilon > 0$, then passing to the limit as $\varepsilon \to 0$ to get the estimate (i) from (ii), first for P_n of finite type (for which the identity (ii) can be proved directly) and then for all P_n in $P_N(^nE')$ by the density therein of the polynomials of finite type; cf [Dw 8], Prop. 1.6.1.

Given now $g \in \mathcal{F}^p_{N,\infty}(E')$ and letting $g_n := \frac{1}{n!}\ \hat{d}^n g(0)$ we define $g(d)$ acting on $f' \in \mathcal{F}_0^{p'}(E)$ by $g(d)f' := \Sigma_{n=0}^{\infty}\ g_n(d)f'$ wherever the series converges.

PROPOSITION 2.2': *Given* $\rho > 0$ *and choosing* $\sigma \geq \rho$ *such that* $f' \in \mathcal{F}^{p'}_{\rho'}(E)$ *for each* m *we have*

$$|||\Sigma^m_{n=0}\, g_n(d)\ f'|||_{\sigma',p'} \leq 2^{1/p'}\,|||g|||_{N,2\sigma,p}|||f'|||_{\rho',p'}\ ,$$

hence $g(d)$ *is a continuous linear operator on* $\mathcal{F}^{p'}_0(E;F')$.

PROOF: Follows from estimating $|||g_n(d)f'|||_{\sigma',\rho'}$ by the termwise application of the estimate (i) in Proposition 2.1' and then the termwise application of the resulting estimate to $\Sigma^m_{n=0}|||g_n(d)f'|||_{\sigma,p'}$: cf [Dw 8], Prop. 1.6.2, 1.7.1 and 1.7.2, as well as the proof of Proposition 2.2 above.

Given $y \in F$ and $\vec{f}' \in \mathcal{F}^{p'}_{\rho'}(E;F')$, letting $\vec{f}'_y(x) := \langle y, \vec{f}'(x)\rangle$ it follows immediately that $|||\vec{f}'_y|||_{\rho',p'} = |||\vec{f}'|||_{\rho',p'}\|y\|$. Given $v \in E$ we have $\tau_v\ \vec{f}'_y = (\tau_v\ \vec{f}')_y$, hence $|||\tau_v\vec{f}'|||_{\rho',p'} = \frac{1}{\|y\|}\,|||\tau_v f'_y|||_{\rho',p'}$ for $y \neq 0$, so that $|||\tau_v\ \vec{f}'|||_{\rho',p'} < \infty$ whenever $|||\tau_v\ \vec{f}'_y|||_{\rho',p'} < \infty$. Setting now $g = e^{-v} = \exp\circ(-\langle v, \rangle)$ we get $g(d) = \tau_v$, hence from Proposition 2.2' we conclude:

COROLLARY: $\mathcal{F}^{p'}_0(E;F')$ *is translation-invariant.*

PROPOSITION 2.3': *Given* $f' \in \mathcal{F}^{p'}_0(E)$ *and* $h \in \mathcal{F}^p_{N,\infty}(E')$ *we have* $\langle\langle h, g(d)f'\rangle\rangle = \langle\langle g\cdot h, f'\rangle\rangle$.

PROOF: Follows from the termwise application of the identity (ii) in Proposition 2.1' to the expansion of $\langle\langle , \rangle\rangle$ in terms of the bilinear forms $\langle\ ,\ \rangle_n$ in Proposition 1.1: cf [Dw 8], Prop. 1.9.1.

In particular, given $u, v \neq 0$ in E and $y \in F$, by setting $g = v^n = \langle v, \rangle^n$, $\vec{g}' = \frac{\partial^n}{\partial v^n}\vec{f}'$ and recalling that

$$\langle y, \vec{g}'(u)\rangle = \langle\langle e^u\cdot y, \vec{g}'\rangle\rangle_F\ ,$$

as well as $\langle\langle \vec{f}\cdot y, \vec{g}'\rangle\rangle_F = \langle\langle \vec{f}, \vec{g}'_y\rangle\rangle$ by [Dw 9], Prop. II. 4, we conclude:

COROLLARY: *Given* $f' \in \mathcal{F}_0^{p'}(E)$, $\vec{f}' \in \mathcal{F}_0^{p'}(E;F')$, $y \in F$, u *and* $v \neq 0$ *in* E, *for each* $n = 0,1,2,\ldots$ *we have:*

(i) $(\partial/\partial v)^n \vec{f}'(u) = \langle\langle e^u \cdot v^n, \vec{f}' \rangle\rangle$

(ii) $\langle y, (\partial/\partial v)^n \vec{f}'(u) \rangle = \langle\langle e^u \cdot v^n \cdot y, \vec{f}' \rangle\rangle_F$.

REMARKS: 1. The analogue of the corollary above holds on $\mathcal{F}_{N,\infty}^p(E';F)$, as follows from Proposition 2.3, but will not be used. A direct proof for $p=1$ is given in [Dw 10], Lemma 3.1.

2. The analogues of the estimate in Proposition 2.2' (with the factor $\|A\|$ as in Proposition 2.2), as well as those of Proposition 2.3' and 2.4', hold for g(d) on $\mathcal{F}_{N,0}^{p'}(E;F')$ with $g \in \mathcal{F}_\infty^p(E')$, and are likewise derived from Proposition 2.1.': cf [Dw 8], Sec. 1.6 through 1.9 when $F = \mathbb{C}$.

A family of solutions of homogeneous equations for $g'(d) \otimes A$ is given by the following proposition.

PROPOSITION 2.4: *Given* u *and* $v \neq 0$ *in* E *as well as* y *in* F, *the function* $\vec{f} = e^u \cdot v^n \cdot y$ *is a solution of* $g'(d) \otimes A\vec{f} = 0$ *if and only if either* $Ay = 0$ *or* u *is a zero of* g' *with order higher than* n *in the direction of* v. *Moreover, such functions are linearly independent for distinct exponents* u *and arbitrary* n, v, y.

PROOF: The argument is the same as for the case $p=1$ in [Dw 10], Props. 3.1 and 3.2: the conditions for $e^u \cdot v^n \cdot y$ to be a solution follow from considering the identity

$$g'(d) \otimes A\,(e^u \cdot v^n \cdot y) = \{\Sigma_{k=0}^n \binom{n}{k} \hat{d}^k g'(u)(v) e^u \cdot v^{n+k}\} \cdot Ay ,$$

the linear independence of $(v^{n-k})_k$ and the non-vanishing of e^u. The linear independence of $\{e^{u_j} \cdot P_j\}_j$, in fact for arbitrary continuous polynomials $\vec{P}_j : E' \to F$ and distinct u_j's, follows from deriving by induction the identity

$$\vec{P}_j = \langle u_{k+1} - u_k, u' \rangle^{-n} \Sigma_{i=1}^n \binom{n}{i} \langle u_j - u_{k+1}, u' \rangle^{n-i} (\partial/\partial u')^i \vec{P}_j$$

from the hypothesis $\Sigma_{j=1}^{k+1} e^{u_j} \cdot \vec{P}_j = 0$ through differentiation along $u' \in E'$ chosen so that $(\partial/\partial u')^n \vec{P}_{k+1} = 0$ and $\langle u_j - u_{k+1}, u' \rangle \neq 0$ for $j \leq k$.

REMARKS: 1. In the preceding proposition, linear independence holds for functions $e^u \cdot \vec{P}$ with distinct exponents $u \in E$ and arbitrary continuous polynomials $\vec{P} : E' \to F$ as shown in the proof.

2. The analogues of all propositions up to this point are true for very general holomorphy types and their dual types (including the compact, current and Hilbert-Schmidt types in locally convex spaces) at least if $F = \mathbb{C}$: cf [Dw 5,6,7,8]. The results in the next two sections are completely known only for the current type-nuclear type pairing (and partially for the Hilbert-Schmidt type: cf [Dw 1,2,3,4], [Bon], [K 1,2,3,4,5].

3. VECTOR - VALUED DIVISION THEOREMS.

The division theorem for the operator $\vec{h}' \to A' \circ g' \cdot \vec{h}'$, given in [Dw 10], Th. 4.1 on Exp(E;F') (entire mappings of exponential type), is now extended to $F_0^{p'}(E;F')$. The analogous result on $F_\infty^{p'}(E;F')$ will also be described. We begin with a strengthened version of [Dw 10], Prop. 4.1:

PROPOSITION 3.1. *Given* $\vec{f}' \in H(E;F')$, $g' \in H(E)$ *such that* $g' \neq 0$, $A \in L(F;F)$ *such that* $AF = F$ *and a total subset* Y *of* $A^{-1}(0)$, *let the "scalar components"* $\vec{f}'_y$, $y \in F$ *of* $\vec{f}'$ *have the following properties:*

(i) *If* $y \notin Y$ *then* $\vec{f}'_y$ *is divisible by* g' *as an entire function along all complex lines in* E *where* g' *does not vanish.*

(ii) *If* $y \in Y$ *then* $\vec{f}'_y = 0$.

Then there is some $\vec{h}' \in H(E;F')$ *satisfying* $A' \circ g' \cdot \vec{h}' = \vec{f}'$.

PROOF: From the case $F = \mathbb{C}$ of [G 2], §8, Prop. 2, or [Dw 8], Lemma 2.3.1 with $E_r = E$, it follows from the hypothesis (i) that for each $y \notin Y$ there is some $h'_{(y)} \in H(E)$ such that $g' \cdot h'_{(y)} = \vec{f}'_y$. By the hypothesis (ii), if $y \in Y$ then $\vec{f}'_y = 0$ and we may set $\vec{h}'_{(y)} = 0$. We now observe:

(a) <u>If $Ay_1 = Ay_2$ then $h'_{(y_1)} = h'_{(y_2)}$</u>: indeed, it follows from the hypothesis (ii) that $\vec{f}'_{y_1} - \vec{f}'_{y_2} = \vec{f}'_{y_1 - y_2} = 0$ (because $y_1 - y_2$ can be approximated by linear combinations of elements of Y), so that $g' \cdot h'_{(y_1)} = g' \cdot h'_{(y_2)}$. Since $g' \neq 0$, there is neighborhood $U \subset E$ such that $g'|_U \neq 0$, so that $h'_{(y_1)}|_U = h'_{(y_2)}|_U$, hence $h'_{(y_1)} = h'_{(y_2)}$ by [H], III. 1.3, th. 3(b).

Following [Dw 10], Th. 4.1, we may then define $\vec{h}': E \to F^*$ (algebraic dual) by $\langle z, \vec{h}'(x) \rangle := h'_{(y)}(x)$ for every $x \in E$ and $z = Ay \in F$. As in [Dw 10], loc. cit., we get:

(b) <u>$A' \circ g' \cdot \vec{h}' = \vec{f}'$</u> (from the definition of $\vec{h}'$).

(c) <u>$\vec{h}'(E) \subset F'$</u> (by considering $\vec{h}'(x) = \lim_n g'(x_n)^{-1} \vec{f}'(x_n) \in F'$ on a sequence $x_n \to x$ where $g'(x_n) \neq 0$, using [H], loc. cit. and the uniform boundedness principle).

(d) <u>$\vec{h}'$ is Gâteaux-analytic</u> (by the analyticity of $\langle z, \rangle \circ \vec{h}' = h'_y$ for each $z = Ay$ in F).

(e) <u>$\vec{h}'$ is bounded on compact sets</u> (by showing that $|(z'' \circ \vec{h}')(K)| \leq 1 + \max\{|h'_{(y)}(x)|: x \in K\}$ for each $z'' \in F''$ and each compact $K \subset E$, where $z = Ay \in F$ is chosen in the unit ball with center z'' in F'' by Alaoglu's theorem.

It follows from (d), (e) and [H], III. 2.2, Prop. 1 (ii)

that $\vec{h}' \in H(E;F')$.

The next proposition differs from [Dw 10], Lemma 4.1 in that the Malgrange-Gupta estimate on quotients of exponential growth estimates in [G 2] is replaced by the Taylor-Boland estimate in [Bol 2] on maximum moduli of quotients of entire functions of bounded type.

PROPOSITION 3.2. With A, g', $\vec{f}'$ and $\vec{h}'$ as in Proposition 3.1, let $g'(0) \neq 0$, $|||g'|||_{\rho',p'} < \infty$, $|||f'|||_{\rho',p'} < \infty$ and $\nu > 6 \cdot 2^{p}_{\rho}$: then there are constants $C_{p,\rho,\nu} > 0$ (depending only on p and ν/ρ) and $\delta_A > 0$ (depending only on A) such that

$$|||\vec{h}'|||_{\nu,p'} \leq C_{p\,\rho,\nu}\delta_A^{-1}|g'(0)|^{-3}\{1+|||\vec{f}'|||_{\rho',p'}\}^3\{1+|||g'|||_{\rho',p'}\}^3.$$

The proof requires the estimates in the lemma below, where $M(R,f') := \text{Max}\{|f'(x)| : \|x\| \leq R\}$ and $H_b(E;F')$ is the space of entire functions from E to F' which are bounded on bounded sets (same for $F = \mathbb{C}$):

LEMMA: (i) *If* $f' \in F^{p'}_{\rho'}(E)$ *then for each* $R > 0$ *we have*

$$M(R^{-1/p}, f') \leq |||f'|||_{\rho',p'} \exp\left(\frac{1}{p} R^{p}\right).$$

(ii) *Given* f' *and* g' *in* $H_b(E)$, *if* $f'/g' \in H(E)$ *and* $g'(0) \neq 0$ *then for each* $R > 0$ *we have*

$$M(R,f'/g') \leq |g'(0)|^{-3}\{1+M(2R,f')\}^3\{1+M(2R,g')\}^3.$$

(iii) *If* $h' \in H_b(E;F')$ *then for each sequence of* $R_n > 0$ *and* $\nu \geq \rho > 0$ *we have*

$$|||\vec{h}'|||^{p'}_{\nu',p'} \leq \Sigma_{n=0}^{\infty}\{n!^{1/p}\frac{1}{R_n}\left(\frac{\rho}{\nu}\right)^{1/p}\}^{np'} M(R_n\rho^{-1/p},\vec{h}')\}^{p'}$$

PROOF: (i) follows from the definition of $M(\ ,\ )$ and of $|||\ \ |||_{\rho',p'}$: cf [Dw 8], Lemma 2.3.3 = [Dw 6], Lemma 2.6. The estimate (ii) is due to Taylor when $E = \mathbb{C}$ and Boland when E is a Banach space: cf [Ta], p. 456 and [Bol 2], Lemma 4.4. Finally, (iii) follows from the Cauchy estimates of [N 1], §6, Prop. 3: cf [Dw 8], Lemma

2.3.5 = [Dw 6], Lemma 2.5.

REMARK: The estimate (iii) in the lemma is the only point in this entire article where the Cauchy estimates are used, hence the only reason for the restriction of the division theorem to the current holomorphy type; cf the same difficulty in [G 1,2,3], [N 2,3], [Bol 1,2,3], [Mat 1,2] and [Dw 5,6,7,8,9,10]. (But we refer to [Dw 1,2,3] for the division of entire functions of Hilbert-Schmidt type by Hilbert-Schmidt polynomials, and to [Di 1,2] for the division of polynomials by polynomials for more general holomorphy types.)

PROOF OF PROPOSITION 3.2. Since AF = F, the image under A of the unit ball in F has non-empty interior by the open mapping theorem, i.e., there is some $\delta_A > 0$ such that

$$\{y \in F: \|Ay\| \leq 1\} \supset \{z_1 \in F: \|z_1\| \leq \delta_A\},$$

hence for each $z \in F$ with $\|z\| \leq 1$ there is some $y \in F$ with $\|y\| \leq 1$ such that $Ay = z_1 := \delta_A z$, thus for each $x \in E$ we have $|\langle z, \vec{h}'(x)\rangle| = \delta_A^{-1} |\vec{h}'_{Ay}(x)|$, so that

$$M(R_n \rho^{-1/p}, \vec{h}'_z) = \delta_A^{-1} M(R_n \rho^{-1/p}, \vec{h}'_{Ay})$$

for any sequence of $R_n > 0$ (to be specified below). Setting $R = 2R_n \rho^{1/p}$ and applying (i) in the lemma to $f' := \vec{f}'_y$ as well as $f' := g'$ we get

$$\{1 + M(2R_n \rho^{-1/p}, \vec{f}'_y)\} \quad \{1 + M(2R_n \rho^{1/p}, g')\} \leq$$
$$\{1 + |||\vec{f}'|||_{\rho',p'}\} \{1 + |||g'|||_{\rho',p'}\} \exp(2 \cdot \frac{1}{p} \cdot 2^p \cdot R_n^p)$$

(where we used $|||\vec{f}'_y|||_{\rho',p'} = |||\vec{f}'|||_{\rho',p'} \|y\| \leq |||\vec{f}'|||_{\rho',p'}$ when $\|y\| \leq 1$). Applying then the estimate (ii) in the lemma, now with $R := R_n \rho^{-1/p}$, we get

$$M(R_n \rho^{-1/p}, \vec{h}'_z) \leq \delta_A^{-1} |g'(0)|^{-3} \{1 + |||\vec{f}'|||_{\rho',p'}\}^3 \{1 + |||g'|||_{\rho',p'}\}^3 \exp(\frac{1}{p} \cdot 6 \cdot 2^p \cdot R_n^p)$$

for every $z \in F$ with $\|z\| \leq 1$ and any sequence of $R_n > 0$, hence $M(R_n\rho^{-1/p},\vec{h}') := \sup\{\|\vec{h}'(x)\| : \|x\| \leq R_n\rho^{-1/p}\} = \sup\{M(R_n\rho^{-1/p},\vec{h}'_z) : \|z\| \leq 1\} \leq$

$$\delta_A^{-1} |g'(0)|^{-3} \{1 + |||\vec{f}'|||_{\rho',p'}\}^3 \{1 + |||g'|||_{\rho',p'}\}^3 \exp(\tfrac{1}{p} \cdot 6 \cdot 2^p \cdot R_n^p).$$

From (iii) in the lemma we then obtain

$$|||\vec{h}'|||^{p'}_{\nu',p'} \leq \Sigma_{n=0}^{\infty} n!^{p'/p} \left[\frac{1}{R_n} (\frac{\rho}{\nu})^{1/p}\right]^{np'} \delta_A^{-p'} |g'(0)|^{-3p'} \cdot$$

$$\cdot \{1 + |||\vec{f}'|||_{\rho',p'}\}^{3p'} \{1 + |||g'|||_{\rho',p'}\}^{3p'} \cdot \exp(\frac{p'}{p} \cdot 6 \cdot 2^p \cdot R_n^p),$$

that is,

$$|||\vec{h}'|||^{p'}_{\nu',p'} \leq C^{p'}_{p,\rho,\nu} \delta_A^{-p'} |g'(0)|^{-3p'} \{1 + |||\vec{f}'|||_{\rho',p'}\}^{3p'} \{1 + |||g'|||_{\rho',p'}\}^{3p'},$$

where

$$C^{p'}_{p,\rho,\nu} := \Sigma_{n=0}^{\infty} \{n!\, n^{-n} (\frac{\rho}{\nu} \cdot n \cdot R_n^{-p})^n \exp(6 \cdot 2^p \cdot R_n^p)\}^{p'/p}.$$

The series defining $C_{p,\rho,\nu}$ converges when $\frac{\nu}{\rho} > 6 \cdot 2^p$ (and diverges when $\frac{\nu}{\rho} < 6 \cdot 2^p$) if we choose the radii of the maximum modulus estimates to be $R_n := (\frac{n}{6 \cdot 2^p})^{1/p}$, as follows from an application of Stirling's estimate and the root test for series: cf [Dw 8], Proof of Lemma 2.3.2 = [Dw 6], Lemma 2.7.

We now combine Propositions 2.4, 3.1 and 3.2 (by use of Proposition 2.3 and the corollaries of Propositions 2.2' and 2.3'), to obtain the division theorem for $\vec{h}' \to A' \circ g' \cdot \vec{h}'$ on $F_0^{p'}(E;F')$:

THEOREM 3.1 (Division Theorem). *Given* $f' \in F_0^{p'}(E;F'), g' \in F_0^{p'}(E)$ *such that* $g' \neq 0$, $A \in L(F;F)$ *such that* $AF = F$ *and a total subet* Y *of* $A^{-1}(0)$, *then the following conditions are equivalent:*

(a) *The equation* $A' \circ g' \cdot \vec{h}' = \vec{f}'$ *has a solution* $\vec{h}'$ *in* $F_0^{p'}(E;F')$.

(b) $\ll \vec{f}, \vec{f}' \gg_F = 0$ *whenever* $\vec{f}$ *is a solution of* $g'(d) \otimes A\vec{f} = 0$ *in* $F^p_{N,\infty}(E';F)$.

(c) $\ll e^u \cdot v^n \cdot y, \vec{f}' \gg_F = 0$ *whenever either* u *is a zero of* g' *with finite order higher than* n *in the direction of* $v \neq 0$ *in* E *or* $y \in Y$.

(d) f'_y *is divisible by* g' *as an entire function on all complex lines in* E *where* $g' \neq 0$ *whenever* $y \notin Y$, *and* $\vec{f}'_y = 0$ *whenever* $y \in Y$.

PROOF: (a) implies (b) by Proposition 2.3 and (b) implies (c) by Proposition 2.4. Let now (c) be satisfied: if $y \notin Y$ and u is a zero of g' with order m in the direction of $v \neq 0$ in E then we conclude by the corollary to Proposition 2.3' that

$$(\partial/\partial v)^n \vec{f}'_y(u) = \ll e^u \cdot v^n \cdot y, \vec{f}' \gg_F = 0$$

for every $n < m$, hence u is also a zero of $\vec{f}'_y$ with order $\geq m$ on $u + \mathbb{C}v$, i.e., $\vec{f}'_y/g'$ has an analytic continuation on all complex lines on which the zeros of g' are of finite order. If $y \in Y$ then for each $x \in E$ we may set $u = x$, $n = 0$ (thus v arbitrary), to get $\vec{f}'_y(x) = \ll e^x \cdot y, \vec{f}' \gg_F = 0$, that is, (d) is satisfied. Finally, let (d) hold true: it follows from Proposition 3.1 that $A' \circ g' \cdot \vec{h}' = \vec{f}'$ for some $\vec{h}' \in H(E;F')$. Let first $g'(0) \neq 0$: by hypothesis we have $g' \in \mathcal{F}^{p'}_{\rho'}(E)$ and $f' \in \mathcal{F}^{p'}_{\rho'}(E;F')$ for some $\rho > 0$, and we choose $\nu > 6 \cdot 2^{p}\rho$ to get $\vec{h}' \in \mathcal{F}^{p'}_{\nu'}(E;F') \subset \mathcal{F}^{p'}_{0}(E;F')$ by Proposition 3.2. If $g'(0) = 0$ then $g'(u) \neq 0$ for some $u \in E$ (since $g' \neq 0$ by hypothesis). Since $\mathcal{F}^{p'}_{0}(E;F')$ is translation-invariant by the corollary of Proposition 2.2', we may apply the same argument as when $g'(0) \neq 0$ to $\tau_{-u}\vec{f}'$, $\tau_{-u}g'$ and $\tau_{-u}\vec{h}'$, getting $\tau_{-u}\vec{h}' \in \mathcal{F}^{p'}_{\nu'}(E;F') \subset \mathcal{F}^{p'}_{0}(E;F')$ for a sufficiently large weight ν, so we still have $\vec{h}' = \tau_u\tau_{-u}\vec{h}' \in \mathcal{F}^{p'}_{0}(E;F')$, that is (a) is satisfied.

REMARK: The division theorem holds verbatim for $\mathcal{F}^{p'}_{\infty}(E;F')$, when $g' \in \mathcal{F}^{p'}_{\infty}(E)$, by use of the analogues of Propositions 2.3 and 2.4, as well as those of the corollaries to Propositions 2.2' and 2.3', for the dual pair consisting of $F^{\mathbf{p}}_{N,0}(E';F)$ and $F^{\mathbf{p}'}_{\infty}(E;F')$: cf [Dw 8], Sec. 2.2 and 2.5, or [Dw 6], Th. 2.1 (ii) when $F = \mathbb{C}$.

4. VECTOR-VALUED EXISTENCE AND APPROXIMATION THEOREMS.

We shall now extend to $F^{\mathbf{p}}_{N,\infty}(E';F)$ the existence and approximation theorems on the solutions $\vec{f}$ of $g'(d) \otimes A\,\vec{f} = \vec{g}$ treated in [Dw 9,10] for $H_{Nb}(E';F)$, and indicate their analogues for $F^{\mathbf{p}}_{N,0}(E';F)$. Improving on the approximation theorem in [Dw 10], Th. 5.1, the vector values y of the exponential-polynomial functions in the kernel of $g'(d) \otimes A$ now need only be in a total subset of the kernel of A:

THEOREM 4.1 (Approximation Theorem). *Let there be given* $g' \in F^{\mathbf{p}'}_{0}(E)$, $A \in L(F;F)$ *with* $AF = F$ *and a total subset* Y *of* $A^{-1}(0)$. *The functions* $\vec{f}_i = e^{u_i} \Sigma'_j \Sigma^{n_{ij}}_{k=0} v^k_{ij} \cdot y_{ij}$ (Σ'_j *denoting a finite sum) with distinct* u_i *for distinct indices* i , *where either* $y_{ij} \in Y$ *or* u_i *is a zero of* g' *with finite order higher than* n_{ij} *in the direction of* $v_{ij} \neq 0$ *in* E, *form a basis for a dense subspace of the space of all solutions* $\vec{f}$ *of* $g'(d) \otimes A\,\vec{f} = 0$ *in* $F^{\mathbf{p}}_{N,\infty}(E';F)$.

PROOF: The argument is the same as in [Dw 9], Th. IV. 2 and [Dw 10] Th. 5.1: the linear independence of the f_i follows from Proposition 2.4, while the vanishing on the kernel of $g'(d) \otimes A$ of the analytic functionals which vanish on the f_i follows from the implication (c) → (a) in Theorem 3.1 together with Propositions 2.3 and 2.4, leading to the totality of the f_i by the Hahn-Banach theorem.

THEOREM 4.2 (Existence Theorem). *Given* $g' \in F^{\mathbf{p}'}_{0}(E)$ *such that* $g' \neq 0$ *and* $A \in L(F;F)$ *such that* $AF = F$ *every equation*

$g'(d) \otimes A\,\vec{f} = \vec{g}$ *with* $\vec{g}$ *in* $\mathcal{F}^{\mathbf{p}}_{N,\infty}(E';F)$ *has a solution* $\vec{f}$ *in the same space.*

PROOF: As with [Dw 10], Th. 5.2, the vanishing of $A'(\vec{h}'(x))$ wherever $A' \circ g' \cdot h'(x) = 0$ and $g'(x) \neq 0$ (by the injectivity of A' as the transpose of a surjective operator) guarantees the injectivity of the $\ll , \gg_F$ -adjoint map $h' \to A' \circ g' \cdot \vec{h}'$ ([H], III. 1.3, th. 3(b)). The $\ll , \gg_F$ -weakly closed character of the range $A' \circ g' \cdot \mathcal{F}^{\mathbf{p}'}_0(E;F')$ follows from its representation as the intersection of the sets $\{f' \in \mathcal{F}^{\mathbf{p}'}_0(E:F') : \ll f, f' \gg_F = 0\}$ for all solutions $\vec{f}$ of $g'(d) \otimes A\,\vec{f} = 0$ in $\mathcal{F}^{\mathbf{p}}_{N,\infty}(E';F)$ (by the implication (b) $\to$ (a) in Theorem 3.1 together with Proposition 2.4). The surjectivity of $g'(d) \otimes A$ follows as usual from the Dieudonné-Schwartz theorem on surjections in Fréchet spaces: cf [Tr 3], Prop. 5.1 (i), p. 25.

REMARKS: 1. The approximation theorem holds on $\mathcal{F}^{\mathbf{p}}_{N,0}(E';F)$ (with $g' \in \mathcal{F}^{\mathbf{p}'}_{\infty}(E)$), as follows from the division theorem on $\mathcal{F}^{\mathbf{p}'}_{\infty}(E;F')$ and the analogue of Propositions 2.3 and 2.4 on $\mathcal{F}^{\mathbf{p}}_{N,0}(E';F) \times \mathcal{F}^{\mathbf{p}'}_{\infty}(E;F')$ (see the remarks following Proposition 2.4 and Theorem 3.1). The existence theorem likewise holds on $\mathcal{F}^{\mathbf{p}}_{N,0}(E';F)$ by the theorem on surjections in spaces with Fréchet duals ([Tr 3], Prop. 5.1(ii), p. 25), provided the bounded sets of $\mathcal{F}^{\mathbf{p}}_{N,0}(E';F)$ are regular (i.e., coincide with the bounded sets of the spaces $\mathcal{F}^{\mathbf{p}}_{N,\rho}(E';F)$), by [FW], §23, no. 5, p. 123 and [RR], Ch. 4, §4, Prop. 15. This regularity condition is known to hold for $\mathcal{F}^{\mathbf{p}'}_{N,0}(E)$ when E is a Hilbert space, a projective limit of a sequence of Hilbert spaces or a Fréchet-Schwartz space (with an inductive limit definition for $\mathcal{F}^{\mathbf{p}'}_{N,0}(E)$ on locally convex domains E which coincides with the one used here for the Banach space case): cf [Dw 5], Prop. 2.3(i), (ii) = [Dw 7], corol. to Prop. 1.4.1 and Prop. 1.5.1, [Dw 6], th. 3.1 = [Dw 8], Th. 2.6.2.

2. Since $g'(d) \otimes A$ maps $F^{p}_{N,4\sigma}(E';F)$ only into $F^{p}_{N,\rho}(E';F)$ with $\sigma \geq \rho$ (not $\sigma < \rho$ even if g' is a polynomial and $F = \mathbb{C}$, as shown in [Dw 4] and [Bon] for $p = 2$, approximation and existence theorems on $F^{p}_{N,\rho}(E';F)$ when $0 < \rho < \infty$ would require a strengthening of Proposition 3.2 to permit choosing $\nu > 4\rho$ instead of $\nu > 6 \cdot 2^{p}\rho$, via perhaps a finer choice of the radii R_n (but see the existence theorem for the Hilbert-Schmidt type with $p = 2$ and $0 < \rho < \infty$ for polynomials g' in [Dw 1, 2, 3]).

5. APPLICATION TO ENTIRE FUNCTIONS WITH ENTIRE FUNCTION VALUES.

We begin by extending the preceding theory to more general range spaces: if F is a Fréchet space with a continuous norm then its topology is determined by a family of norms $\|\ \|_r$ indexed by an ordered set with a countable cofinal subset (it is enough to add the continuous norm to sufficienty many continuous seminorms determining the topology of F). F is then a complete countably normed space in the sense of Gelfand. Letting F_r denote the completion of F with respect to $\|\ \|_r$ it follows that the natural mappings $F \to F_r$ are injective and F has the projective limit topology induced by these mappings. We may then define $F^{p}_{N,\infty}(E';F)$ as $\bigcap_r F^{p}_{N,\infty}(E';F_r)$ and $F^{p'}_{0}(E;F')$ as $\bigcup_r F^{p'}_{0}(E;F'_r)$, with the obvious identification of mappings into F (resp. F'_r) with mappings into F_r (resp. F'). $F^{p}_{N,\infty}(E';F)$ is then still a Fréchet space when regarded as a projective limit of the Fréchet spaces $F^{p}_{N,\infty}(E';F_r)$, and the pairing of $F^{p}_{N,\infty}(E';F)$ with $F^{p'}_{0}(E;F')$ given by the corollary to Proposition 1.1 still holds: now $|\langle\langle \vec{f}, \vec{f}' \rangle\rangle| \leq \|\|\vec{f}\|\|_{N,\rho,r,p} \|\|\vec{f}'\|\|_{\rho',r,p'}$ if $\vec{f} \in F^{p}_{N,\rho}(E';F_r)$ and $\vec{f}' \in F^{p'}_{\rho'}(E;F'_r)$, where $|||\vec{f}|||_{N,\rho,r,p}$ and $|||\vec{f}'|||_{\rho',r,p'}$ are the norms in the corresponding spaces of maps $E' \to F_r$ and $E \to F'_r$ respectively. Given $A \in L(F,F)$, the estimates of Proposition 2.2 and 2.2' still hold with $\|A\|$ replaced by the appropriate growth estimate on A regarded as a map from F_s into F_r for each r and an appropriate s (given by A) satisfying $\| \ \|_r \leq \| \ \|_s$. The duality between $g'(d) \otimes A$ and $A' \circ g' \cdot$ with $g' \in F_0^{p'}(E)$ of Proposition 2.4 clearly remains valid. Since the open mapping theorem remains true for the Fréchet spaces, Proposition 3.2 also remains true. It follows that the division theorem 3.1, hence the approximation theorem 4.1 and the existence theorem 4.2, hold unchanged when F is a complete countably normed space.

Let now E_i, i = 1,2 be complex Banach spaces, and let there be given $p_i \geq 1$ and $g'_i \in F_0^{p'_i}(E_i)$, i = 1,2 (with $F_0^{p'_i}(E_i) := \mathrm{Exp}(E_i)$ if $p_i = 1$). Since, unlike [Dw 10], the vector values of solution functions now need be only in a total subset of the kernel of A, we obtain the following approximation theorem for $g'_1 \otimes g'_2(d)$:

THEOREM 5.1. *The functions* $\vec{f}_i : E'_1 \to F_{N,\infty}^{p_2}(E'_2)$ *of the form* $\vec{f}_i = e^{u_i} \Sigma'_j \Sigma_{k=0}^{n_{ij}} v_{ij}^k \cdot e^{\mu_{ij}} \cdot \nu_{ij}^{m_{ij}}$, *with distinct* u_i, *where either* u_i *is a zero of* g'_1 *with order higher than* n_{ij} *along* $v_{ij} \neq 0$ *in* E_1 *or* μ_{ij} *is a zero of* g'_2 *with order higher than* m_{ij} *along* $\nu_{ij} \neq 0$ *in* E_2, *form a basis for a dense subspace of the solution space of* $g'_1(d) \otimes g'_2(d)\vec{f} = 0$ *in* $F_{N,\infty}^{p_1}(E'_1; F_{N,\infty}^{p_2}(E'_2))$.

PROOF: By the approximation theorem when $F = \mathbb{C}$ we know that the functions $e^{\mu_{ij}} \nu_{ij}^{m_i}$ form a total subset of the kernel of $g'_2(d)$ in $F_{N,\infty}^{p_2}(E'_2)$. Moreover, $||| \ |||_{N,\rho,p_2}$ is a true norm on $F_{N,\infty}^{p_2}(E'_2)$ for each $\rho > 0$. It is enough to let $F = F_{N,\infty}^{p_2}(E'_2)$ and let Y be the set of functions $e^{\mu_{ij}} \cdot \nu_{ij}^{m_{ij}}$ in Theorem 4.1.

We also have an existence theorem for $g_1'(d) \otimes g_2'(d)$:

THEOREM 5.2. *If* $g_1' \neq 0$ *and* $g_2' \neq 0$ *then every equation* $g_1'(d) \otimes g_2'(d)\vec{f} = \vec{g}$ *with* $\vec{g}$ *in* $\mathcal{F}^{p_1}_{N,\infty}(E_1'; \mathcal{F}^{p_2}_{N,\infty}(E_2'))$ *has a solution* $\vec{f}$ *in the same space.*

PROOF: By the existence theorem when $F = \mathbb{C}$ we know that $g_2'(d)$ is surjective on $\mathcal{F}^{p_2}_{N,\infty}(E_2')$. It is enough then to set $F = \mathcal{F}^{p_2}_{N,\infty}(E_2')$ and $A = g_2'(d)$ in theorem 4.2.

REMARKS: 1. The relationship with the operators $(g_1' \otimes g_2')(d,d)$, or the more general operators $G'(d,d)$ with $G' \in \mathcal{F}^{p'}_{\theta;\rho'}(E_1 \times E_2)$, acting on the spaces $\mathcal{F}^{p}_{\theta,\rho}(E_1' \times E_2')$, is still to be clarified, even for the nuclear type $\theta = N$. All results here do hold on $\mathcal{F}^{p}_{N,\infty}(E_1' \times E_2')$ with $G' \in \mathcal{F}^{p'}_{0}(E_1 \times E_2)$, however, as follows from setting $E = E_1 \times E_2$ and $f = \mathbb{C}$ in sections 1 through 4.

2. If $E_1 = \mathbb{C}$, $E_2 = E$ and $p_1 = p_2 = 1$ we are reduced to the spaces $\mathcal{H}(\mathbb{C}, \mathcal{H}_{Nb}(E'))$ (of $\mathcal{H}_{Nb}(E')$-valued entire functions on $\mathbb{C}$). The relationship with $\mathcal{H}_{Nb}(\mathbb{C} \times E')$ would be a particularly interesting topic of investigation, as its study leads to initial value problems in infinite dimension: indeed, given $\vec{f}$ and $\vec{g}$ in $\mathcal{H}(\mathbb{C}; \mathcal{H}_{Nb}(E'))$ as well as g' in $\mathrm{Exp}(E)$, the equation $\partial/\partial t\, \vec{f}(t,x') = g'(d)\vec{f}(t,x') + \vec{g}(t,x')$ is equivalent to the equation $G'(d,d)\vec{f} = \vec{g}$, where the function $G' : \mathbb{C} \times E \to \mathbb{C}$ is defined by $G'(t,x) := t - g'(x)$, provided $\vec{f}$ and $\vec{g}$ are regarded as scalar-valued functions on $\mathbb{C} \times E'$. Similar considerations hold regarding $\mathcal{F}^{p}_{N,\infty}(\mathbb{C} \times E')$, as well as for other holomorphy types.

6. APPLICATION TO VECTOR-VALUED VARIATIONAL EQUATIONS.

The results in the preceding sections apply to holomorphic

mappings defined on spaces of essentially bounded measurable functions, with integrable variational kernels. More precisely, let M be a measure space with a positive measure m and set $E = L^1(M)$ (all function spaces on M are relative to the measure m). A function $\vec{f}$ on $L^\infty(M)$ with values in a (Banach or Fréchet) space F is said to have an n-th variational derivative at $x' \in L^\infty(M)$ if its n-th Fréchet derivative at x' has a kernel in $L^1_{s(M^n;F)}$ (symmetric integrable functions $M \times \ldots \times M \to F$) denoted by

$$\delta^n \vec{f}/\delta x'^n: (t_1,\ldots,t_n) \to \delta^n \vec{f}(x')/\delta x'(t_1)\ldots\delta x'(t_n)$$

so that

$$\hat{d}^n \vec{f}(x')(v') = \int \ldots \int \delta^n \vec{f}(x')/\delta x'(t_1)\ldots\delta x'(t_n) v'(t_1)\ldots v'(t_n) dm(t_1)\ldots dm(t_n)$$

for every $v' \in L^\infty(M)$. This kernel is then the variational derivative of f. We have:

PROPOSITION 6.1. *A function* $\vec{P}_n: L^\infty(M) \to F$ *is an n-homogeneous nuclear polynomial if and only if there is a kernel* $\vec{x}_n \in L^1_s(M^n:F)$ *such that*

$$\vec{P}_n(x') = \int \ldots \int \vec{x}_n(t_1,\ldots,t_n)\, x'(t_1) \ldots x'(t_n) dm(t_1)\ldots dm(t_n)$$

for every $x' \in L^\infty(M)$. *In this case we have*

$$\| \vec{P}_n \|_N = \int \ldots \int \| \vec{x}_n(t_1,\ldots,t_n) \| \, dm(t_1)\ldots dm(t_n).$$

PROOF: $P_N({}^nL^\infty(M);F)$ can be identified with the nuclear completion of $L^1(M) \odot \ldots \odot L^1(M) \otimes F$ (where $\odot$ denotes the symmetric tensor product). The repeated application of the Dunford-Pettis theorem (on the isometry between $L^1(M;F)$ and the nuclear completion of $L^1(M) \otimes F$) then leads to the identification of $L^1_s(M^n;F)$ with $P_N({}^nL^\infty(M);F)$: cf. [Tr 2], Th. 46.2.

Together with the translation-invariance of $F^p_{N,\infty}(E';F)$ corollary to Proposition 2.2), by application of the root test to the norms $|||\vec{f}|||_{N,\rho,p}$ regarded as power series in ρ we the have:

COROLLARY: *A function* $\vec{f} : L^{\infty}(M) \to F$ *is in* $F^{p}_{N,\infty}(L^{\infty}(M) ; F)$ (*or* $H_{Nb}(L^{\infty}(M) ; F)$ *when* $p = 1$) *if and only if* $\vec{f}$ *has variational derivatives of all orders at some* $x' \in L^{\infty}(M)$ *such that*

$$(*) \qquad \lim_n \{n!^{-1/p} \int \cdots \int \| \delta^n \vec{f}(x')/\delta x'(t_1) \ldots \delta x'(t_n) \| dm(t_1) \ldots dm(t_n)\}^{1/n} = 0$$

In this case we have

$$|||\vec{f}|||_{N,\rho,p} = \{\Sigma_{n=0}^{\infty} \rho^n \frac{1}{n!} (\int \cdots \int \| \delta^n \vec{f}(x')/\delta x'(t_1) \ldots \delta x'(t_n) \| dm(t_1) \ldots dm(t_n))^p\}^{1/p}$$

for each $\rho > 0$, *and* $x' = 0$.

REMARK. The analogue of the condition (*) above applied to the norms $\| \hat{d}^n \vec{f}(0) \|_{\theta}$ in place of the integrals clearly characterizes $F^{p}_{\theta,\infty}(E';F)$ for general holomorphy types θ.

We can also characterize polynomials and holomorphic mappings of current type on $L^1(M)$:

PROPOSITION 6.2. *A function* $\vec{P}'_n : L^1(M) \to F'$ *is a continuous* n-*homogeneous polynomial on* $L^1(M)$ *if and only if there is a kernel* $x'_n \in L^{\infty}_s(M^n; F')$ *such that*

$$\vec{P}'_n(x) = \int \cdots \int \vec{x}'_n(t_1, \ldots, t_n) x(t_1) \ldots x(t_n) dm(t_1) \ldots dm(t_n)$$

for every $x \in L^1(M)$. *We then have*

$$\| \vec{P}'_n \| = \text{ess. sup } \{ \| \vec{x}'_n(t_1, \ldots, t_n) \| : t_i \in M \}.$$

PROOF: Follows from the isometry between $P_N({}^nE';F)'$ and $P({}^nE;F')$ [Dw 9], that between $P_N({}^nL^{\infty}(M);F)$ and $L^1_s(M^n;F)$ (Proposition 6.1), and finally that between $L^1_s(M^n;F)'$ and $L^{\infty}_s(M^n;F')$.

The translation - invariance of $F^{p'}_0(E;F')$ (corollary to Proposition 2.2') and the application of the root test to the norms $|||\vec{f}'|||_{\rho' p'}$ regarded as power series in ρ' then give us:

COROLLARY: *A function* $\vec{f}' : L^1(M) \to F'$ *is in* $F^{p'}_0(L^1(M) ; F')$

(or $Exp(L^1(M);F')$ *when* $p' = \infty$ *if and only if* $\vec{f}'$ *has variational derivatives of all orders at some* $x \in L^1(M)$ *such that*

$$(+) \quad \limsup (n!^{-1/p'} \text{ ess. sup.}\{\| \delta^n \vec{f}'(x)/\delta x(t_1)\ldots\delta x(t_n)\| : t_i \in M\})^{1/n} < \infty$$

In this case we have

$$|||f'|||_{\rho',p'} = \{ \Sigma_{n=0}^{\infty} \frac{\rho'^n}{n!} (\text{ess. sup}\{\| \delta^n \vec{f}'(x)/\delta x(t_1)\ldots \delta x(t_n)\| : t_i \in M\})^{p'}\}^{1/p'}$$

for some $\rho' > 0$ *if* $p' < \infty$, *and* $x = 0$.

REMARK: The analogue of condition (+) applied to the norms $\| \hat{d}^n \vec{f}'(0) \|_\theta$ characterizes $F^{p'}_{\theta,0}(E;F')$ for general holomorphy types θ.

We may then treat variational equations on $L^\infty(M)$:

PROPOSITION 6.3: *The existence and approximation theorems hold for variational equations of the form*

$$(a) \quad \Sigma_{n=0}^{\infty} \frac{1}{n!} \int\ldots\int A\delta^n \vec{f}(x')/\delta x'(t_1)\ldots\delta x'(t_n) x'_n(t_1,\ldots,t_n) dm(t_1)\ldots dm(t_n) = \vec{g}(x'),$$

where the variational derivatives $\delta^n \vec{f}/\delta x'^n$ *and* $\delta^n \vec{g}/\delta x'^n$ *satisfy* (*) *in the corollary to Proposition 6.1 and the kernels* x'_n *satisfy* (+) *in the corollary to Proposition 6.2.*

In particular, if $\vec{g} = 0$ *then linear combinations of functions* $L^\infty(M) \to F$ *of the from*

$$x' \mapsto \{\exp \{\int u(t)x'(t)dm(t)\}\cdot\{\int v(t)x'(t)dm(t)\}^n\}y,$$

where either y *lies in a total subset of* $A^{-1}(0)$ *or* u *is a zero with order higher than* n *in the direction of* v *of the function* g' *given by*

$$x \mapsto \Sigma_{n=0}^{\infty} \int\ldots\int x'_n(t_1,\ldots,t_n)x(t_1)\ldots x(t_n) dm(t_1)\ldots dm(t_n) \text{ in } L^1(M),$$

are dense in the space of solutions $\vec{f}$.

PROOF: It is enough to set $F = \mathbb{C}, \vec{f}' = g'$ and $\frac{1}{n!}(\delta^n g'/\delta x^n)_{x=0} =: x'_n$ in the corollary to Proposition 6.2, and observe that the variational equation (a) is the equation $g'(d) \otimes A\vec{f} = \vec{g}$ with $\vec{f}, \vec{g}$ in $F^p_{N,\infty}(L^\infty(M);F)$ and $g' \in F^{p'}_0(L^1(M))$.

REMARK: Propositions 6.1 and 6.2, hence the proof of Proposition 6.3, do not hold on $L^p(M)$ with arbitrary p: e.g., if p = 2 then the functions

$$x' \mapsto \int\cdots\int \vec{x}_n(t_1,\ldots,t_n)x'(t_1)\ldots x'(t_n)dm(t_1)\ldots dm(t_n) \qquad \text{for}$$

$x' \in L^2(M)$ and $\vec{x}_n \in L^2_s(M^n;F)$ are the n-homogeneous Hilbert-Schmidt polynomial mappings from $L^2(M)$ to F, so that it is the treatment in [Dw 1,2,3] that applies.

REFERENCES

[A 1] N. U. AHMED, On the $L^p(p \geq 1)$ stability of a class of non-linear systems, Proc. IEEE 57 (1969), 1795-1797, MR 47, 8169.

[A 2] N. U. AHMED, Closure and completeness of Wiener's orthogonal set G_n in the class $L^2(\Omega,B,M)$ and its application to stochastic hereditary differential systems, Inform. and control 17 (1970), 161-174.

[A 3] N. U. AHMED, Strong and weak synthesis of non-linear systems with constraints in the system space G_λ, Inform. and Control 23 (1973), 71-85.

[Ar] R. ARON, Tensor products of holomorphic functions, Nederl. Akad. Wetensch. = Indag. Math. 35 (76) (1973), Fasc. 3, 192-202.

[Bol 1] Ph. J. BOLAND, Espaces pondérés de fonctions entières nucléaires sur un espace de Banach, C. R. Acad. Sci. Paris Sér. A-B 275 (1972), A587-A590.

[Bol 2] Ph. J. BOLAND, Some spaces of entire and nuclearly entire functions on a Banach space, J. Reine Angew. Math. Part I: *270* (1974), 38-60. Part II: *271* (1974), 8-27.

[Bol 3] Ph. J. BOLAND, Malgrange theorem for entire functions on nuclear spaces, Proc. on Infinite-Dimensional Holomorphy, Lecture Notes in Math., Nº 346, Springer Verlag, New York, 1974.

[Bol 4] Ph. J. BOLAND, Holomorphic functions on *DFN* (dual of Fréchet nuclear) spaces, Séminaire Pierre Lelong 14ème année, 1973/74, Lecture Notes in Math. Nº 474, Springer-Verlag, New York, 1975.

[Bol 5] Ph. J. BOLAND, Holomorphic functions on nuclear spaces, Trans. Amer. Math. Soc., 209, (1975).

[BD] PH. J. BOLAND AND S. H. DINEEN, Convolution operators on G-holomorphic functions in infinite dimension, Trans. Amer. Math. Soc. *190*(1974), 313-323.

[Bon] O. BONNIN, Représentation holomorphe des distributions tempérées. Transformation de Fourier-Borel. Opérateurs de dérivations partielles de type Hilbert-Schmidt en dimension infinie (d'après Thomas A. W. Dwyer, III), Séminaire Pierre Lelong 14ème année, 1973/74, Lecture Notes in Math. Nº 474, Springer-Verlag, New York, 1975, pp. 123-141.

[Br 1] R. W. BROCKETT, Nonlinear systems and differential geometry, Proc. IEEE 64 (1976).

[Br 2] R. W. BROCKETT, Volterra series and geometric control theory, Automatica 12 (1976).

[Br 3] R. W. BROCKETT, Finite and infinite dimensional bilinear realizations, J.Franklin Inst., 301, Nº6 (1976).

[Di 1] S. H. DINEEN, Holomorphic functions on a Banach space, Bull. Amer. Math. Soc. *76* (1970), 883-886. MR *41*, 4216.

[Di 2] S. H. DINEEN, Holomorphy types on a Banach space, Studia Math. *39* (1971), 241-288.

[DL] M. D. DONSKER AND J. L. LIONS, Fréchet-Volterra variational equations, boundary value problems and function space integrals, Acta Math. *108* (1962), 147-228. MR *27*, 1701.

[Dw 1] T. A. W. DWYER, III, Partial differential equations in Fischer-Fock spaces for the Hilbert-Schmidt holomorphy type, Bull. Amer. Math. Soc. *77* (1971), 725-730. MR *44*, 7288.

[Dw 2] T. A. W. DWYER, Holomorphic Fock representations and partial differential equations on countably Hilbert spaces, Bull. Amer. Math. Soc. *79* (1973), 1045-1050. MR *47*, 9283.

[Dw 3] T. A. W. DWYER, Partial differential equations in holomorphic Fock spaces, Functional Analysis and Applications, Lecture Notes in Math. Nº 384, Springer-Verlag, New York, 1974, pp. 252-259.

[Dw 4] T. A. W. DWYER, Holomorphic representation of tempered distributions and weighted Fock spaces, Analyse

Fonctionnelle et Applications, Actualites Sci.et Industr. Nº 1367, Hermann, Paris, 1975, pp. 95-118.

[Dw 5] T. A. W. DWYER, Dualité des espaces de fonctions entières en dimension infinie, C. R. Acad. Sci. Paris Sér. A-B *280* (1975), A1439-A1442.

[Dw 6] T. A. W. DWYER, Équations différentielles d'ordre infini dans des espaces localement convexes, C.R. Acad. Sci. Paris *281* (1975), A163-A166.

[Dw 7] T. A. W. DWYER, Dualité des espaces de fonctions entières en dimension infinie, Ann. Inst. Fourier (Grenoble) *26* (1976), Nº 4.

[Dw 8] T. A. W. DWYER, Differential operators of infinite order in locally convex spaces, to appear.

[Dw 9] T. A. W. DWYER, Convolution equations for vector-valued entire functions of nuclear bounded type, Proc. Symposium on Infinite-Dimensional Function Theory, Northern Illinois University, 1973 = Trans. Amer. Math. Soc. *217* (1976).

[Dw 10] T. A. W. DWYER, Vector-valued convolution equations for the nuclear holomorphy type, Proc. Royal Irish Acad., 76, Sec A, Nº 11, (1976).

[FW] K. FLORET AND J. WLOKA, Einfhrüng in die Theorie der lokalkonvexen Raüme, Lecture Notes in Math. Nº 56, Springer-Verlag, New York, 1968.

[G 1] C. P. GUPTA, Malgrange theorem for nuclearly entire functions of bounded type on a Banach space, Notas de Matemática Nº 37, Instituto de Matemática Pura e

Aplicada, Rio de Janeiro, Brasil, 1968.

[G 2] C. P. GUPTA, Convolution operators and holomorphic mappings on a Banach space, Séminaire d'Analyse Moderne Nº 2. Université de Sherbrooke, Canada, 1969.

[G 3] C. P. GUPTA, On the Malgrange theorem for nuclearly entire functions of bounded type on a Banach space, Nederl. Akad. Wetensch. Proc. Ser. A *73* = Indag. Math. *32* (1970), 356-358.

[H] M. HERVÉ, Analytic and plurisubharmonic functions in finite and infinite dimensions, Lecture Notes in Math. Nº 198, Springer-Verlag. New York, 1970.

[K 1] P. KRÉE, Solutions faibles d'équations aux dérivées fonctionnelles I, Séminaire Pierre Lelong, 13ème année, 1972/73, Lecture Notes in Math. Nº 410, Springer-Verlag, New York, 1974.

[K 2] P. KRÉE, II, Séminaire Pierre Lelong, 14ème année, 1973/74, Lecture Notes in Math. Nº 474, Springer-Verlag, New York, 1975.

[K 3] P. KRÉE, Applications des méthodes variationelles aux équations aux dérivées partielles sur un espace de Hilbert, C. R. Acad. Sci. Paris Sér. A-B *278* (1974), A753-A755.

[K 4] P. KRÉE, Extension du calcul différentiel en dimension infinie, C. R. Acad. Sci. Paris Sér. A - B *280* (1975), A7-A9.

[K 5] M. KRÉE, Proprieté de trace en dimension infinie, d'espaces du type de Sobolev, C. R. Acad. Sci. Paris

Sér. A-B *279*, A157-A160.

[Mar 1] A. MARTINEAU, Équations différentielles d'ordre infini, 2ème Colloque sur l'Analyse Fonctionnelle, CBRM, Liège, 1964, pp. 37-47.

[Mar 2] A. MARTINEAU, Équations différentielles d'ordre infini, Bull, Soc. Math. France *95* (1967), 109-154.

[Mat 1] M. DE MATOS, Sur le théorème d'approximation et d'existence de Malgrange-Gupta, C. R. Acad. Sci. Paris Sér. A-B *271* (1970), A1258-A1259. MR *44*,3105.

[Mat 2] M. DE MATOS, Holomorphic mappings and domains of holomorphy, Monografias do Centro Brasileiro de Pesquisas Físicas nº 27, Rio de Janeiro, Brasil,1970.

[N 1] L. NACHBIN, Topology on spaces of holomorphic mappings, Ergebnisse de Mathematik und ihrer Grenzgebiete vol. 47, Springer-Verlag, New York, 1969. MR *40*, 7787.

[N 2] L. NACHBIN, Convolution operators in spaces of nuclearly entire functions on a Banach space, Functional Analysis and Related Fields (Chicago, 1968), Springer-Verlag, New York, 1970, pp. 167-171.

[N 3] L. NACHBIN, Concerning holomorphy types for Banach spaces, Proc. International Colloquium on Nuclear spaces and Ideals in Operator Algebras (Warsaw, 1969), studia Math. *38*(1970),407-412. MR 43, 3787.

[N 4] L. NACHBIN, Convoluções em funções inteiras nucleares, Atas da 2ª Quinzena de Análise Funcional e Equações Diferenciais, Sociedade Brasileira de Matemática, 1972.

[N 5] L. NACHBIN, Recent developments in infinite-dimensional holomorphy, Bull. Amer. Math. Soc. *79* (1973), 625-640. Zbl *279*. 32017.

[RR] A. ROBERTSON AND W. ROBERTSON, Topological vector spaces, Cambridge University Press, 1963.

[Ta] B. A. TAYLOR, Some locally convex spaces of entire functions, Proc. Symposia in Pure Math., Vol 11, Amer. Math. Soc., 1968.

[Tr 1] F. TRÈVES, Linear partial differential equations with constant coefficients, Gordon and Breach, New York, 1966.

[Tr 2] F. TRÈVES, Topological vector spaces, distributions and kernels, Academic Press, New York, 1967.

[Tr 3] F. TRÈVES, Linear partial differential operators, Gordon and Breach, New York, 1970.

[W] N. WIENER, Non-linear problems in random theory, Technology Press of M.I.T. and John Wiley, New York, 1958.

DEPARTMENT OF MATHEMATICAL SCIENCES
NORTHERN ILLINOIS UNIVERSITY,
DEKALB, ILLINOIS, 60115, U.S.A.

Infinite Dimensional Holomorphy and Applications, Matos (ed.)

ON THE RANGE OF ANALYTIC FUNCTIONS INTO A BANACH SPACE

By J. *Globevnik*

Let Δ be the open unit disc in C and let X be a separable complex Banach space. Does there exist an analytic function $f : \Delta \to X$ such that the convex hull of $f(\Delta)$ is contained and dense in the unit ball of X ? This problem, raised by D. Patil at the International Conference on Infinite Dimensional Holomorphy, University of Kentucky, 1973, was the motivation to start studying certain range properties of analytic functions into a Banach space.

Throughout, Δ is the open unit disc in C and $\partial\Delta$ is its boundary. Let X be a complex Banach space. The closure of a set $S \subset X$ is denoted by $\bar{S}$. If $r > 0$ we denote by $B_r(X)$ the open ball in X, of radius r, centered at the origin. By $H(\Delta,X)$ we denote the space of all analytic functions from Δ to X, by $H^\infty(\Delta,X)$ we denote the Banach space of all bounded analytic functions from Δ to X with sup norm and by $A(\Delta,X)$ we denote its closed subspace of functions having continuous extension to all $\bar{\Delta}$. We write H^∞ for $H^\infty(\Delta,C)$. If K is a compact Hausdorff space we denote by $C(K,X)$ the Banach space of all continuous functions from

K to X with sup norm.

Let X be a separable complex Banach space. Does there exist an $f \in H(\Delta,X)$ whose range is dense in some ball? The positive answer is provided by

THEOREM 1 (see [10], p.157, [15], p.298) *Let X be a complex Banach space. Given any injective sequence* $\{z_n\} \subset \Delta$ *with no cluster point in* Δ *and any sequence* $\{w_n\} \subset X$ *there exists* $f \in H(\Delta,X)$ *satisfying* $f(z_n) = w_n$ *for all* n.

Theorem 1 does not provide the answer to the question whether there exists an $f \in H^\infty(\Delta,X)$ whose range is dense in some ball. A natural way to show that such a function exists is the interpolation of bounded sequences of vectors in Banach spaces by bounded analytic functions. Call a sequence $\{z_n\} \subset \Delta$ *interpolating sequence* (see [11]) if given any bounded sequence $\{w_n\} \subset C$ there exists an $f \in H^\infty$ such that $f(z_n) = w_n$ for all n. The well known theorem of L. Carleson gives a necessary and sufficient condition for a sequence to be interpolating (see [11]). Call a sequence $\{z_n\} \subset \Delta$ a *general interpolating sequence* if given any bounded sequence $\{w_n\}$ in any complex Banach space X there exists $f \in H^\infty(\Delta,X)$ satisfying $f(z_n) = w_n$ for all n. Using the theory of H^p-spaces of vector-valued functions it was shown in [2] that the proof of the Carleson interpolation theorem due to Shapiro and Shields (see [11]) can be modified to obtain

THEOREM 2 (see [2]) *A sequence* $\{z_n\} \subset \Delta$ *is a general interpolating sequence if and only if it is an interpolating sequence.*

Note that Theorem 2 can be deduced also from a result of P. Beurling (see [2], [3]).

It was proved in [2] that given an interpolating sequence $\{z_n\} \subset \Delta$ there exists a universal bound for the norms of the interpolating functions, i.e. there exists $M < \infty$ such that for any sequence $\{w_n\}$ in any complex Banach space X there exists $f \in H^\infty(\Delta, X)$ satisfying $f(z_n) = w_n$ for all n and $||f(z)|| \leq M.\sup_n ||w_n||$ $(z \in \Delta)$. Even in the scalar case it seems to be an open problem what is the least lower bound for such M. There exist sequences $\{z_n\} \subset \Delta$ for which this bound is close to 1:

THEOREM 3 (see [2]) *Given* $\varepsilon > 0$ *there exists a sequence* $\{z_n\} \subset \Delta$ *such that for any complex Banach space* X *and for any bounded sequence* $\{w_n\} \subset X$ *there exists* $f \in H^\infty(\Delta, X)$ *satisfying*

$$(i) \quad f(z_n) = w_n \quad \text{for all } n$$

$$(ii) \quad ||f(z)|| \leq (1+\varepsilon).\sup_n ||w_n|| \quad (z \in \Delta).$$

Observe that given a separable complex Banach space X Theorem 3 gives an approximate solution to Patil's problem since given $\varepsilon > 0$ it gives an $f \in H^\infty(\Delta, X)$ whose range is contained in $B_{1+\varepsilon}(X)$ and dense in $B_1(X)$. However, as observed in [2] such a method does not give the exact soution of Patil's problem.

The solution of Patil's problem in the finite dimensional case was found in [5] as an application of the following generalization of the well known Rudin-Carleson theorem.

THEOREM 4 (see [5], [13], [16]) *Let* X *be a complex Banach space Given a closed set* $F \subset \partial\Delta$ *of Lebesgue measure* 0 *and* $f \in C(F, X)$ *there exists* $\tilde{f} \in A(\Delta, X)$ *satisfying* $\tilde{f}|F = f$ *and* $||\tilde{f}|| = ||f||$. *Consequently if* X *is finite dimensional there exists* $\tilde{f} \in A(\Delta, X)$ *such that* $\tilde{f}(\Delta)$ *is contained and dense in* $B_1(X)$.

REMARK 1 To deduce the second part of the above theorem from its first part, let $F \subset \partial\Delta$ be a Cantor set of Lebesgue measure 0 and let f be a continuous surjection from F onto $\overline{B_1(X)}$ which exists by the compactness of $\overline{B_1(X)}$ (see [12], p.166).

In the infinite dimensional case Patil's problem was solved independently in [1] and [6]. In [1] this was done in a typically infinite dimensional way: by constructing first a function that solves the problem for c_0 and then composing it by an analytic function mapping $B_1(c_0)$ into $B_1(X)$ densely. In [6] the solution was found as an application of the following generalization of an interpolation result of [9]:

THEOREM 5 (see [6]) *Let* $E \subset \partial\Delta$ *be a closed set and let* $F \subset \partial\Delta - E$ *be a relatively closed set of Lebesgue measure* 0.*Given a complex Banach space* X *and a bounded continuous function* $f: F \to X$ *there exists a bounded continuous function* $\tilde{f}: \overline{\Delta} - E \to X$, analytic on Δ and satisfying

$$\text{(i)} \quad \tilde{f}|F = f$$

$$\text{(ii)} \quad \sup_{z \in \overline{\Delta} - E} ||\tilde{f}(z)|| = \sup_{s \in F} ||f(s)||$$

Consequently if X *is separable then there exists* $\tilde{f} \in H(\Delta, X)$ *whose range is contained and dense in* $B_1(X)$.

REMARK 2 To deduce the second part of the above theorem from its first part, let $E = \{1\}$, let $F = \{z_n\} \subset \partial\Delta - \{1\}$ be an injective sequence converging to 1 and let $f(z_n) = w_n$ where $\{w_n\} \subset B_1(X)$ is a sequence, dense in $B_1(X)$.

An interesting consequence of Theorem 5 is that the space H^∞ is isometrically universal for all complex Banach spaces possessing countable determining sets, i.e. that every such space is isometrically isomorphic to a subspace of H^∞ (recall that a determining set for a Banach space X is any subset S of the dual

X' such that $||x|| = \sup_{u \in S} |u(x)|$ for all $x \in X$) (see [7]).

The results presented above lead naturally to various problems. First, let us ask when does Theorem 5 hold if $B_1(X)$ is replaced by some other set. Let us say that Δ is *analytically dense* in a subset P of a separable complex Banach space X if there exists $f \in H(\Delta, X)$ whose range is contained and dense in P.

PROBLEM 1 *Obtain a (geometrical, topological) characterization of the sets in which Δ is analytically dense.*

This problem seems to be hard. Recently a partial solution was found for the case when the sets in question are open:

THEOREM 6 (see [7]) *Δ is analytically dense in an open subset P of a separable complex Banach space if and only if P is connected.*

COROLLARY 1 *Let X be a complex Banach space and let Y be a separable complex Banach space. Given any open connected set Ω in Y there exists an analytic mapping $F : B_1(X) \to Y$ whose range is contained and dense in Ω.*

The proof of Corollary 1 is trivial. One applies $u \in X'$, $u \neq 0$ to map $B_1(X)$ onto an open disc in C and then by Theorem 6 one maps this disc analytically and densely into Ω. Since the range of $f \in H(\Delta, Y)$ is always separable such a construction is not possible if the space Y is nonseparable. Consequently we have

PROBLEM 2 *Let Y be a nonseparable complex Banach space. Determine the class of all complex Banach spaces X having the following property: given any open connected set Ω in Y there exists an analytic mapping $F : B_1(X) \to Y$ whose range is contained and dense in Ω.*

CONJECTURE *If P is an open connected set in a complex Banach space X then there exists an analytic mapping $F : B_1(X) \to X$ whose range is contained and dense in P.*

Since it is easier to construct the continuous functions having some prescribed range properties than the analytic ones it is obvious that various generalizations of the Rudin-Carleson theorem are an efficient tool when proving the existence of an analytic function whose range has certain density properties (see Remarks 1, 2). In particular, when proving the existence of an analytic function whose range is dense in some ball we first constructed a continuous function on a suitable subset of $\partial\Delta$ whose range was dense in this ball, and then we extended this function to a function analytic on Δ whose range was contained in the same ball. It is a natural question whether such an extension is possible for some other classes of sets in place of the balls. Let us say that a subset P of a complex Banach space X has the *analytic extension property* if given any closed set $E \subset \partial\Delta$, any relatively closed subset $F \subset \partial\Delta - E$ of Lebesgue measure 0 and any continuous function $f : F \to X$ satisfying $f(F) \subset P$ there exists a continuous extension $\tilde{f} : \bar{\Delta} - E \to X$, $\tilde{f}|F = f$, analytic on Δ and satisfying $\tilde{f}(\bar{\Delta} - E) \subset P$.

PROBLEM 3 *Obtain a (geometrical, topological) characterization of the sets having the analytic extension property.*

This problem seems to be very hard. Recently a partial solution was found for the case when the sets in question are open:

THEOREM 7 (see [8]) *An open subset of a complex Banach space has the analytic extension property if and only if it is con-*

nected.

By Theorem 5, any closed ball in a complex Banach space X has the analytic extension property. To prove the same for more general closed sets seems much more difficult. For example, does the closed shell $\{x \in X : 1/2 \leq ||x|| \leq 1\}$ have the analytic extension property?

ACKNOWLEDGEMENT The author acknowledges gratefully the support from the Boris Kidric Fund, Ljubljana, Yugoslavia.

REFERENCES

[1] R.M.ARON: The range of vector-valued holomorphic mappings. Proc.Conf.Anal.Funct., Krakow 1974, Ann.Polon.Math. 33 (1976).

[2] R.M.ARON, J.GLOBEVNIK, M. SCHOTTENLOHER: Interpolation by vector-valued Analytic functions. Rendiconti di Mat.(2) V.9, serie VI (1976).

[3] L.CARLESON: Interpolations by bounded analytic functions and the corona problem. International Congress of Mathematicians, Stockholm 1962.

[4] ---------: Representations of continuous functions. Math. Zeit. 66(1957) 447-451.

[5] J.GLOBEVNIK: The Rudin-Carleson theorem for vector- valued functions. Proc. Am. Math. Soc. 53(1975).

[6] ---------: Analytic functions whose range is dense in a ball. Journ. Funct. Anal. 22(1976).

[7] ---------: The range of vector-valued analytic functions. To appear in Ark. för Math. The range of vector-

valued analytic functions II. Ark. för Mat.14(1976)

[8] ————: Analytic extensions of vector - valued functions. Pac. Journ. Math. 63 (1976).

[9] E.A.HEARD, J.H.WELLS: An interpolation theorem for subalgebras of H^∞. Pac.Journ.Math. 28(1969) 543-553.

[10] Y.HERVIER: On the Weierstrass problem in Banach spaces. Proc. on Infinite Dimensional Holomorphy. Lect. Notes in Math. 364, Springer 1974, pp. 157-167.

[11] K.HOFFMAN: Banach spaces of analytic functions. Prentice - Hall 1962.

[12] J.L.KELLEY: General topology. Van Nostrand 1955.

[13] D.H.OBERLIN: Interpolation and vector-valued functions. Journ. Funct. Anal. 15(1974) 428-439.

[14] W.RUDIN: Boundary values of continuous analytic functions. Proc. Amer. Math. Soc. 7(1956) 808-811.

[15] W.RUDIN: Real and complex analysis. McGraw Hill 1966.

[16] E.L.STOUT: On some restriction algebras.Functions Algebras. Scott, Foresman 1966, pp. 6 - 11.

ADDED IN PROOF: B. Josefson (Some remarks on Banach valued polynomials on $c_o(A)$, to appear in these Proc.) proved that the above conjecture is false by constructing a counterexample in $X = c_o(A)$ with A uncountable. On the other hand the author (on the ranges of analytics maps in infinite dimensions, to appear) showed that the assertion in the conjecture is true for $X = l^p(A)$ $(1 \leq p < \infty)$. Further, the author (The range of analytic extensions, to appear in Pac. Journ.Math.) obtained partial solutions of Problem 3; in particular, he gave a complete topological description of the subsets of C having the analytic extension property.

INSTITUTE OF MATHEMATICS,
PHYSICS AND MECHANICS
UNIVERSITY OF LJUBLJANA
LJUBLJANA, YUGOSLAVIA

Infinite Dimensional Holomorphy and Applications, Matos (ed.)

ω-SPACES AND σ-CONVEX SPACES

By ERIK GRUSELL

In this paper we give an example of a locally convex topological vector space E, such that E is an ω-space, but is not σ-convex with respect to the open sets. However, E is σ-convex with respect to the open and connected pseudo-convex sets. (For definitions see below). To prove this we show that E is not σ-convex with respect to the open sets, and that every plurisubharmonic (and in fact also every continuous) function in E depends only on countably many coordinates.

DEFINITION 1 [Dineen, Noverraz, Schottenloher 3] Let PCVX(F) denote the family of open and connected pseudo-convex subsets of the locally convex space F. Let $\alpha = \{p_j\}_{j=1}^{\infty}$ be a countable set of continuous seminorms on F, and let F_α denote the space F endowed with the topology generated by the seminorms in α. F is called *σ-convex with respect to the pseudoconvex domains* (or *σ-convex*) if:

$$PCVX(F) = \bigcup_{\alpha} PCVX(F_\alpha)$$

where the union is taken over all countable sets α of continuous seminorms on F.

DEFINITION 2 Let F, α and F_α be as in Definition 1, and let OPEN(F) denote the family of open subsets of F. F is called *σ-convex with respect to the open sets* if:

$$\mathrm{OPEN}(F) = \bigcup_\alpha \mathrm{OPEN}(F_\alpha),$$

where the union has the same meaning as in Definition 1.

DEFINITION 3 [Dineen 1, 2] Let F be a locally convex space, and let $\mathcal{H}(U)$ be the set of complex valued analytic functions defined on U, where U is an open and connected subset of F. This means that $f \in \mathcal{H}(U)$ if f is continuous and $\lambda \to f(x + \lambda y)$ is analytic in a neighbourhood of zero for every $x \in U$ and $y \in F$. F is called an *ω-space* if for every $f \in \mathcal{H}(U)$ where $U \subset F$ is open and connected, there exists a sequence $\{p_i\}_{i=1}^\infty$ of continuous seminorms such that f is also continuous in the topology generated by the sequence $\{p_i\}_{i=1}^\infty$.

Since $\mathrm{OPEN}(F) \supset \mathrm{PCVX}(F)$ it is trivial that if F is σ-convex with respect to OPEN(F), then F is also σ-convex with respect to PCVX(F). It is also clear from the definitions that if F is σ-convex with respect to OPEN(F), then F is an ω-space.

The converses of these implications are not true. We show this by giving an example of an ω-space E, which is not σ-convex with respect to OPEN(E), but with respect to PCVX(E).

Let $E = \{(x_r)_{r \in \mathbb{R}};\ \Sigma|x_r| < \infty\}$ and let the topology on E be defined by the set of seminorms of the form

$$p_A(x) = \sum_{r \in A} |x_r|$$

where $A \subset \mathbb{R}$ is a countable set. We first show that E is not σ-convex with respect to OPEN(E) and then give a lemma which implies that E is σ-convex with respect to PCVX(E), and also that E is an ω-space.

PROPOSITION E *is* σ-*convex with respect to* PCVX(E) *but not with respect to* OPEN(E) *and* E *is an* ω-*space.*

PROOF To prove that E is not σ-convex with respect to OPEN(E) it is sufficient to find one open set $U \subset E$ such that $U \notin \text{OPEN}(E_\alpha)$ for any countable set α of continuous seminorms on E.

Let $a_r = \{r + k,\ k \in \mathbb{Z}\}$, for $r \in [0,1[\ = I$. Then $\mathbb{R} = \bigcup_{r \in I} a_r$, each a_r is denumerable, and $a_r \cap a_s = \emptyset$ if $r \neq s$. Further, for $r \in I$ let $x^r = (x_t^r)_{t \in \mathbb{R}} \in E$, where

$$x_t^r = \begin{cases} 0 & \text{if } r \neq t \\ 1 & \text{if } r = t. \end{cases}$$

Define $U_r = x^r + \{x;\ p_{a_r}(x) < 1\}$, and $U = \bigcup_{r \in I} U_r$. Then $U \in \text{OPEN}(E)$. Suppose $U \in \text{OPEN}(E_\alpha)$ for some α. Then $U \in \text{OPEN}(E_\beta)$ where $\beta = \{p_A\}$ and $A = \bigcup_{p_B \in \alpha} B$. Since A is countable, there is an $r_0 \in I$ such that $A \cap a_{r_0} = \emptyset$. $x^{r_0} \in U$ by construction, but since $p_A(x^{r_0}) = 0$, every p_A-open neighbourhood of x^{r_0} contains zero. Since $0 \notin U$ this shows that x^{r_0} has no p_A-open neighbourhood contained in U, so that U cannot be p_A-open, which contradicts the assumption. We have thus shown that $U \notin \text{OPEN}(E_\alpha)$ for any α.

To prove that E is an ω-space and that E is σ-convex with respect to PCVX(E) we give a lemma (Lemma 1), which says that a plurisubharmonic function in E only depends on a countable number of coordinates. This shows immediately that E is an ω-space, and as a corollary it follows that E is σ-convex with respect to PCVX(E). Lemma 2 says that the continuous functions on E also have this property.

LEMMA 1 *If* $f \in \text{PSH}(U)$, *where* U *is an open and connected subset of* E, *then* f *depends only on countably many variables in* U, *i.e. there exists a countable set* $A \subset \mathbb{R}$ *such that* $\operatorname{supp} y \subset \complement A \Longrightarrow f(x+y) = f(x)$ *for every* $x \in U$, *where* $\operatorname{supp} y = \{r;\ y_r \neq 0\}$.

PROOF Choose $x_0 \in U$ so that $f(x_0) > -\infty$. Because f is upper semicontinuous in x_0, there is a countable set $A \subset \mathbb{R}$ and a constant $\varepsilon > 0$ such that $p_A(x-x_0) < \varepsilon \Longrightarrow x \in U$ and $f(x) < f(x_0)+1$. If $\operatorname{supp} y \subset \complement A$ and $p_A(x-x_0) < \varepsilon$ this means that $f(x+\lambda y) < f(x_0)+1$ for every $\lambda \in \mathbb{C}$, and since $\lambda \to f(x+\lambda y)$ is a subharmonic function it cannot be bounded without reducing to a constant. Thus $f(x+y) = f(x)$ if $p_A(x-x_0) < \varepsilon$ and $\operatorname{supp} y \subset \complement A$. Assume that there exists a point $x_1 \in U$ such that $p(x_1-x_0) > \varepsilon$, and such that $x_0 + \lambda(x_1-x_0) \in U$ for $|\lambda| < \alpha$ for some $\alpha > 1$. Let $K > 1$ be a constant such that $f(x_1) < f(x_0) + K$. The set $V=\{x; f(x) < f(x_0)+K\}$ is a pseudoconvex set, so that $-\log d_V(x, y)$, where $d_V(x,y) = \inf\{|\lambda|;\ x+\lambda y \notin V\}$, is a plurisubharmonic function in $V \times (E \setminus \{0\})$. If we choose y with $\operatorname{supp} y \subset \complement A$, then $\lambda \to -\log d_V(x_0+ \lambda(x_1-x_0), y)$ is a subharmonic function which is equal to $-\infty$ in a neighbourhood of zero. It must be equal to $-\infty$ everywhere so that $-\log d_V(x_1, y) = -\infty$, which means that $\lambda \to f(x_1 + \lambda y)$ is bounded by $f(x_0) + K$ and therefore constant. Since U is connected, every point in U can be reached in this manner in a finite number of steps, starting from x_0. It is thus proved that $f(x+y) = f(x)$ if $x \in U$ and $\operatorname{supp} y \subset \complement A$.

This lemma can also be proved in a way somewhat similar to the way in which Lemma 2 is proved, but the proof given above, which was suggested by C.O. Kiselman, is simpler.

COROLLARY E *is* σ*-convex with respect to* PCVX(E).

PROOF If $U \in$ PCVX(E), then $-\log d_U(x,y)$ is a plurisubharmonic function in $U \times (E\setminus\{0\})$ where $d_U(x,y) = \inf\{|\lambda|;\ x+\lambda y \notin U\}$. For the space E defined above, it holds that $E \times E$ is topologically homomorphic to E. Thus we can apply Lemma 1 also to the space $E \times E$ so that $-\log d_U(x,y)$ depends only on countably many coordinates. But then there exists a continuous seminorm p_A on $E \times E$, so that $-\log d_U(x,y)$ is p_A-continuous. Hence U must be p_B-open, if $B\times B \supset A$, since $U = \bigcup_{k=1}^{\infty} \bigcup_{y\in E\setminus\{0\}} \{x \in U; -\log d_U(x,y)<k\}$.

It is interesting to note that the lemma remains true if "$f \in PSH(U)$" is replaced by "f is continuous in U". We state this as a separate lemma, because the proof is different.

LEMMA 2 *If f is a continuous, complex valued function in U, and U is an open subset of E, then f depends only on countably many variables in U.*

PROOF Let $x_0 \in U$. Let $A \subset \mathbb{R}$ be a countable set and $\varepsilon > 0$ be a constant such that $V = \{x;\ p_A(x-x_0) < \varepsilon\} \subset U$. For every $\alpha > 0$ there is a countable set $A_\alpha \supset A$ and a constant δ, $\varepsilon > \delta > 0$, such that $p_{A_\alpha}(x'-x_0) < \delta \Longrightarrow |f(x')-f(x_0)| < \delta$. Let $A_{x_0} = \bigcup_{k=1}^{\infty} A_{1/k}$. Then $\operatorname{supp} y \subset \complement A_{x_0} \Longrightarrow f(x_0+y) = f(x_0)$. Let B be a countable set, $B \subset \mathbb{R}$. Then there exists a countable set $B_1 \supset B$ such that $x \in U$, $\operatorname{supp} x \subset B$ and $\operatorname{supp} y \subset \complement B_1 \Longrightarrow f(x+y) = f(x)$. To construct B_1, let $E_B = \{x \in E;\ \operatorname{supp} x \subset B\}$. Then E_B is topologically homomorphic to ℓ^1 and thus is separable. Let $\{x_j\}_1^\infty \subset E_B \cap U$ be dense in the closure of $E_B \cap U$, and for every x_j let $C_j \subset \mathbb{R}$ be a countable set such that $\operatorname{supp} y \subset \complement C_j \Longrightarrow f(x_j + y) = f(x_j)$. Then $B_1 = \bigcup_{j=1}^{\infty} C_j$ has the desired properties, because if $\operatorname{supp} y \subset \complement B_1$ then

$$f(x) = \lim_{j\to\infty} f(x_{kj}) = \lim_{j\to\infty} f(x_{kj} + y) = f(x + y)$$

for some subsequence $\{x_{kj}\}$ of the sequence $\{x_j\}_1^\infty$.

Let B_1 be the empty set and define B_{j+1} so that

$$x \in U,\ \text{supp } x \subset B_j \text{ and supp } y \subset \complement B_{j+1} \implies f(x+Y) = f(x).$$

Let $B = \bigcup_{j=1}^{\infty} B_j$. Assume $x \in U$, supp $x \subset B$ and supp $y \subset \complement B$. Then for every $\alpha > 0$ there is an element $x_\alpha \in U$ with supp $x_\alpha \subset B_j$ for some j and such that $|f(x) - f(x_\alpha)| < \alpha$ and $|f(x+y) - f(x_\alpha + y)| < \alpha$. supp $y \subset \complement B$ implies supp $y \subset \complement B_j$ so that $f(x_\alpha + y) = f(x_\alpha)$. Thus

$$|f(x + y) - f(x)| \leq |f(x + y) - f(x_\alpha + y)| + |f(x_\alpha) - f(x)| < 2\alpha .$$

Since α was arbitrary this shows $f(x + y) = f(x)$.

Since for every $x \in U$, $x = x' + x''$ where supp $x' \subset B$, supp $x'' \subset \complement B$, $f(x+y) = f(x)$ whenever $x \in U$ and supp $y \subset \complement B$.

REFERENCES

[1] Dineen, S., Holomorphically complete locally convex vector spaces, Sém. P. Lelong, 1971/72, *Lecture notes in mathematics* 332 (1973), 77-111.

[2] Dineen, S., Surjective limits of locally convex spaces and their application to infinite dimensional holomorphy. To appear in *Jour. Math. Pures et Appl.*

[3] Dineen, S., Noverraz, Ph., and Schottenloher, M., Le problème de Levi dans certains espaces vectoriels topologiques localement convexes. Bull. Soc. Math.France 104 (1976).

ERIC GRUSELL
Dept. of MATHEMATICS
SYSSLOMANSGATAN 8
S - 752 23 UPPSALA
SWEDEN

Infinite Dimensional Holomorphy and Applications, Matos (ed.)

SOME PROPERTIES OF THE IMAGES OF ANALYTIC MAPS

By *MICHEL HERVÉ*

Throughout the paper: X denotes a complex Banach space; letters such as U or V, open sets in X; $\mathcal{A}(U,V)$ the class of (Fréchet-) analytic maps of U into V; $\mathcal{O}(U,V)$ the class of one-one maps f of U onto V such that $f \in \mathcal{A}(U,V)$ and $f^{-1} \in \mathcal{A}(V,U)$; finally d denotes the distance in X associated with the norm.

1. A NEW PROOF OF EARLE-HAMILTON'S [1] THEOREM, which reads as follows.

THEOREM 1 *Let* U *be open bounded in* X *and* $f \in \mathcal{A}(U,U)$ *be such that the image* f(U) *lies at a positive distance from* $\complement U$*: then* f *has a unique fixed point in* U.

PROOF Let $\Phi = \{\phi \in \mathcal{A}(U,\mathbb{C}) : |\phi| < 1\}$; for $a \in U$, $b \in X$, set

(1.1) $$\alpha(a;b) = \sup_{\phi \in \Phi} |\phi'(a).b| .$$

This supremum is finite: in fact, for any $\phi \in \Phi$, $\phi(a+tb)$ is an analytic function of $t \in \mathbb{C}$, $|t| < d(a,\complement U)/||b||$, whose derivative for $t = 0$ is $\phi'(a).b$, hence

(1.2) $$\alpha(a;b) \leq ||b||/d(a,\complement U).$$

On the other hand, if $b \neq 0$ the same supremum is positive: in fact, if $u \in X'$ (the adjoint space to X) with $|u(b)|/||b||_X = ||u||_{X'} = 1/\text{Diam } U$, then $\phi = u - u(a) \in \Phi$, hence $\alpha(a;b) \geq |u(b)|$ and

$$(1.3) \qquad \alpha(a;b) \geq ||b||/\text{Diam } U \ .$$

If $f \in \mathcal{A}(U,U)$, then $\{\phi \circ f: \phi \in \Phi\} \subset \Phi$; Since $(\phi \circ f)'(a).b = \phi'[f(a)].[f'(a).b]$, we get

$$(1.4) \qquad \alpha[f(a);\ f'(a).b] \leq \alpha(a;b) \ .$$

Now we use the assumption made on $f(U)$ in the same way as EARLE HAMILTON did, namely: since U is bounded, there exists $\mu > 0$ such that

$$(1.5) \qquad g(x) = f(x) + \mu[f(x) - f(a)] \in U \quad \forall\, x \in U.$$

Then $g \in \mathcal{A}(U,U)$ with $g(a) = f(a)$, $g'(a) = (1+\mu)\, f'(a)$; therefore

$$(1+\mu)\ \alpha[f(a);\ f'(a).b] = \alpha[g(a):\ g'(a).b] \leq \alpha(a;b)$$

sharpens (1.4) into

$$(1.6) \qquad \alpha[f(a);\ f'(a).b] \leq \frac{\alpha(a;b)}{1+\mu} \quad \forall\, a \in U,\ b \in X.$$

Let $f_0 = \text{id}$, $f_n = f \circ f_{n-1}$: since

$f'_n(a).b = f'[f_{n-1}(a)].[f'_{n-1}(a).b]$, the iteration of (1.6) yields

$$(1.7) \qquad \alpha[f_n(a);\ f'_n(a).b] \leq \frac{\alpha(a;b)}{(1+\mu)^n} \quad \forall\, a \in U,\ b \in X,\ n \in N.$$

Putting (1.2),(1.3) and (1.7) together, we get

$$(1.8) \qquad ||f'_n(a).b|| \leq \frac{\text{Diam } U}{d(a,\complement U)}\ \frac{||b||}{(1+\mu)^n} \quad \forall\, a \in U,\ b \in X,\ n \in N.$$

If $U \supset S = \{a + t\,(b-a): 0 \leq t \leq 1\}$, from (1.8) follows

$$(1.9) \qquad ||f_n(b) - f_n(a)|| \leq \frac{\text{Diam } U}{d(S,\complement U)}\ \frac{||b-a||}{(1+\mu)^n} \ ,$$

and for arbitrary $a, b \in U$ we have

(1.10) $$\sup_{n\in N} (1+\mu)^n \|f_n(b) - f_n(a)\| < +\infty.$$

With $b = f(a)$, from (1.10) follows that $(f_n(a))_{n\in N}$ is a Cauchy sequence; with arbitrary a and b, that $l = \lim f_n(a)$ does not depend on a. By the assumption made on $f(U)$, $l \in U$ and l is a unique fixed point of f.

2. THE DEPENDENCE OF THE FIXED POINT ON A PARAMETER Let U be open bounded in X and, for each $t \in T$, let $f_t \in \mathcal{A}(U,U)$ be such that $f_t(U)$ lies at a positive distance (depending on t) from $\complement U$; then f_t has a unique fixed point $l(t) \in U$.

PROPOSITION 2.1 *Let T be a topological space,* $t_0 \in T$*: if each point* $a \in U$ *has a neighborhood* A *such that* $f_t(x) \to f_{t_0}(x)$ *as* $t \to t_0$ *uniformly for* $x \in A$*, then* $l(t) \to l(t_0)$ *as* $t \to t_0$.

PROOF Take $a = l(t_0)$ in the proof of Theorem 1: since each map $X \ni b \mapsto \phi'(a).b$, $\phi \in \Phi$, is linear, the map $b \to \alpha(a;b)$ defined by (1.1) is a seminorm; by (1.2) and (1.3) it is a norm equivalent to the given one, and we may assume it is the given one.

Then, if $\mu_0 > 0$ satisfies (1.5) for f_{t_0}, (1.6) means that

$$\|f'_{t_0}(a)\| \leq \frac{1}{1+\mu_0} \quad ;$$

since f_{t_0} is continuously differentiable, A in the assumption of 2.1 may be chosen as an open ball with center a, such that

$$\|f'_{t_0}(x)\| \leq \frac{1}{1+\mu_0/2} = \lambda \ \forall\, x \in A;$$

then $f_{t_0}(A) \subset a + \lambda A$ and, for t sufficiently near t_0, $f_t(A) \subset a + \frac{1+\lambda}{2} A$, with the consequences:

a) $f_t|_A$ has a fixed point, namely $l(t)$;

b) (1.9) can be written for $f_t|_A$ with $b = f_t(a)$ and a number μ depending only on λ, therefore $l(t)$ is the uniform limit of the sequence $b_n(t)$ defined by

(2.2) $$b_1(t) = f_t(a),\ b_n(t) = f_t[b_{n-1}(t)].$$

PROPOSITION 2.3 *Let* T *be an open set in a normed linear space: if* $t \mapsto f_t(x)$ *is analytic for each* $x \in U$, *then* $t \mapsto l(t)$ *is analytic.*

PROOF Since the analytic maps $t \to f_t(x)$ are bounded together, they are equicontinuous ([4], III.2.2), which entails that the assumption of 2.1 holds, and that $(t,x) \mapsto f_t(x)$ is a continuous map of $T \times U$ into U. Since this continuous map is separately analytic, it is analytic, the composed maps b_n defined by (2.2) are analytic, and so is the map l by the last statement in the proof of 2.1.

3. SOME EXTENSIONS OF THE THEOREM Given a real Banach space X_0, let $X = X_0 + i\,X_0$ be the complexified space, a complex Banach space with the norm

$$\|x_0+ix_1\|_X = \sup_{\theta \in \mathbb{R}} (\|x_0\cos\theta - x_1\sin\theta\|_{X_0} + \|x_0\sin\theta + x_1\cos\theta\|_{X_0}),$$

which satisfies

(3.1) $$\|x_0\|_{X_0} + \|x_1\|_{X_0} \le \|x_0+ix_1\|_X \le (\|x_0\|_{X_0} + \|x_1\|_{X_0})\sqrt{2}.$$

THEOREM 3.2 *Let* V *be open bounded in* X_0, $U = V + iX_0$, *and* $f \in \mathcal{A}(U,U)$ *be such that* $f(U)$ *lies at a positive distance from* $\complement U$: *then* f *has a unique fixed point in* U.

PROOF In the proof of Th. 1, (1.2) and (1.4) remain unaltered; so does (1.5) since, by (3.1), the projection of $f(U)$ into V lies at a positive distance from $\complement V$; thus we only have to find a substitute for (1.3).

Given $b = b_0 + ib_1 \neq 0$, choose $k = 0$ or 1 such that $||b_k||_{X_0} \geq ||b||_X/2\sqrt{2}$, then $v \in X_0'$ such that

$$|v(b_k)|/||b_k||_{X_0} = ||v||_{X_0'} = 1/\text{Diam } V ;$$

setting $u(x_0+ix_1) = v(x_0) + iv(x_1)$, we have $u \in X'$, $||u||_{X'} \leq ||v||_{X_0'}$, $|u(b)| \geq |v(b_k)| \geq ||v||_{X_0'} ||b||_X/2\sqrt{2}$, $\psi = u - u(a) \in \mathcal{A}(U,\mathbb{C})$ with $|\text{Re } \psi(x)| = |v(x_0) - v(a_0)| < 1$, therefore $\phi = (\psi/2-\psi) \in \Phi, \phi'(a) = \frac{u}{4}$: in (1.3),(1.8),(1.9), Diam U is replaced by $8\sqrt{2}$.Diam V.

REMARK 3.3 In the statement of Th. 1, the boundedness of $f(U)$ may be assumed instead of the boundedness of U: in fact, if $0 < \rho \leq d[f(U), \complement U]$, $U' = \{x \in X: d[x,f(U)] < \rho\}$ is an open bounded subset of U, and $f(U')$ lies at a positive distance from $\complement U'$. Similarly, in the statement of Th. 3.2, the boundedness of the projection of $f(U)$ into V may be assumed instead of the boundedness of V.

REMARK 3.4 In the classical case when X has a finite dimension, a proof of Th. 1, under the corresponding assumption that $f(U)$ is relatively compact in U, can be found in [3], chap. IV, 1; Th. 3.2 does not seem to have been stated before.

PROPOSITION 3.5 *Let* U *and* U' *be open sets in* X *such that:* $U \supset U'$; $U \setminus U'$ *has an empty interior; for any* $a \in U$, $b \in X$, *if* ω *is the connected component containing* 0 *of* $\{\zeta \in \mathbb{C}: a + \zeta b \in U\}$, *the set* $\{\zeta \in \omega: a + \zeta b \in U \setminus U'\}$ *is either polar or* ω *itself. Let* $f \in \mathcal{A}(U',U)$ *be such that* $f(U')$ *is bounded and lies at a positive distance from* $\complement U$: *then* f *can be continued into a map* $\in \mathcal{A}(U,U)$, *which has a unique fixed point in* U.

PROOF By the argument used in Remark 3.3, U may be assumed bounded; then Th. 6.9 in [5] proves the existence of a continuation $\in \mathcal{A}(U,U)$, for which Th. 1 above holds.

4. ONE-ONE RESTRICTIONS TO OPEN BALLS WITH A GIVEN CENTER Let $B = \{x \in X: ||x|| < 1\}$ and $\mathcal{F} = \{f \in \mathcal{A}(B,X): f(0)=0, f'(0) = id\}$; given $\mu > 0$, let $\mathcal{F}(\mu) = \{f \in \mathcal{F} : ||f(x) - x|| \leq \mu \ \forall\, x \in B\}$.

The classical SCHWARZ lemma for the scalar function $\zeta \to u[f(\zeta a) - \zeta a]$ $(u \in X', a \in X, ||a||=1)$ shows that $||f(x)-x|| \leq \mu||x||^2 \ \forall\, x \in B,\ f \in \mathcal{F}(\mu)$.

PROBLEM 4.1 *Find a radius ρ depending only on μ such that each $f \in \mathcal{F}(\mu)$ can be restricted to some open subset $U_f \ni 0$ of B into a map $\in \mathcal{O}(U_f, \rho B)$. Note that SCHWARZ's lemma for the inverse map yields $\rho||x|| \leq ||f(x)|| \ \forall\, x \in U_f$ and $\rho \leq 1$.*

PROBLEM 4.2 *Find a radius r depending only on μ such that $f|_{rB} \in \mathcal{O}(rB, V_f)$ for each $f \in \mathcal{F}(\mu)$ and some open set V_f in X.*

THEOREM 4.3 (partly in [2]). *The best value of ρ answering Problem 4.1 is $\rho = 1 - \mu$ if $\mu \leq \frac{1}{2}$, $\rho = 1/4\mu$ if $\mu \geq \frac{1}{2}$.*

PROOF Let $g_y(x) = x - f(x)+y$: if $||y|| < 1 - \mu$, $g_y(B)$ lies at a positive distance from $\complement B$, and by Prop. 2.3 the unique fixed point of g_y in B depends analytically on y, therefore we answer Problem 4.1 with $\rho = 1 - \mu$, $U_f = B \cap f^{-1}(\rho B)$; if $r \leq 1$ and $||y|| < r - \mu r^2$, $g_y(rB)$ lies at a positive distance from $\complement(rB)$, therefore we answer Problem 4.1 with $\rho = r - \mu r^2$, $U_f = (rB) \cap f^{-1}(\rho B)$, and may take $r = 1/2\mu$ if $\mu \geq \frac{1}{2}$.

In order to show that the result is sharp, choose $a \in X$ with $||a||_X = 1$, $u \in X'$ with $u(a) = ||u||_{X'} = 1$, and consider the example

$$f(x) = x - \mu a\, u^2(x) \quad : \tag{4.4}$$

if $\mu \leq \frac{1}{2}$, $f(x) = (1-\mu)a$ or $x - a = \mu a\, u(x-a)\, u(x+a)$ implies $|u(x+a)| \geq \frac{1}{\mu} \geq 2$ and $||x|| \geq 1$; if $\mu \geq \frac{1}{2}$, $f(x) = a/4\mu$ implies $u(x) = 1/2\mu$, $f'(x) = id - au$, $f'(x).a = 0$, $f'(x)$ non invertible.

REMARK 4.5 If $\mu \geq \frac{1}{2}$, a radius r answering Problem 4.2 cannot exceed $1/2\mu$. In fact, with (4.4) we have $f'(a/2\mu) = id - au$ non invertible.

THEOREM 4.6 *Let* $\mu \geq \frac{1}{2}$ *: then the best value of* r *answering Problem 4.2 in all Banach spaces* X *is* $r = 1/4\mu$, *and* $f(B/4\mu) \supset \frac{3}{16\mu} B$ $\forall f \in \mathcal{F}(\mu)$.

PROOF Set $g(x) = f(x) - x$, take x and $y \in \frac{1}{2}B$: $g'(x).y$ is ([4], III.1.1 and III.1.3) the mean value for $0 \leq \theta \leq 2\pi$ [denoted by MV in (4.7)] of $e^{-i\theta} g(x + e^{i\theta} y)$, hence

(4.7) $||g'(x).y|| \leq MV\, ||g(x+e^{i\theta}y)|| \leq \mu.MV\, ||x+e^{i\theta}y||^2 \leq \mu\,(||x||+||y||)^2.$

If moreover $||y|| = ||x||$, we obtain $||g'(x).y|| \leq 4\mu\, ||x||\, ||y||$ and

$$(4.8) \qquad ||g'(x)|| \leq 4\mu\, ||x|| \quad \forall\, x \in \tfrac{1}{2}B.$$

Now $x \in B/4\mu$ implies $||g'(x)|| < 1$, $f'(x) = id + g'(x)$ invertible; x and $y \in B/4\mu$, $x \neq y$, imply $||g(x)-g(y)|| < ||x-y||$, $f(x) \neq f(y)$: by the classical inversion theorem for continuously differentiable maps, $f|_{B/4\mu}$ has an open image V_f and an analytic inverse map.

If $y \in \frac{3}{16\mu}B$, the recursion formula

$$(4.9) \qquad x_0 = y,\ x_n = y - g(x_{n-1})$$

defines a sequence of points $x_n \in B/4\mu$, for it is majorized by the recursion formula

$$t_0 = 3/16\mu,\ t_n = (3/16\mu) + \mu t_{n-1}^2 .$$

The more precise inequalities $||x_n|| \leq ||y|| + \mu t_{n-1}^2 \leq ||y|| + 1/16\mu$ and

$$\frac{||x_{n+1} - x_n||}{||x_n - x_{n-1}||} \leq \sup \{||g'(x)|| : ||x|| \leq ||y|| + 1/16\mu\} < 1$$

prove the existence of $x = \lim x_n \in B/4\mu$, and $f(x) = y$.

As an example of a Banach space where the results are sharp, we choose $X = \mathbb{C}^2$ with the following notation. The norm of $x = (x_1, x_2) \in X$ is $||x|| = \frac{1}{2}(|x_1 - ix_2| + |x_1 + ix_2|)$, f maps $x = (x_1, x_2)$ onto $y = (y_1, y_2)$ such that

$$y_1 = x_1 - e^{-i\pi/4}\mu(x_1^2 + x_2^2), \quad y_2 = x_2 - e^{i\pi/4}\mu(x_1^2 + x_2^2)$$

or $y_1 + iy_2 = x_1 + ix_2$, $y_1 - iy_2 = \left[1 - \mu\sqrt{2}(1-i)(x_1 + ix_2)\right](x_1 - ix_2)$;

setting $\quad U = \{(x_1, x_2) \in \mathbb{C}^2 : x_1 + ix_2 \neq 1/\mu\sqrt{2}(1-i)\}$,

we have $\quad f|_U \in \mathcal{O}(U, U)$.

Since 1 is the upper bound of $|x_1^2 + x_2^2| = |x_1 - ix_2|\,|x_1 + ix_2|$ for $||x|| < 1$, μ is the upper bound of $||f(x) - x||$;

$$J_f(x) = 1 - \mu\sqrt{2}\left[(1-i)x_1 + (1+i)x_2\right]$$

vanishes at some point x such that $||x|| = 1/4\mu$, namely $(1-i)x_1 = (1+i)x_2 = 1/(2\mu\sqrt{2})$; finally $||f^{-1}(y)|| = 1/4\mu$ for some point $y \in U$ such that $||y|| = 3/16\mu$, namely $y_1 - iy_2 = 1/8\mu$, $y_1 + iy_2 = e^{i\pi/4}/4\mu$.

5. THE CASE OF HILBERT SPACES

THEOREM 5.1 *Let* X *be a Hilbert space and* $\mu \geq 1$: *then the best value of* r *answering Problem 4.2 is* $r = 1/2\mu$, *and* $f(B/2\mu) \supset B/4\mu$ $\forall f \in \mathcal{F}(\mu)$.

PROOF Since X is a Hilbert space, we have the special formula

$$||x+e^{i\theta}y||^2 + ||x-e^{i\theta}y||^2 = 2(||x||^2+||y||^2),$$

which turns the end of (4.7) into MV $||x+e^{i\theta}y||^2 = ||x||^2+||y||^2$ and (4.8) into $||g'(x)|| \leq 2\mu||x|| \forall x \in \frac{1}{2} B$. So the inversion theorem can be used for $f|_{B/2\mu}$ instead of $f|_{B/4\mu}$.

If $y \in B/4\mu$, (4.9) is majorized by the recursion formula

$$t_0 = 1/4\mu,\ t_n = (1/4\mu) + \mu t_{n-1}^2,$$

with the same consequences as in the proof of Th. 4.6. Finally the results are sharp by Example (4.4), as it was pointed out in Remark 4.5.

REMARK 5.2 *The simple example* $X = \mathbb{C}$, $f(x) = x - \mu x^3$, *where* $f'(1/\sqrt{3\mu}) = 0$, *shows that a radius* r *answering Problem* 4.2 *in Hilbert spaces cannot exceed* $1/\sqrt{3\mu}$, *and this is smaller than* $1/2\mu$ *if* $\mu < 3/4$. *Then a substitute for* $1/2\mu$ *has to be found if* $\mu < 1$.

If $\frac{1}{2} \leq ||x|| < 1$, the inequality $||g'(x).y|| \leq \mu(||x||^2+||y||^2)$ proceeding from (4.7) may be written for $||y|| < 1 - ||x||$ only, and therefore only yields

$$||g'(x)|| \leq \mu \frac{||x||^2 + (1-||x||)^2}{1-||x||} =$$

$$= \mu\left(\frac{1}{1-||x||} - 2||x||\right) \ \forall x \in B ;$$

so the radius $r = \frac{1}{4\mu}(2\mu - 1 + \sqrt{1+4\mu-4\mu^2})$, which is smaller than $1/2\mu$ and decreases from 1 to $\frac{1}{2}$ as μ increases from 0 to 1, can be substituted for $1/2\mu$ in Th. 5.1, but the author does not know if this value of r is sharp.

6. THE CASE OF HAUSFORFF, LOCALLY CONVEX, SEQUENTIALLY COMPLETE SPACES

As an analogue of closed balls in such a space E, let $\bar{B}$ be bounded, closed, balanced and convex, let X be the linear subspace of E spanned by $\bar{B}$, q the gauge of $\bar{B}$ in X: $q(x) = \inf\{\lambda > 0 : x \in \lambda\bar{B}\}$, $x \in X$.

Since $\bar{B}$ is closed, $\bar{B} = \{x \in X : q(x) \leq 1\}$; since $\bar{B}$ is bounded, any continuous seminorm p on E satisfies $p \leq cq$ on X for some constant c; since E is Hausdorff, q is a norm on X, which from now on defines the topology of X. This is finer than the topology induced on X by the given one, and the following example shows that it may be actually finer.

EXAMPLE 6.1 Let ω be the open unit disc in $\mathbb{C}$, $E = \mathcal{A}(\omega,\mathbb{C})$ with the usual topology of compact convergence, $\bar{B} = \{x \in E : |x(\zeta)| \leq 1 \ \forall\ \zeta \in \omega\}$: then $X = \{x \in E : x \text{ bounded}\}$, with $q(x) = \sup_{\zeta\in\omega} |x(\zeta)|$ if $x \in X$.

PROPOSITION 6.2 *X is a Banach space.*

PROOF Let (x_n) be a Cauchy sequence in X for the norm q, $x = \lim x_n$ in E, $\beta = \sup q(x_n)$: since $\beta\bar{B}$ is closed in E, $x \in \beta\bar{B}$; now let n and $n' \geq n_0$ imply $q(x_n - x_{n'}) \leq \varepsilon$: since $x_n - \varepsilon\bar{B}$ is closed in E, $n \geq n_0$ implies $x \in x_n - \varepsilon\bar{B}$ or $q(x_n - x) \leq \varepsilon$.

THEOREM 6.3 *Let* $B = \{x \in X : q(x) < 1\}$ *and g be a Gâteaux-analytic map of* B *into* E, *which only means the existence in* E *of* $\lim_{t\to 0} \frac{1}{t}[g(a+tb)-g(a)]$ $\forall\ a \in B,\ b \in X$: *if* $g(B) \subset \mu B$ *for some number* μ, *then* $g \in \mathcal{A}(B;X)$ *with the topology of* X.

PROOF Let $q(a) + q(b) < 1$; since the closed convex set $\mu\overline{B}$ contains $e^{-in\theta}g(a+e^{i\theta}b)$ $\forall\ \theta \in \mathbb{R}$, it also contains, with the notation used in [4], III.1.3, and in the proof of Th. 4.6 above:

$$g_n(a;b) = MV\left[e^{-in\theta}g(a+e^{i\theta}b)\right].$$

Now the expansion $g(a) + \sum_{n\geq 1} t^n g_n(a;b)$ converges in E to $g(a+tb)$ for $|t| < 1$; since the closed set $\frac{\mu|t|^2}{1-|t|}$ contains $t^2 g_2(a;b)+\ldots+t^n g_n(a;b)$ for each $n \geq 2$, it also contains $g(a+tb) - g(a) - t\,g_1(a;b)$, which yields the inequalities

$$q\left[\frac{g(a+tb) - g(a)}{t} - g_1(a;b)\right] \leq \frac{\mu|t|}{1-|t|},$$

showing that $g_1(a;b) = \lim_{t\to 0} \frac{1}{t}\left[g(a+tb)-g(a)\right]$ in X, and

$$q\left[g(a+b) - g(a)\right] \leq \frac{\mu q(b)}{1-q(a)-q(b)},$$

proving that g is continuous, for the topology of X, at the point a.

COROLLARY 6.4 *Let* f *be a Gâteaux-analytic map of* B *into* E, *with the properties:* $f(0) = 0$; $\lim_{t\to 0} f(tx)/t = x$ in E $\forall\ x \in X$; $f(x) - x \in \mu B$ $\forall\ x \in B$; $\mu \geq \frac{1}{2}$. *Then, with the topology of* X, $f|_{B/4\mu} \in \mathcal{O}(B/4\mu, V_f)$ *for some open subset* V_f *of* X, *containing* $\frac{3}{16\mu}B$.

EXAMPLE 6.5 Let ω be a bounded simply connected open set in $\mathbb{C}$; assume that, for some real number δ, any two points in ω can be linked by a path in ω whose length does not exceed δ. Then, for any $\phi \in \mathcal{A}(\omega,\mathbb{C})$ with $|\phi| < 3/(16\,\delta^2)$, the unique solution vanishing at a given point $\alpha \in \omega$ of the RICCATI differential equation $u' + u^2 = \phi$ is analytic on ω, with $|u| < 1/4\delta$, and the map

$\phi \longmapsto u$ is analytic, with the topology of uniform convergence on $X = \{x \in \mathcal{A}(\omega, \mathbb{C}) : x \text{ bounded}\}$.

PROOF Let $E = \mathcal{A}(\omega, \mathbb{C})$ with the usual topology of compact convergence, $\overline{B} = \{x \in E : |x(\zeta)| \leq 1/2\delta^2 \ \forall\ \zeta \in \omega\}$; for $x \in E$, define $g(x)$ as the function $\omega \ni \zeta \longmapsto \left[\int_\alpha^\zeta x(\tau)\, d\tau\right]^2$ (where the integral is taken along any path in ω with a length $\leq \delta$), which has a modulus $< 1/4\delta^2 \ \forall\ x \in B$. Thus $g(B) \subset \frac{1}{2}B$, and on the other hand $g \in \mathcal{A}(E,E)$.

By Corollary 6.4 with $\mu = \frac{1}{2}$: for any $\phi \in \frac{3}{8}B$, the equation $x + g(x) = \phi$ has a unique solution $x \in \frac{1}{2}B$, depending analytically on ϕ for the topology of X; $u(\zeta) = \int_\alpha^\zeta x(\tau)\, d\tau$ is the required solution of the differential equation, and $|x(\zeta)| < 1/4\delta^2 \ \forall\ \zeta \in \omega$ implies $|u(\zeta)| < 1/4\delta \ \forall\ \zeta \in \omega$.

BIBLIOGRAPHY

[1] C.EARLE and R.HAMILTON-A fixed point theorem for holomorphic mappings (Proc. of Symposia in pure Math., XVI,A.M.S., 1970).

[2] L.HARRIS - On the size of balls covered by analytic transformations (Preprint, University of Kentucky).

[3] M.HERVÉ - Several complex variables, local theory (Oxford University press, 1963).

[4] M.HERVÉ - Analytic and plurisubharmonic functions in finite and infinite dimensional spaces (Lecture notes in Math., 198,1971).

[5] M.HERVÉ - Lindelöf's principle in infinite dimensions (Lecture notes in Math., 364, 1974, p. 41-57).

UNIVERSITÉ PIERRE ET MARIE CURIE
4, PLACE JUSSIEN, PARIS 5
et
ÉCOLE NORMALE SUPÉRIEURE,
45 rue d'ULM, PARIS 5,
FRANCE

Infinite Dimensional Holomorphy and Applications, Matos (ed.)

SOME REMARKS ON BANACH VALUED POLYNOMIALS ON $C_o(A)$

By *BENGT JOSEFSON* *

1. J. Globevnik [5] has proved that for every open connected set U in a separable complex Banach space E there exists a holomorphic function f from the open unit ball B in $\mathbb{C}$, the complex line, into U such that f(B) is dense in U. In [5] even sharper results are obtained and in the case U is a ball this is also proved in [2]. In this note we shall show that this result cannot be generalized to arbitrary, non-separable Banach spaces.More precisely we shall prove, Proposition 2, that every holomorphic function from an open connected set in $C_o(A)$ into $\ell^p(\Gamma)$, $P > 0$, has separable range for all index sets A and Γ. This proposition follows more or less obviously from results due to A.Pelczynski and Z. Semadeni. We also show, Theorem 3, that there exists an open set $U \subset C_o(A)$ and a constant $\varepsilon > 0$ so that, when A is un-

---

(*) Supported by the Swedish Natural Science Research council contract Nº F 3435-004 and Dublin Institute for Advanced Studies.

countable, every holomorphic function f from the open unit ball B_o in $C_o(A)$ into $U + \varepsilon \cdot B_o$ has not a range which is dense in U.

NOTATION: Let $U \subset E$ be an open set of a Banach space E and let F be a Banach space. Then $\mathcal{H}(U,F)$ denotes the set of all holomorphic functions from U into F, $P(^nE,F)$ is the set of all continuous, n-homogeneous polynomials on E into F and P(E) denotes all complex valued, continuous polynomials on E. We note that if U is balanced then $f \in \mathcal{H}(U,F)$ has a Taylor series expansion $f(z) = \sum_0^\infty P_n(z)$ where $P_n \in P(^nE,F)$. See [8] for details. $C_o(A)$ is the Banach space of all complex valued functions on the indexset A which are arbitrarily small outside finite subsets of A. Finally $\ell^P(A)$ denotes the Banach space $\ell^P(A) = \{Z = \{Z_\alpha\}_{\alpha\in A};\ \Sigma|Z_\alpha|^P < \infty\}$. If $V \subset A$ and $Z=\{Z_\alpha\}_{\alpha\in A} \in C_o(A)$ or $\ell^P(A)$ supp $Z = \{\alpha \in A;\ Z_\alpha \neq 0\}$ and $\mathrm{Proj}_{[V]} Z=\{Z'_\alpha\}_{\alpha\in A}$ where $Z'_\alpha = Z_\alpha$ if $\alpha \in V$ and $Z'_\alpha = 0$ elsewhere. We note that supp Z always is countable.

2. We recall the following result which is due to A. Pelczynski and Z. Semadeni. See [7], [8].

THEOREM 1: *If* F *is a Banach space which does not contain* C_o, *then for every* $n \in \mathbb{N}$, $P(^nC_o(A),F) = P_k(^nC_o(A),F)$ *where* P_k *denotes the polynomials which maps the unit ball of* $C_o(A)$ *into a relatively compact set of* F.

We need also the following result of R. Aron, see [1].

THEOREM 2. $P(^nC_o(A))$ = *the closed span of the collection* ϕ^n *where* $\phi \in \ell^1(A)$ *the dual space of* $C_o(A)$.

REMARK: Theorem 2 implies that every function $f \in \mathcal{H}(B_o,\mathbb{C})$, where B_o is the open unit ball in $C_o(A)$, can be approximated on every strictly smaller ball by polynomials on $C_o(A)$ with finite spectrum. The spectrum Spf, of a function f on $C_o(A)$ is the intersection of all subsets S of A such that f factors through S and we will denote with $P_F(C_o(A))$ the continuous polynomials on $C_o(A)$ which have finite spectrum. We note that a function is determined by its spectrum.

If $P \in P({}^nC_o(A),\mathbb{C})$ then P can be written

$$P(z) = a^{r_1\cdots r_n}_{\alpha_1\cdots\alpha_n} z^{r_1}_{\alpha_n} \cdots z^{r_n}_{\alpha_n} \quad \text{where} \quad a^{r_1,\ldots,r_n}_{\alpha_1,\ldots,\alpha_n} \in \mathbb{C}.$$

In [3] and [7] the following is proved

PROPOSITION 1. $(\Sigma \mid a^{r_1,\ldots,r_n}_{\alpha_1,\ldots,\alpha_n} \mid^2)^{1/2} \leq \|P\| = \sup_{\|z\|\leq 1} \mid P(z)\mid.$

PROPOSITION 2. *Every* $f \in \mathcal{H}(B_o,\ell^P(\Gamma))$, *where* $P > 0$ *and* Γ *is arbitrary, factors through a countable subset of* A.

PROOF: It is enough to prove the proposition for every

$$P \in P({}^nC_o(A),\ \ell^P(\Gamma))$$

because of the Taylor series expansion. $P = \{P_\gamma\}_{\gamma\in\Gamma}$ where $P_\gamma \in P({}^nC_o(A),\mathbb{C})$. According to Theorem 1 $P(B_o)$ is a relatively compact set in $\ell^P(\Gamma)$ since it is wellknown that C_o is not contained in $\ell^P(\Gamma)$. But it is also wellknown that every relatively compact set of $\ell^P(\Gamma)$ has its support on a countable subset of Γ. Hence there is a countable set $\{\gamma_j\}_{j\in\mathbb{N}}$ such that $P_\gamma \equiv 0$ if $\gamma \notin \{\gamma_j\}_{j\in\mathbb{N}}$. But then the proposition follows from proposition 1, which gives that each P_γ factors through a countable set, and from the fact that a countable union of countable sets is a countable set. Q.E.D.

REMARK 1. Theorems 1,2 and Proposition 2 are also true if we replace $C_o(A)$ by $C(K)$, the Banach space of continuous functions on a dispersed, compact Hausdorff space.

REMARK 2. For every $P \in P({}^nC_o(A), \ell^1(\Gamma))$ the coefficient estimate corresponding to Proposition 1 holds which can be seen from the fact that $Q(t,z) = \sum_{\gamma\in\Gamma} t_\gamma \cdot P_\gamma(z)$ can be regarded as a n+1-homogeneous polynomial on $C_o(A \cup \Gamma)$ into $\mathbb{C}$ such that $\|Q\| = \|P\|$ where $P = \{P_\gamma\}_{\gamma\in\Gamma}$ and $t = \{t_\gamma\}_{\gamma\in\Gamma} \in C_o(\Gamma)$.

REMARK 3. If $f \in \mathcal{H}(U, \ell^p(\Gamma)$ where U is an open connected set in $C_o(A)$ (or $C(K)$) it follows from the proof of Proposition 2 that f has a separable range because only countably many $f_\gamma \not\equiv 0$ ($f_\gamma = \mathrm{Proj}_{[\gamma]} f$). On the other hand a modification of Hirschowitz example [6] shows that there exist an open, connected and bounded set $U \subset C_o(A)$ and a function $f \in \mathcal{H}(U,\mathbb{C})$ such that f depends on all variables.

Let e_α be the α-th unit vector in $C_o(A)$. Put

$$U_\alpha = \{z \in C_o(A);\ \inf_{x\in B_{\mathbb{C}}} \|z - xe_\alpha\| < 1/10\}$$

where $B_{\mathbb{C}}$ is the unit ball in $\mathbb{C}$. Put $U = \bigcup_{\alpha\in A} U_\alpha$. U is an open bounded connected set in $C_o(A)$. Finally let B_o be the open unit ball in $C_o(A)$ and $\varepsilon \cdot B_o$ be the open ball with radius ε. The set U was suggested to me by R. Aron.

THEOREM 3. *Let* $f \in \mathcal{H}(B_o, C_o(A))$, *where* A *is uncountable, be such that* $f(B_o)$ *is dense in* U. *Then* $f(B_o)$ *is not contained in* $U + 1/10 \cdot B_o$.

PROOF: Let $f \in \mathcal{H}(B_o, C_o(A))$ have a range which is dense in U. There exist a constant $k > 0$ and an uncountable set $H \subset A$ such that

$$\sup_{z \in (1-k)B_o} |f_\alpha(z)| > 1/2 + 1/10 \text{ if } \alpha \in H \text{ where } f_\alpha = \mathrm{Proj}_{[\alpha]} f.$$

This follows because if we assume it is false there exists, for every sequence $(k_j)_j$ of positive numbers tending to zero, a sequence $(H_j)_j$ of countable subsets of A such that

$$\sup_{z \in (1-k_j)B_o} |f_\alpha(a)| \leq 1/2 + 1/10 \text{ if } \alpha \notin A \setminus H_j .$$

Hence

$$\sup_{z \in B_o} |f_\alpha(z)| \leq 1/2 + 1/10 \quad \text{if} \quad \alpha \notin A \setminus \bigcup_1^\infty H_j$$

which contradicts the assumption that $f(B_o)$ is dense in U since $x \cdot e_\alpha \in U$ if $x \in B_{\mathbb{C}}$ for all $\alpha \in A$.

From Theorem 2 it follows that we can take, for every $\alpha \in A$, a $g_\alpha \in P_F(C_o(A))$ such that

$$\sup_{z \in (1-k)B_o} |f_\alpha(z) - g_\alpha(z)| = \|f_\alpha - g_\alpha\|_{(1-k)B_o} < 1/10.$$

hence

$$\|g_\alpha\|_{(1-k)B_o} > 1/2.$$

We need now the following lemma.

LEMMA: *Let* $\{g_\alpha\}_{\alpha \in H}$ *be an uncountable collection of complex-valued, continuous polynomials on* $C_o(A)$ *with finite spectrum and let* $\varepsilon > 0$. *Then there exist an uncountable set* $H_1 \subset H$, *an integer* s *and* $g'_\alpha \in P_F(C_o(A))$ *such that* $\|g_\alpha - g'_\alpha\|_{B_o} < \varepsilon$ *and* $g'_\alpha(z) = \sum_{i=1}^{s} g_i(z) \cdot g_i^\alpha(z)$ *if* $\alpha \in H_1$, *where* $g_i, g_i^\alpha \in P_F(C_o(A))$ *and* $\mathrm{Sp}\, g_i^\alpha \cap \mathrm{Sp}(g_j^\gamma + g_\ell) = \emptyset$ *if* $H_1 \ni \alpha \neq \gamma \in H_1$ *for all* i,j *and* $\ell \leq s$. *Furthermore, there exist, for every* i, *an integer* q_i, *an isomorphism* $h_i^\alpha : C_o(\mathrm{Sp}\, g_i^\alpha) \to \mathbb{C}^{q_i}$ *and a polynomial* $g^i \in P(\mathbb{C}^{q_i})$ *such that* $g^i(h_i^\alpha(z)) = g_i^\alpha(z)$ *for all* $\alpha \in H_1$.

PROOF OF THE THEOREM CONTINUED. Take an uncountable set $H_1 \subset H$

and g'_α as in the Lemma and take α and $\gamma \in H_1$. Take also $z \in (1-K)B_o$ such that $|g'_\alpha(z)| > 1/2 - \varepsilon$ which is possible if $|g_\alpha(z)| > 1/2$. Since g_j and g^α_i have disjoint spectrum we may assume that z is chosen so that $z = z^0 + z^1$ where $\mathrm{supp}\, z^0 \subset \bigcup_1^s \mathrm{Sp}\, g_1$ and $\mathrm{supp}\, z^1 \subset \bigcup_1^s \mathrm{Sp}\, g^\alpha_i$. Put

$$z^{2,i} = (h^\gamma_i)^{-1} \circ h^\alpha_i(\mathrm{Proj}_{[\mathrm{Sp}\, g^\alpha_i]} z^1) \text{ and } z^2 = \sum_{i=1}^{s} z^{2,i},$$

where $(h^\gamma_i)^{-1}$ is the natural inverse of h^γ_i. $\mathrm{Supp}\, z^2 \subset \bigcup_1^s \mathrm{Sp}\, g^\gamma_i$ and $g'_\gamma(z^2+z^0) = g'_\alpha(z^1+z^0)$ since $g^\gamma_i(z^{2,i}) = g^i(h^\alpha_i(z^{1,i})) = g^\alpha_i(z^{1,i})$, where

$$z^{1,i} = \mathrm{Proj}_{[\mathrm{Sp}\, g^\alpha_i]} z^1 .$$

Further $g'_\alpha(z^0 + z^1 + z^2) = g'_\gamma(z^0 + z^1 + z^2) = g'_\alpha(z^0 + z^1)$ since $\mathrm{supp}\, z^2 \cap \mathrm{Sp}\,(g^\alpha_i + g_i) = \emptyset$ and $\mathrm{supp}\, z^1 \cap \mathrm{Sp}(g^\gamma_i + g_i) = \emptyset$ according to the assumptions. But $z' = z^0 + z^1 + z^2 \in (1 - K)B_o$ because $\| h^\alpha_i \| = 1$, $\| z^0+z^1 \| \leq 1+k$ and $\mathrm{supp}\, z^2 \cap \mathrm{supp}(z^0+z^1) = \emptyset$. Now $|g'_\alpha(z')| = |g'_\gamma(z')| > 1/2 - \varepsilon$ hence $|f_\alpha(z')| > 1/2 - 1/10 - 2\varepsilon$ and $|f_\gamma(z')| > 1/2 - 1/10 - 2\varepsilon$. Thus $f(z') \notin U + 1/10\, B_o$ if ε is small enough because $x e_\alpha + y e_\gamma \in U$ if $|x|$ and $|y| \geq 1/10$. Q.E.D.

PROOF OF THE LEMMA. There exist an uncountable set $H^0 \subset H$ and an integer ℓ such that g_α has its spectrum on at most ℓ variables and such that the degree of g_α is less than ℓ for all $\alpha \in H_o$. Let $V \subset A$ be a set such that there exists an uncountable set $H^V \subset H^0$ so that every g_α depends on the variable z_γ if $\gamma \in V$ when $\alpha \in H^V$. V contains at most ℓ elements since each g_α depends on at most ℓ variables. Let V^0 be a maximal set with the properties above. V^0 exists but can be the empty set. Also there exists an integer s such that $g_\alpha = \sum_1^s g^{\alpha,1}_i \cdot g^\alpha_i$ where $g^{\alpha,1}_s = 1$, $g^{\alpha,1}_i$ are polynomials on $C_o(A)$ with spectrum in V^o

if $i < s$ and g_i^α are polynomials with spectrum outside V^0.

Now there are, for every fixed $\alpha \in H$, at most countably many $\gamma \in H^{V_0}$ such that $V_\alpha \cap V_\gamma \neq \emptyset$, where $V_\alpha = \bigcup_{i=1}^{s} \operatorname{supp} g_i^\alpha$, because V^0 is maximal and V_α is finite. Hence there is an uncountable set $H^1 \subset H^{V^0}$ such that $V_\alpha \cap V_\gamma = \emptyset$ if $H^1 \ni \alpha \neq \gamma \in H^1$.

$\{g_1^{\alpha,1}\}_{\alpha \in H^1}$ is an uncountable set of polynomials in $C_o(V^0)$, where V^0 is finite, with degree less than ℓ. Hence there exist, for every $\varepsilon_0 > 0$, an uncountable set $H^2 \subset H^1$ and $g_1 \in P(C_o(V^0))$ such that $\|g_1^{\alpha,1} - g_1\|_{B_o} < \varepsilon_0$ if $\alpha \in H^2$. We can repeat this argument $s - 1$ times and get that there are an uncountable set $H^s \subset H^{s-1} \subset \ldots \subset H^1$ and $g_i \in P(C_o(V^0))$ such that

$$\|g_i^{\alpha,1} - g_i\|_{B_o} < \varepsilon_0 \quad \text{if} \quad \alpha \in H^s.$$

Put $V_i^\alpha = \operatorname{supp} g_i^\alpha$. There exists an uncountable set $H^{s+1} \subset H^s$ such that the number of elements q_i in V_i^α is independent of $\alpha \in H^{s+1}$. Let $\ell_i^\alpha : V_i^\alpha \to (1,\ldots,q_i)$ be a one-to-one map and let $h_i^\alpha : C_o(V_i^\alpha) \to \mathbb{C}^{q_i}$ be the mapping induced by ℓ_i. Put

$$G_i^\alpha(z) = g_i^\alpha((h_i^\alpha)^{-1}(z))$$

where $z \in \mathbb{C}^{q_i}$ and $(h_i^\alpha)^{-1}$ is the natural inverse of $h_i^\alpha \cdot G_i^\alpha \in P(\mathbb{C}^{q_i})$. Now by the same reasons as above there exist an uncountable set $H^{s+2} \subset H^{s+1}$ and $G_i \in P(\mathbb{C}^{q_i})$ such that $\|G_i^\alpha - G_i\|_{B_{\mathbb{C}^{q_i}}} < \varepsilon_o$ if $\alpha \in H^{s+2}$ where $B_{\mathbb{C}^{q_i}}$ is the unit ball in $\mathbb{C}^{q_i}$ with the C_o-norm. Put $g_i^{\alpha,0} = G_i(h_i^\alpha(z))$. Then $\|g_i^{\alpha,0} - g_i^\alpha\|_{B_o} < \varepsilon_0$ and

$$g_i^{\alpha,0} \in P(C_o(A)).$$

But then the lemma follows if we put $H_1 = H^{s+2}$, $G_i = g^i$, $g^{\alpha,0} = g_i^\alpha$ and $g_\alpha = \sum_1^s g_i \cdot g_i^\alpha$ if ε_0 is chosen so small that

$$s \cdot \operatorname*{Max}_{\substack{\alpha \in A \\ i \leq s}} (\|g_i^\alpha\|_{B_o} + \|g_i^{\alpha,1}\|_{B_o} + \|g_i^\alpha g_i^{\alpha,1}\|_{B_0})\varepsilon_o < \varepsilon. \text{ Q.E.D.}$$

The author is grateful to Professors R. Aron and J. Globevnik for useful discussions on this paper.

REFERENCES

[1] R. ARON, Compact polynomials and compact differentiable mappings between Banach spaces. To appear.

[2] ———, The range of a vector valued holomorphic mappings. To appear.

[3] S. DINEEN, Unpublished manuscript.

[4] J. GLOBEVNIK, The range of vector valued analytic functions. To appear in Arkiv för Matematik.

[5] ——— ' The range of vector valued analytic functions II. To appear.

[6] A. HIRSCHOWITZ, Remarques sur les ouverts d'holomorphic d'un produit denombrable de droites. Annales de l'Institut Fourier 19 (1969).

[7] B. JOSEFSON, A counterexample in the Levi problem. Lecture Notes in Mathematics 364 (1974).

[8] L. NACHBIN, Topology on spaces of holomorphic mappings. Springer-Verlag, Ergebnisse der Math. 47 (1969).

[9] A. PELCZYNSKI, A theorem of Dunford-Pettiš type for polynomial operators. Bull. Acad. Pol. Sc. XI, 6, (1963).

[10] A. PELCZYNSKI AND Z. SEMADENI, Spaces of continuous functions III. St. Math. XVIII (1959).

Uppsala University
Department of Mathematics
Sysslomansgatan 8
S-752 23 Uppsala, Sweden

Infinite Dimensional Holomorphy and Applications, Matos (ed.)

DOMAINS OF EXISTENCE IN INFINITE DIMENSION

By *GETULIO KATZ*

INTRODUCTION

Among the subjects studied in infinite dimensional holomorphy the one of analytic continuation is perhaps the most developed. (See [6] for a survey and a fairly complete bibliography).

The Levi problem asks whether it is true that a domain is pseudo-convex if and only if it is a domain of existence.In the first part of this paper we give a positive answer to the Levi problem in the case of Riemann domains over Banach spaces with Banach approximation property (B.A.P.). This result follows from investigations about properties of permanence for domains of existence under elementary set operations.

In the second part we prove that a convex set in $\ell_p(A)$ (A any index set) is a domain of existence. This result stresses that the behavior of the space of polynomials is directly related to the answer to the Levi problem.

This paper is based an the author's doctoral thesis written under the guidance of Leopoldo Nachbin at the University of

Rochester.

1. Let E and F be locally convex Hansdorff spaces (l.c.s.); $U \subset E$ be an open set; $\mathcal{P}(^mE;F)$ the set of all continuous m-homogeneous polynomials from E into F; C.S.(F) the set of all continuous seminorms in F. Denote by $\mathcal{H}(U;F)$ the set of holomorphic mappings from U into F as defined below.

DEFINITION *A function* $f : U \to F$ *is holomorphic if for all* $\xi \in U$ *and* $m = 0,1,2,\ldots$ *there exist* $P_m \in \mathcal{P}(^mE;F)$ *such that for all* $\beta \in$ C.S.(F) *there exists* $V(\xi)$ U *such that*

$$\lim_{m\to\infty} \beta\left[f(x) - \sum_{n=0}^{m} P_n(x-\xi)\right] = 0 \quad \text{uniformly on } V.$$

DEFINITION *Let* E *be a* l.c.s., *the pair* (X,p) *is called a Riemann domain over* E *if* $X \neq \phi$, X *is a connected Hausdorff space and* $p : X \to E$ *is a local homeomorphism.*

Holomorphic functions; morphism between Riemann domains over the same basic space; $d^\alpha : X \to \mathbb{R}^+$ the distance in X relatively to $\alpha \in$ C.S.(E); The concept of A-domain of holomorphy (where $A \subset \mathcal{H}(X)$); etc., are all defined in the obvious way.

Let (X_1,p_1) and (X_2,p_2) be two Riemann domains over E_1 and E_2 respectively. Let $X = X_1 x X_2$ and $p = (p_1,p_2) : X \to E_1 x E_2$, then (X,p) is a Riemann domain over $E_1 x E_2$.

Assume that E_1, E_2 are metrizable l.c.s. whose topologies are generated by seminorms $\alpha_1 < \alpha_2 \ldots$ and $\beta_1 < \beta_2 \ldots$ respectively. Then $E = E_1 x E_2$ is a metrizable l.c.s. whose topology is generated by $\gamma_1 < \gamma_2 < \ldots$ where $\gamma_i(x_1,x_2) = \sup\{\alpha_i(x_1), \beta_i(x_2)\}$.

Let $x_2 \in X_2$; $Y = X_1 x\{x_2\}$; $q: \begin{array}{l} Y \to E_1 \\ (x_1,x_2) \to p_1(x_1) \end{array}$. Then (Y,q) is a Riemann domain over E_1 such that $Y \simeq X_1$.

THEOREM 1 *Let* (X_i,p_i) $i = 1,2$ *and* (Y,q) *as above. Then if* (X,p) *is a domain of existence so is* (Y,q).

PROOF We use some results of [5] and all definitions required in the proof can be found in this paper.

Suppose (X,p) is a domain of existence. Then by theorem 4.3 of [5], there is an admissible covering $\mathcal{V}$ of X, such that X is an $A_{\overline{\mathcal{V}}}$-domain of holomorphy. $\mathcal{V}$ may be taken countable.

Let $\mathcal{U} = \{V \cap Y \mid V \in \mathcal{V}\}$. $\mathcal{U}$ is an open covering of Y. We claim that $\mathcal{U}$ is admissible. By proposition 4.2 of [5] we have that X is $A_{\mathcal{V}}$-convex and $A_{\mathcal{V}}$-separated. We shall prove that Y is $A_{\mathcal{U}}$-convex and $A_{\mathcal{U}}$-separated.

Assume $\{u,\omega\} \subset Y \subset X$, $u \neq \omega$. As X is $A_{\mathcal{V}}$-separated, there exists $f \in A_{\mathcal{V}}$ such that $f(u) \neq f(\omega)$. Let $f_1 \in \mathcal{H}(Y)$ defined by $f_1(y) = f(y)$. As $f \in A_{\mathcal{V}}$ then $\|f\|_V < \infty$ for all $V \in \mathcal{V}$. So $\|f_1\|_{V\cap Y} \leq \|f_1\|_V < \infty$; it follows that $f_1 \in A_{\mathcal{U}}$ and $f_1(u) = f(u) \neq f(w) = f_1(w)$. Therefore Y is $A_{\mathcal{U}}$-separated.

Now as X is $A_{\mathcal{V}}$-convex, then by theorem 3.6 of [5] we have that for all sequences (x_n) in X such that $d_X^{\gamma_m}(x_n) \to 0$ for all γ_m, there exists $f \in A_{\mathcal{V}}$ such that $\sup|f(x_n)| = \infty$.

Let $(y_n) \in Y$ such that $d_Y^{\alpha_m}(y_n) \to 0$ for all α_m. Then $d_X^{\gamma_m}(y_n) \to 0$ for all γ_m. We have that $f_1 \in A$ and $\sup|f_1(y_n)| = \infty$. Again by theorem 3.6 of [5] we get that Y is $A_{\mathcal{U}}$-convex and by proposition 4.2 and theorem 4.3 that Y is a domain of existence. q.e.d.

DEFINITION *A Banach space* E *is said to have the Banach approximation property (B.A.P.) if* E *is separable and there exists a sequence of operators of finite rank* $(u_n)_{n\in\mathbb{N}}$ *such that* $u_n(x) \to x$ *for all* $x \in E$.

COROLLARY *Let* (X_1, p_1) *be a Riemann domain over a Banach space* E_1 *with B.A.P. Then* X_1 *is pseudo-convex if and only if* X_1 *is a domain of existence.*

PROOF It is known that a separable Banach space has B.A.P. if and only if it is a direct subspace of a Banach space with basis. (see [4]).

Let $E = E_1 \times G$ where E is a separable Banach space with basis. Assume (X,p) is a Riemann domain over E such that $X = X_1 \times G$ and $p = (p_1, id): X \to E$. As X_1 and G are pseudo-convex so is (X,p).

Now by [1] (X,p) is a domain of existence, hence by theorem 1, $X_1 \times \{0\}$ is also a domain of existence. Finally as $X_1 \times \{0\} \simeq X_1$ it follows that (X_1, p_1) is a domain of existence.q.e.d.

In contrast with theorem 1 where separability was crucial, we shall state.

THEOREM 2 *Let* $i = 1,2$, $\mathcal{U}_i \subset E_i$ *a domain of existence in a l.c.s.* E_i. *Then* $\mathcal{U}_1 \times \mathcal{U}_2$ *is a domain of existence.*

PROOF The proof is based on two technical lemmas.

LEMMA 1 *Suppose* E *is a l.c.s. and* $\mathcal{U} \subset E$. $\mathcal{U}$ *is a domain of existence if and only if there exists* $g \in \mathcal{H}(\mathcal{U})$, *such taht for all* $Z \in \partial\mathcal{U}$ *there does not exist a neighborhood* W *(Z-balanced and convex) such that there exists* $\tilde{g} \in \mathcal{H}(W)$ *and* $\tilde{g}$ *agrees with* g *in a connected component* Ω *of* $W \cap \mathcal{U}$ *and* $Z \in \partial\mathcal{U} \cap \partial\Omega$;

LEMMA 2 *Suppose* E *and* G *are two l.c.s., then* $\mathcal{U} \times G$ *is a domain of existence provided* $\mathcal{U}$ E *is a domain of existence.*

PROOF OF THE THEOREM $\mathcal{U}_1 \times \mathcal{U}_2 = (\mathcal{U}_1 \times E_2) \cap (E_1 \times \mathcal{U}_2)$. Let $f_1 \in \mathcal{H}(\mathcal{U}_1 \times E_2)$ and $f_2 \in \mathcal{H}(E_1 \times \mathcal{U}_2)$ such that: $\mathcal{U}_1 \times E_2$

is the domain of existence of f_1 and $E_1 \times \mathcal{U}_2$ the domain of existence of f_2.

We have that $\partial(\mathcal{U}_1 \times \mathcal{U}_2) = (\partial\mathcal{U}_1 \times \partial\mathcal{U}_2) \cup (\partial\mathcal{U}_1 \times \mathcal{U}_2) \cup (\mathcal{U}_1 \times \partial\mathcal{U}_2)$. It is easy to prove that for all $x \in \partial\mathcal{U}_1 \times \partial\mathcal{U}_2$ and $x \in A^{\text{open}} \subset E_1 \times A_2$ then $A \cap (\partial\mathcal{U}_1 \times \mathcal{U}_2) \neq \emptyset$. We claim that $f_1 + f_2 \in \mathcal{H}(\mathcal{U}_1 \times \mathcal{U}_2)$ has $\mathcal{U}_1 \times \mathcal{U}_2$ as natural domain.

If not, there exists $Z \in \partial(\mathcal{U}_1 \times \mathcal{U}_2)$, $\mathcal{W}(Z)$ and $\tilde{g} \in \mathcal{H}(\mathcal{W})$ such that $\tilde{g} = f_1 + f_2$ in a connected component Ω of $\mathcal{W} \cap (\mathcal{U}_1 \times \mathcal{U}_2)$ such that $Z \in \partial(\Omega)$.

a) Suppose $Z \in \partial(\mathcal{U}_1) \times \mathcal{U}_2$. Then there exists $B_2 \overset{\text{domain}}{\subset} \mathcal{U}_2$, $B_1 \overset{\text{domain}}{} E_1$, such that $Z \in B_1 \times B_2 = \mathcal{W}_1 \subset \mathcal{W}$; $\mathcal{W}_1 \subset \mathcal{U}_1 \times E_2$. Besides as $Z \in \partial(\Omega)$ then $\Omega \cap \mathcal{W}_1 \neq \phi$. Let θ be a connected component of $\Omega \cap \mathcal{W}_1 \subset \mathcal{W}_1 \cap (\mathcal{U}_1 \times \mathcal{U}_2)$.

We have $\tilde{g} = f_1 + f_2$ in θ; $\tilde{g} - f_2 \in \mathcal{H}(\mathcal{W}_1)$ and $\tilde{g} - f_2 = f_1$ in θ. This contradicts the fact that f_1 has $\mathcal{U}_1 \times E_2$ as domain of existence.

b) Suppose $Z \in \mathcal{U}_1 \times \partial\mathcal{U}_2$. Same proof as in (a).

c) If $Z \in (\partial\mathcal{U}_1) \times (\partial\mathcal{U}_2)$, let $\mathcal{W}(Z)$ as in lemma 1. Suppose A_1 and A_2 convex, Z_i - equilibrated open neighborhoods of A_i and $Z \in A_1 \times A_2 \subset \mathcal{W}$, where $Z = (Z_1, Z_2)$. As $Z \in \partial(\Omega)$, there exists $(x_1, x_2) \in \Omega \cap (A_1 \times A_2)$. So $(Z_1, x_2) \in A_1 \times A_2$; $s = (Z_1, x_2) \in (\partial\mathcal{U}_1) \times \mathcal{U}_2$. Let A_3 = connected component of $\mathcal{U}_2 \cap A_2$ such that $x_2 \in A_3$. Then $s \in A_1 \times A_3 = \mathcal{W}_1$; $\mathcal{W}_1 \mathcal{W}$; $\mathcal{W}_1 \subset E_1 \times \mathcal{U}_2$; $\mathcal{W}_1 \subset \mathcal{U}_1 \times E_2$ and $(x_1, x_2) \in \Omega \cap (A_1 \times A_2)$. Thus reasoning as in part (a) we would get that $\mathcal{U}_1 \times E_2$ is not the domain of existence of f_1. It follows that $\mathcal{U}_1 \times \mathcal{U}_2$ is a domain of existence. q.e.d.

2. In this section we shall follow the techniques of [2] to

prove that in $\ell_p(A)$ the open convex sets are domains of existence.

Let F be a Banach space, $\mathcal{U} \subset F$ be a convex open subset of F containing the origin and $\bar{B}$ the closed unit ball in F.

Let $\varepsilon > 0$ such that $\varepsilon\bar{B} \subset U$; using the open-mapping and Hahn-Banach theorems, one can prove that for all $x \in \partial U$ there exists $\phi_x \in F'$ satisfying: $\|\phi_x\| < 2/\varepsilon$, $\operatorname{Re} \phi_x(x) = 1$ and $\operatorname{Re}(\phi_x(Z)) < 1$ whenever $Z \in U$.

Let $f_x(Z) = \dfrac{1}{[\phi_x(Z)-\phi_x(x)]^i}$, where $i \in \mathbb{N}$ is fixed. We have that $f_x \in \mathcal{H}(U)$ since $\operatorname{Re}(\phi_x(Z)- \phi_x(x)) \neq 0$ for all $Z \in U$.

PROPOSITION 1 *For all* $Z \in U$, $\gamma \in (0, \infty)$ *such that* $Z + 2\gamma\bar{B} \subset U$, *we have that* $\sup\limits_{x\in\partial U} \sup\limits_{Y\in\gamma\bar{B}} |f_x(Z+Y)| \leq M_{Z,\gamma} < \infty$.

PROOF* Let $Z \in U$ and fix $\gamma \in (0,\infty)$ such that $Z + 2\gamma\bar{B} \subset U$. Define $a = \sup \{r>0 | rZ\in U\}$ and $M_{Z,\gamma} = (1/2\ (1 - \frac{1}{a}))^{-i}$. If $Y\in\gamma \bar{B}$; $\operatorname{Re}(\phi_x(Z+2Y)-\phi_x(x)) \leq 0$, then $\operatorname{Re} \phi(Y) \leq \frac{1}{2}(1-\operatorname{Re}(\phi_x(Z))$. Therefore $\operatorname{Re}(\phi_x(Z+Y) - \phi_x(x)) \leq \operatorname{Re} (\phi_x(Z)) + \frac{1}{2}(1 - \operatorname{Re}(\phi_x(Z)) - 1 = \frac{1}{2}(\operatorname{Re}(\phi_x(Z))-1)$. But $\operatorname{Re}(\phi_x(Z)) \leq \frac{1}{a}$, therefore $|\operatorname{Re}(\phi_x(Z+Y) - \phi_x(x)| \geq 1/2\ (1- \frac{1}{a})$. So $|f_x(Z+Y)| \leq M_{Z,\gamma}$ for all $Y\in\gamma\bar{B}$. q.e.d.

Let $\ell_p(A), p \in [1,\infty)$, be the set $E(A)=\{f:A\to\mathbb{C} | \sum\limits_{\alpha\in A} |f(\alpha)|^p<\infty\}$ endowed with the norm $\|f\| = (\sum\limits_{\alpha\in A} |f(\alpha)|^p)^{1/p}$. $\ell_p(A)$ is a Banach space. Let $\{\ell_\alpha\}_{\alpha\in A}$ given by $\ell_\alpha(\beta) = \begin{cases} 1 \text{ if } \alpha = \beta \\ 0 \text{ otherwise} \end{cases}$

THEOREM 3 *Let* $0 \in U \subset \ell_p(A)$ *be an open convex set. Then* U *is*

---

* This elegant proof is due to David Prill, mine was considerably more complicated.

a domain of existence.

PROOF If $U = \ell_p(A)$ the result is clear. So we may consider $U \neq \ell_p(A)$, and we may also assume that A is well ordered, say $A = [0,\psi)$ where ψ is the cardinal number of A. Finally it is possible to prove that there exists a dense subset $\{x_\alpha\}_{\alpha \in A}$ in ∂U.

The proof of the theorem is based in three claims.

CLAIM 1 *There exists $\ell : A \to A$ such that ℓ is injective and* $(x_\gamma)_{\ell(\beta)} = g_{\ell(\beta)} = 0$ if $\begin{cases} \gamma < \beta \\ \gamma \in A \end{cases}$ where $g_{\ell(\beta)}$ is the corresponding associated family of coefficients functionals.

Let $q = \begin{cases} p \text{ if } p \text{ is integer,} \\ [p] + 1 \text{ otherwise, where } [p] = \text{integer part of } p \end{cases}$

CLAIM 2 *The function* $Z \to Z^q_{\ell(\alpha)} \cdot f_{x_\alpha}(Z)$, *cannot be continued over* x_α. *(In the definition of* f_{x_α} *take* $i = q + 1$).

Set $g(Z) = \sum_{\alpha \in A} C_\alpha \cdot Z^q_{\ell(\alpha)} \cdot f_{x_\alpha}(Z)$ where C_α is defined as follows

$$C_\alpha = \begin{cases} 0 \text{ if } \sum_{\gamma<\alpha} C_\gamma \left[(1-r)x_\alpha + r \in \ell_{\ell(\alpha)}\right]^q_{\ell(\alpha)} \cdot f_{x\gamma}\left[(1-r)x_\alpha + r \in \ell_{\ell(\alpha)}\right] \longrightarrow \infty \text{ as } r \longrightarrow 0 \\ 1 \text{ otherwise} \end{cases}$$

CLAIM 3 *We have that $g \in \mathcal{H}(\mathcal{U})$, where g is defined as above.*

We shall conclude the proof showing that g cannot be continued over any point x_α, $\alpha \in A$; and therefore g cannot be continued over any point of ∂U since $\{x_\alpha\}_{\alpha \in A}$ is dense in ∂U. Indeed, $\sum_{\gamma>\beta} C_\gamma (\xi x_\beta + y\ell_{\ell(\beta)})^q \cdot f_{x\gamma}(\xi x_\beta + y\ell_{\ell(\beta)}) = 0$ for all $\xi, y \in \mathbf{C}$ such that $\xi x_\beta + y\ell_{\ell(\beta)} \in U$, because the way we defined ℓ. Now if

$$\sum_{\gamma<\beta} \left[C_\gamma (1-r)\, x_\beta + r \in \ell_{\ell(\beta)}\right]^q_{\ell(\gamma)} \cdot f_{x\gamma}\left[(1-r)x_\beta + r \in \ell_{\ell(\beta)}\right] \to \infty$$

as $r \to 0$, then $g[(1-r)x_\beta + r\, \epsilon\, \ell_{\ell(\beta)}] =$

$$= \sum_{\gamma<\beta} \left[C_\gamma((1-r)x_\beta + \epsilon\, \ell_{\ell(\beta)})^q\right]_{\ell(\gamma)} \cdot f_{x\beta}((1-r)x_\beta + \epsilon\, \ell_{\ell(\beta)})\Big] \to \infty$$

as $r \to 0$.

If as $r \to 0$

$$\sum_{\gamma<\beta}\left[C_\gamma((1-r))x_\beta + r\epsilon\ell_{\ell(\beta)}\right]^q_{\ell(\gamma)} \cdot \left[f_{x\gamma}((1-r))x_\beta + r\, \epsilon\, \ell_{\ell(\beta)}\right] \not\longrightarrow \infty$$

then $\underline{\lim} \left|\sum_{\gamma<\beta} [(\ )].[(\ )]\right| < \infty$, so $\left|g\ |(1-r)x_\beta + r\epsilon\ell_{\ell(\beta)}|\right| \geq$

$$\geq \left|\sum_{\gamma<\beta}\right| + \left|(1-r)x_\beta + r\, \epsilon\, \ell_{\ell(\beta)})^q_{\ell(\beta)}\right| \cdot \left|f_{x\beta}\left[(1-r)x_\beta + r\epsilon\ell_{\ell(\beta)}\right]\right| \to \infty$$

on a sequence of points with infimum 0. Indeed in this case $C_\beta = 1$ and the behavior of the sum is given by the last term, which goes to fininity by the claim 2. This finishes the proof.

BIBLIOGRAPHY

|1| HERVIER, Y., Sur la problème de Levi pour les espaces Étalés Banachiques, C.R. Acad. SCI, V.275, ser A.p. 821, 1972.

|2| JOSEFSON, B., Counterexample in the Levi problem. Proceedings on infinite dimensional holomophy, Lecture notes in Mathematics. Springer Verlag, V. 364, 1973, 168-177.

|3| NACHBIN, L., Topology on spaces of holomorphic mappings. Ergebnisse der Mathematik und ihere grenzze biete, Springer-Verlag, Hejt-47, 1969.

|4| PELCZINSKI, A., Any separable. Banach space with the bounded approximation property is a complemented subspace of

a Banach space with basis. Studia Math; E 40, 1971, p. 239-243.

[5] SCHOTTENLOHER, M., Analytic continuation and regular classes in l.c. Hausdorff spaces. Portugaliae Matematica. (to appear).

[6] SCHOTTENLOHER, M., Riemann domains: Basic results and open problems. Proceedings on infinite dimensional holomorphy. Lecture notes in Math, Springer-Verlag V.364, 1973.

DEPARTAMENTO DE MATEMÁTICA
UNIVERSIDADE FEDERAL DE PERNAMBUCO
CIDADE UNIVERSITÁRIA - TEL. 27-2388
RECIFE - BRASIL

Infinite Dimensional Holomorphy and Applications, Matos (ed.)

GEOMETRIC ASPECTS OF THE THEORY OF BOUNDS FOR ENTIRE FUNCTIONS IN NORMED SPACES

By C. O. KISELMAN

CONTENTS:

1. INTRODUCTION

A phenomenon in infinite-dimensional complex analysis which has no counterpart in finite dimensions is that an entire function may be unbounded on a bounded set. For example the series

$$f(x) = \sum_1^\infty x_k^k, \quad x = (x_1, x_2, \ldots) \in \ell^p,$$

converges everywhere in ℓ^p, $1 \leq p < \infty$, and defines an entire function which is unbounded in every ball of radius larger than one. This leads e.g. to the definition of a bounding set, i.e. a set which is mapped onto a bounded set by every entire function, and this concept has been studied by many authors. A fundamental result is that in certain spaces only the relatively compact sets are bounding, and, generally speaking, bounding sets have no interior. If on the other hand we consider sets where an individual entire function is bounded it is clear that these sets may have interior points: they are the subsets of the open sets

$$\omega_{k,f} = \{x \in E;\ |f(x)| < k\}$$

for some k. The geometry of these sets is of interest. However, to describe all possible families $(\omega_{k,f})_{k>0}$ is a formidable task, and one is led to measuring the growth of f in some simplified way. The most fundamental notion is that of the radius of boundedness: if f is entire on E, a normed space, the radius of boundedness at $x \in E$ is the least upper bound $R_f(x)$ of all numbers r such that f is bounded in the ball $\{y;\ ||y - x|| \leq r\}$. How does $R_f(x)$ depend on x? This is a difficult and interesting question, obviously connected to both complex analysis and infinite-dimensional geometry. It turns out that even if one is interested only in $R_f(x)$, several related concepts come into the picture, and the main purpose of this lecture is to propose, in §2, a whole family of related measures of the growth of an entire (or plurisubharmonic) function.

First of all we must broaden the view to include the plurisubharmonic functions. It then becomes natural to define the *radius of boundedness* $R_u(x)$ of any numerical function $u: E \to [-\infty, +\infty[$ as the supremum of all numbers r such that u is bounded above in the ball of radius r and center at x. This, of course, is to allow $u = \log |f|$, f entire, and we are not concerned here with sets where f is small. We remark that the radius of boundedness of $u = \log |f|$ is then equal to the *radius of convergence* of f, i.e. the least upper bound of all r such that the Taylor series of f at x converges uniformly in the ball of center x and radius r (see Nachbin [7, p. 26]).

The radius of boundedness may be regarded as a kind of boundary distance. In fact, let E be a normed space, u a plurisubharmonic function in E, in symbols $u \in PSH(E)$, and put

$$\Omega_k = \{x \in E;\ u(x) < k\}.$$

Then Ω_k is a pseudoconvex open set and this implies that the function $-\log d_k$ is plurisubharmonic in Ω_k, where $d_k(x)$ is the distance from $x \in \Omega_k$ to $\partial\Omega_k$. Hence,

$$R_u = \lim_{k \to \infty} d_k = \sup_k d_k,$$

and by well-known properties of plurisubharmonic functions,

(1.1) $-\log R_u$ *is plurisubharmonic in* E,

for it is locally the limit of $-\log d_k \in PSH(\Omega_k)$. Also, since every d_k is Lipschitz continuous with Lipschitz constant 1, we see that

(1.2) $|R_u(x) - R_u(y)| \leq ||x-y||,\ x,\ y \in E,$

which holds, of course, for any function u, not just for $u \in PSH(E)$. Property (1.1) was first proved by Lelong [5,p.176]. Thus the radius of boundedness appears as the regularized bound ary distance from $(x,0) \in E \times \mathbb{C}$ to $\partial\omega$ in the "direction" $E \times \{0\}$, where

$$\omega = \{(x,t) \in E \times \mathbb{C};\ u(x) + \log |t| < 0\}$$

is an open subset of $E \times \mathbb{C}$, pseudoconvex if $u \in PSH(E)$.

Instead of $R_u(x)$ one may of course define

$$R_{u,A}(x) = \sup\ (r; u \text{ is bounded above in } x + rA)$$

for any set A. However, this is not so interesting unless we manage to impose a structure on the family of sets A. In §2 we shall do precisely that for certain sets A, viz. those which are linear images of the unit ball of some other normed space. We may then study the dependence of $R_{u,A}(x)$ on A as well as on x, and this leads to new problems as well as to some answers. In §3 and §4 we give applications of this-it is hoped that they are not the last ones. In §3 we determine zones where a plurisubharmonic function must be large if its radius of boundedness decays slowly at infinity. This is proved using $R_{u,A}$ for $A = A_\lambda$ a linear image of the unit ball of the same space under a one-parameter family of mappings:

$$A_\lambda = (I + (\lambda-1)\pi)(B) = \{x \in E;\ ||x + (\tfrac{1}{\lambda} - 1)\pi(x)|| \leq 1\}, \lambda > 0.$$

In §4 we study when equality in (1.2) can occur: this turns out to be rather exceptional, and the result already shows that not every function R in c_0 or ℓ^p, $1 < p < \infty$, satisfying (1.1) and (1.2) is a radius of boundedness of an entire (or even plurisubharmonic) function. In §5 we state a precise estimate for the Lipschitz constant of a radius of boundedness which decreases slowly at infinity. Theorem 5.1 was (essentially) proved in [4].

The problem of constructing an entire function with prescribed radius of convergence is discussed in the final Section 6 were we first give the main theorem from Section 4 of [4], and then discuss a special case (see Theorem 6.2) which is not covered by the methods of [4].

My basic conjecture concerning the radius of convergence has been this: any function R: $\ell^1 \to]0, +\infty[$ satisfying (1.1) and (1.2) is the radius of convergence of some entire function on ℓ^1. A test for any new methods would be to prove this first for functions R depending on, say, Σx_j only. A more general conjecture is that Theorem 6.1 holds for ℓ^p, $1 \leq p < \infty$, and c_0 without the finiteness condition. In fact in this geometric setting the natural concept is that of the τ-local radius of boundedness $R_{\tau,u}$, and the conjecture concerning this is that for $\tau = \sigma(E,E')$, E reflexive (say) and infinite-dimensional, $-\log R_{\tau,u}$ is just any plurisubharmonic function. It should however be remarked that if Theorem 6.2 is best possible or close to it, then the conjecture does not hold for ℓ^1 (but may still hold for the reflexive spaces ^{p}, $1 < p < \infty$).

NOTATION With the exception of Theorem 4.1 all spaces are normed linear spaces over the complex numbers. If E and F are normed spaces we let L(E,F), or just L, denote the continuous linear mappings from E into F, and A = A(E,F) the continuous affine mappings, identified in a natural way with F × L. In L we use the topology given by the norm

$$||f||_L = \sup_{||x||_E \leq 1} ||f(x)||_F,$$

and in A the product topology of F X L. The closed unit ball of E is denoted by B_E or just B. We use vector operations also for sets, thus e.g. $\lambda_1 A_1 + \lambda_2 A_2 = \{\lambda_1 a_1 + \lambda_2 a_2; a_1 \in A_1 \text{ and } a_2 \in A_2\}$. We write PSH($\Omega$) and $\mathcal{O}(\Omega)$ for the set of, respectively, all plurisubharmonic functions and all analytic functions in Ω. The star in f^* ("étoile de Lelong" in the terminology of Martineau [6]) means upper regularization, i.e. the operation of taking the interior of the epigraph. By a uniform neighbourhood of a set M in a normed space we mean a set containing $M + \varepsilon B$ for some $\varepsilon > 0$.

2. GEOMETRIC INDICATORS OF THE GROWTH OF A PLURISUBHARMONIC FUNCTION

The definition of the radius of boundedness R_u of a plurisubharmonic function u on a normed space E means that for $(x,t) \in E \times \mathbb{C}$, $|t| < R_u(x)$ if and only if u is bounded above on $x + tB$. Knowing R_u is equivalent to knowing the set

$$\Omega = \{(x,t) \in E \times \mathbb{C};\ |t| < R_u(x)\},$$

and $-\log R_u$ is plurisubharmonic if and only if Ω is pseudoconvex. More detailed information is provided by the function

$$\tilde{u}(x,t) = \sup_{||y|| \leq |t|} u(x+y),$$

and Ω is gotten from $\tilde{u}$ in a simple way: it is the interior of the set where $\tilde{u}$ is less than $+\infty$. We can thus say that $\tilde{u}$ is a "*functional indicator*" and that Ω (or its boundary distance) is a "*geometric indicator*" of the growth of u.

The purpose of this section is to introduce a whole family of "indicators" of the growth of a plurisubharmonic function u which are more general than $\tilde{u}$ and Ω above, but still have enough structure to allow calculations to be made.

The general version is given in the following theorem.

THEOREM 2.1 *Let* E *be a normed space,* X *a set and* H *a linear space of bounded mappings* $h : X \to E$ normed by

$$||h||_H = \sup_{x \in X} ||h(x)||_E$$

and containing the constant mappings. Define, for any numerical function u *on* E,

$$\tilde{u}(h) = \sup_{x \in X} u(h(x)),\ h \in H,$$

and let Ω *be the set of all* $h \in H$ *such that* $\tilde{u}$ *is bounded above in a uniform neighborhood* $h(X) + \varepsilon B_E$ *of* $h(X)$. *Then* Ω *is pseudoconvex and* $\tilde{u}^* \in PSH(\Omega)$ *if* $u \in PSH(E)$. *If* ϕ *is a filter on* X *we put*

$$\tilde{u}_\phi(h) = \lim_{M \in \phi} \sup_{x \in M} u(h(x))$$

and have the analogous conclusion for $\tilde{u}^*_\phi$ *on*

$$\Omega_\phi = \{h \in H;\ u \text{ is bounded above in } h(M) + \varepsilon B_E \text{ for some } \varepsilon > 0 \text{ and some } M \in \phi\}.$$

If E is regarded as a subspace of H (via the constant functions), $\tilde{u}$ and $\tilde{u}^*$ become extensions of u.

This theorem is too general to be of use as it stands: for each particular application one will have to choose X and H to match the problem. It is of special importance to find criteria for $\tilde{u}^*(h)$ to be equal to $\tilde{u}(h)$. The following instance of Theorem 2.1 therefore seems to be at a reasonable level of abstraction.

THEOREM 2.2 *Let* G *and* E *be normed spaces, and let* $A = A(G, E)$ *be the linear space of all affine continuous mappings of* G *into* E. *Let* $u \in PSH(E)$, *let*

$$\tilde{u}(h) = \sup_{\|y\|_G \le 1} u(h(y)), \ h \in A,$$

and let Ω *be the set of all* $h \in A$ *such that* u *is bounded above in* $h(B_G) + \varepsilon B_E$ *for some* $\varepsilon > 0$. *Then* Ω *is pseudoconvex in* A *and* $\tilde{u}^* \in PSH(\Omega)$. *Also* $h \in \Omega$ *and* $\tilde{u}^*(h) = \tilde{u}(h)$ *whenever* G *and* E *are Banach spaces,* h *is surjective and* $\tilde{u}$ *is less than* $+\infty$ *near* h. *In particular,* $\tilde{u}|_{\Omega \cap A_{iso}}$ *is plurisubharmonic in the pseudoconvex open subset* $\Omega \cap A_{iso}$ *of* A, A_{iso} *denoting the isomorphisms of* G *onto* E.

One can of course give a more general instance of Theorem 2.1 by letting H denote, say, a space of holomorphic mappings $h : G \to E$, bounded on the unit ball $X = B_G$ of G. However, the difficulty then is to decide, for a given h, whether $\tilde{u}^*(h)=\tilde{u}(h)$.

Theorem 2.2 generalizes a result which has been proved and applied in [4]: we take $G = E$ and restrict attention to those affine mappings $h \in A(E,E)$ which are of the form

$$h(y) = x + ty, \ y \in E,$$

for some $x \in E$ and some $t \in \mathbb{C}$. This leads to the function $\tilde{u}(x,t)$ considered at the beginning of this section. Letting ϕ denote the filter of neighborhoods of the origin for some topology τ on E (e.g. $\tau = \sigma(E,E')$) we get corresponding "local" objects of which the most important is the τ-*local radius of boundedness* $R_{\tau,u}(x)$ which is the supremum of all numbers r such that u is bounded above in $x + r(B \cap W)$ for some τ-neighborhood W of the origin. We shall use these constructions in the proof of Theorem 4.4. On the other hand, in §3 we shall use Theorem 2.2 with affine mappings not of the form $y \mapsto x + ty$.

From one plurisubharmonic function $u \in PSH(E)$ we thus get a family of others, reflecting in various ways the growth of u. Let for instance, in the notation of Theorem 2.2, $d(h)$ denote the distance from $h \in A(G,E)$ to the boundary of Ω measured in some more or less arbitrary way (we may use any continuous norm in A). Then $-\log d \in PSH(\Omega)$. More generally, if $d(h,f)$ is the distance from h to $\partial\Omega$ in the direction f, i.e. $d(h,f)=\sup\ (r;\ h+tf \in \Omega$ for all complex t with $|t| \leq r)$, then $-\log d(h,f)$ is plurisubharmonic in $\Omega \times A$ (see Noverraz [8, Théorème 2.2.1]; it is no harm to allow $f = 0$). In particular, we may take h constant, $h(y) = x \in E$ for all $y \in G$, and f linear and then $d(h,f) = d(x,f)$ is the least upper bound of all numbers r such that u is bounded above in a uniform neighborhood of $x + rf(B_G)$; we may justly call this number the *radius of boundedness at* x *with respect to* $f(B_G)$. We state this result as a corollary of Theorem 2.2, to be used in the proof of Theorem 3.1:

COROLLARY 2.3 *Let* G *and* E *be normed spaces and* $u \in PSH(E)$.*Then minus the logarithm of the radius of boundedness of* u *at* x *with respect to* $f(B_G)$ *is a plurisubharmonic function of* $(x,f) \in E\times L(G,E)$.

PROOF OF THEOREM 2.1 Consider for a fixed $x \in X$ the continuous linear map

$$\delta_x : H \ni h \to h(x) \in E.$$

Then $u \circ \delta_x$ is plurisubharmonic in H so the upper regularization of $\tilde{u} = \sup_{x\in X} u \circ \delta_x$ is plurisubharmonic in the open set Ω' where the family $(u \circ \delta_x)_{x\in X}$ is locally bounded above. It is easily seen that Ω' is precisely Ω as defined in the statement of the theorem. (The constant mappings are used in proving

$\Omega' \subset \Omega$).

To see that Ω is pseudoconvex we put

$$\omega_{k,x} = \{h \in H;\ u(h(x)) < k\},\ k \in \mathbb{N},\ x \in X,$$

$$\omega_k = \Big(\bigcap_{x\in X} \omega_{k,x}\Big)^{\circ}, \text{ and}$$

$$\omega = \bigcup_{k\in \mathbb{N}} \omega_k .$$

Clearly $\omega_{k,x}$ is pseudoconvex in H if $u \in PSH(E)$ and the operations used to get ω_k and ω preserve pseudoconvexity. Finally, one sees easily that $\omega = \Omega$.

The statement about $\tilde{u}_\phi$ now follows by applying the first part of the theorem to a set $M \in \phi$ and then passing to the limit. (The usual "lim-sup-star" theorem when one takes the limit along a directed set requires this set to possess a denumerable cofinal subset, but this is no longer necessary if, as is the case here, the family of functions decreases. For details in a similar situation, see [4, proof of Proposition 2.2]).

PROOF OF THEOREM 2.2 Only the statement about surjections is not a direct consequence of Theorem 2.1. Let $\varepsilon > 0$, let $h_0 \in A$ be surjective, and assume that $\tilde{u}$ is less than $+\infty$ in some neighborhood of h_0, say for $\|h - h_0\|_A \leq \delta$ where we now use the norm

$$\|h\|_A = \|x\|_E + \|f\|_L$$

in $A = A(G,E)$; $x \in E$ and $f \in L = L(G,E)$ being defined by $h(y) = x + f(y)$, $y \in G$. If $h_0(y) = x_0 + f_0(y)$, $y \in G$, we consider

$$\psi(t) = \tilde{u}(x_0 + tf_0) = \sup_{y\in B_G} u(x_0 + tf_0(y)) \leq +\infty.$$

Obviously ψ depends on $|t|$ only and is a convex function of $\log |t|$. Since $\tilde{u}(h) < +\infty$ for $\|h - h_0\|_A \leq \delta$ we have

$\psi(1+\delta_1) < +\infty$ where $\delta_1 = \delta/||f_0||_L$, and by logarithmic convexity, $\psi(1+\delta_2) < \psi(1) + \varepsilon$ for some $\delta_2 > 0$. Pick $\delta_3 > 0$ such that $2\delta_3 B_E \subset \delta_2 f_0(B_G)$ (we suppose now that G and E are Banach spaces and apply the homomorphism theorem). Then if $||x - x_0|| \leq \delta_3$ and $||f - f_0|| \leq \delta_3$ we obtain

$$x + f(B_G) \subset x_0 + \delta_3 B_E + f_0(B_G) + \delta_3 B_E \subset x_0 + f_0(B_G) + \delta_2 f_0(B_G) =$$

$$= x_0 + (1+\delta_2) f_0(B_G).$$

Hence $\tilde{u}(h) = \tilde{u}(x+f) \leq \psi(1+\delta_2) \leq \tilde{u}(h_0) + \varepsilon$ which proves that $\tilde{u}^*(h_0) = \tilde{u}(h)$ and at the same time that $h_0 \in \Omega$.

We finally remark that A_{iso} is a pseudoconvex open set in A (possibly empty) (for the distance $d(h_0)$ from $h_0 = x_0 + f_0$ to the boundary of A_{iso} is at least $||f_0^{-1}||^{-1}_{L(E,G)}$ by the Neumann series so $-\log ||f_0^{-1}||_{L(E,G)}$ is a plurisubharmonic function in A_{iso} tending to $+\infty$ at the boundary). Therefore $\Omega \cap A_{iso}$ is pseudoconvex, and $\tilde{u}^* = \tilde{u}$ in $\Omega \cap A_{iso}$. This completes the proof of Theorem 2.2.

REMARK 2.4 We do not know whether in general a point $h_0 \in A$ such that $\tilde{u}(h) < +\infty$ for all h near h_0 belongs to Ω. This is true, as we have seen, if h_0 is surjective and the spaces involved are Banach.

REMARK 2.5 Let E and F be Banach spaces. It has been observed above that the isomorphisms in $L(E,F)$ form a pseudoconvex open set. Similarly, the epimorphisms, the monomorphisms with closed range, the direct epimorphisms, the direct monomorphisms, and the direct homomorphisms form open sets in $L(E,F)$. Are these open sets pseudoconvex?

3. ZONES WHERE A GIVEN PLURISUBHARMONIC FUNCTION IS LARGE

By the very definition of the radius of boundedness, a function u assumes arbitrarily large values in every shell

$$\{y;\ R_u(x) - \varepsilon < \|y - x\| < R_u(x) + \varepsilon\}$$

where $\varepsilon > 0$. It is natural to ask if it is possible to describe subsets of this shell where u is also unbounded above. We shall do so assuming that u is plurisubharmonic and that R_u decreases slowly at infinity. The description will be in terms of slabs $\alpha B \cap \beta\pi^{-1}(B)$ where $\pi: E \to E$ is a projection in E.

THEOREM 3.1 *Let* $E = \ell^p(J)$, $1 \leq p < \infty$, *where* J *is an infinite index set, and let* F *be a subspace of* E *spanned by a finite number of coordinates. Assume that the radius of boundedness of a function* $u \in PSH(E)$ *satisfies an estimate of the form*

$$R_u(x) \geq C\,\|x\|^{-\gamma} \text{ for } x \in F,\ \|x\| \geq r_1,$$

where $C > 0$ *and* $\gamma \geq 0$ *denote constants. Let* $\pi: E \to F$ *be the canonical projection onto* F. *Then* u *is unbounded in* $\alpha B \cap \beta\pi^{-1}(B)$ *for every* α *and* β *satisfying*

$$\alpha > R_u(0) \quad \text{and} \quad \beta > R_u(0)\left(\frac{\gamma}{1+\gamma}\right)^{1/p}.$$

Note that the number r_1 may be very large, so no explicit assumption is made on the behavior of $R_u(x)$ for small $\|x\|$. For $p = 2$, E is an arbitrary infinite-dimensional Hilbert space, and F any finite-dimensional subspace. The theorem then says roughly that if R_u decreases slowly in a certain direction x, then u assumes large values close to the hyperplane $\{y;\ \langle y,x\rangle = 0\}$.

PROOF Changing the constant C if necessary we may, and shall, assume that

$$R_u(x) > C\, ||x||^{-\gamma} \quad \text{for} \quad x \in F,\ ||x|| \geq r_1, \text{ and}$$

$$R_u(x) > C\, r_1^{-\gamma} \quad \text{for} \quad x \in F,\ ||x|| \leq r_1.$$

Let $K(a,b)$ denote an "ellipsoid" with half-axes a and b:

$$K(a,b) = \{x \in E;\ (||\pi(x)||/a)^p + (||x - \pi(x)||/b)^p \leq 1\}.$$

We claim that u is bounded above in $K(a,b)$ if $a \geq r_1$ and $b=C\, a^{-\gamma}$. In fact, u is bounded above in every ball $x + b_1 B$, where $x \in F$, $||x|| \leq a$, and $b_1 < R_u(x)$ for all these points x. We can choose such a number $b_1 > b$. Now let

$$\omega_x = x + b_1 B \cap \varepsilon_x \pi^{-1}(B)$$

where ε_x is positive but so small that

$$\omega_x \supset K(a,b) \cap (x + \varepsilon_x \pi^{-1}(B)).$$

Since $K(a,b) \cap F$ is compact, finitely many ω_x with $x \in K(a,b) \cap F$ suffice to cover $K(a,b) \cap F$, and then they cover $K(a,b)$ as well. Hence u is bounded above there.

Letting now $R_u(\lambda;\ x)$ denote the radius of boundedness of u with respect to $f_\lambda(B)$ where f_λ is the linear mapping

$$f_\lambda(x) = x + (\lambda - 1)\ \pi(x),$$

we therefore have, since $f_\lambda(B) = K(|\lambda|,1)$,

$$R_u(\lambda;\ 0) \geq b = C\, a^{-\gamma} \quad \text{for} \quad a \geq r_1,$$

where $\lambda = a/b$, i.e. $a = (C\lambda)^{1/(\gamma+1)}$. Equivalently,

$$R_u(\lambda;0) \geq C^{\frac{1}{\gamma+1}}\, \lambda^{-\frac{\gamma}{\gamma+1}} \quad \text{for } \lambda \geq C^{-1}\, r_1^{\gamma+1}$$

Now the function $\phi(\lambda) = -\log R_u(\lambda;0)$ is subharmonic in $\lambda \in \mathbb{C}$ by Corollary 2.3.Furthermore, and this is most essential, $\phi(\lambda)$ depends only on $|\lambda|$, since, as already noted, $f_\lambda(B) = K(|\lambda|,1)$. When $|\lambda|$ is large we have

$$\phi(\lambda) \leq \frac{\gamma}{\gamma+1} \log |\lambda| - \frac{1}{\gamma+1} \log C,$$

and the logarithmic convexity enables us to extend this estimate to all λ with $|\lambda| \geq 1$:

$$\phi(\lambda) \leq \frac{\gamma}{\gamma+1} \log |\lambda| + \phi(1) \quad \text{for } |\lambda| \geq 1.$$

In particular, going back to $R_u(\lambda;0)$, we obtain for every given number $c > \gamma/(\gamma + 1)$ that

$$R_u(1 + t;\ 0) > R_u(1;0)(1 - ct) = R_u(0)(1 - ct) \text{ for } t > 0.$$

This means that u is bounded above in the "ellipsoid" $K(a,b)$ for $a = (1 + t)b$ and $b = R_u(0)(1 - ct)$. Let now β be fixed with $\beta > R_u(0)(\gamma/(\gamma + 1))^{1/p}$. Since u is unbounded above in $(R_u(0)+ t^2)B$, u must also be unbounded above in $(R_u(0) + t^2)B \setminus K(a,b)$ for these a,b and t. A simple calculation now shows that this set is contained in $\beta\pi^{-1}(B)$ if c is close enough to $\gamma/(\gamma + 1)$ and t is small. (This is only a matter of finding the intersection of the two curves

$$\xi^p + \eta^p = (R_u(0) + t^2)^p,$$

$$(\xi/a)^p + (\eta/b)^p = 1$$

in $\mathbb{R}^2$). This proves the theorem.

When p tends to $+\infty$ the slab $\alpha B \cap \beta\, \pi^{-1}(B)$ increases and tends, roughly speaking, to a ball. This is to be expected since in the limiting case we have the following example.

EXAMPLE 3.2 The entire function

$$f(x) = \sum_{2}^{\infty} (x_1 x_k)^k, \quad x \in c_0,$$

has radius of convergence

$$R_f(x) = 2((|x_1|^2 + 4)^{1/2} + |x_1|)^{-1}.$$

In particular $R_f(0) = 1$ and $R_f(x) \geq C|x_1|^{-1}$ for $|x_1| \geq 1$. It is easily seen that if f is unbounded in the set $\alpha B \cap \{x; |x_1| \leq \beta\}$ for every $\alpha > 1$, then $\beta \geq 1$, i.e. the slab contains the unit ball.

REMARK 3.3 It can be shown that the number $(\gamma/(\gamma + 1))^{1/p}$ is best possible. For $1 < p < \infty$ one can give an alternative proof of Theorem 3.1 using Theorem 5.1. For $p = 1$, however, this is not possible.

4. THE INEQUALITY $|R(x) - R(y)| < ||x - y||$

We shall investigate when equality in the estimate

$$|R_u(x) - R_u(y)| \leq ||x - y||$$

can occur for $x \neq y$: in the classical coordinate spaces ℓ^p, $1 \leq p < +\infty$, and c_0 we can have equality only in ℓ^1 and c_0 for continuous functions u and only in ℓ^1 for plurisubharmonic functions u. The first result of this section therefore belongs to real functional analysis.

THEOREM 4.1 *Let* E *be a (real or complex) locally uniformly convex space, or, more generally, a normed space satisfying the weaker condition (4.2) below. Then for every upper semicontinuous function* $u: E \to [-\infty, +\infty[$ *with finite radius of boundedness* R_u *we have*

$$|R_u(x) - R_u(y)| < ||x - y||, \quad x, y \in E, \quad x \neq y.$$

A normed space is called *locally uniformly convex* if

(4.1) *for every* $x \in E$ *with* $||x|| = 1$ *and every* $\varepsilon > 0$ *there is a* $\delta > 0$ *such that for* $||y|| = 1$, $||x+y|| \geq 2-\delta$ *implies* $||x-y|| \leq \varepsilon$.

If δ can be chosen independently of x, E is called *uniformly convex.* We recall that the Banach spaces $\ell^p(J)$, $1 < p < \infty$, are uniformly convex. In any separable Banach space (Kadec [3]) or, more generally, in any weakly compactly generated Banach space (Troyanski [9]) there is an equivalent locally uniformly convex norm.

Now consider a weakened version of (4.1):

(4.2) *For every* $x \in E$ *with* $||x|| = 1$ *there is a compact set* K *such that for every* $\varepsilon > 0$ *there is a* $\delta > 0$ *such that for* $||y||=1$,

$$||x+y|| \geq 2 - \delta \quad \textit{implies} \quad y \in K + \varepsilon B.$$

Thus (4.1) is obtained from (4.2) by requiring K to be $\{x\}$. On the other hand, uniform c-convexity, a property introduced by Globevnik [1], is weaker than uniform convexity and does not imply (4.2).

PROOF OF THEOREM 4.1 Assume that the conclusion were false. By normalizing we may then assume that $R_u(0) = 1$ and that $R_u(\lambda x)=1-\lambda$ for some x with $||x|| = 1$ and some λ with $0 < \lambda < 1$. This means that u is bounded above in

$$\{y;\ ||y|| < 1 - 1/k\} \cup \{y;\ ||y-\lambda x|| < 1 - \lambda - 1/k\},$$

say $u \leq M_k$ there, and unbounded above in

$$\{y;\ ||y - \lambda x|| < 1 - \lambda + 1/k\}.$$

Take z_k in the last-mentioned ball such that $u(z_k) > M_k$, hence $||z_k|| \geq 1 - 1/k$ and $1 - \lambda - 1/k \leq ||z_k - \lambda x|| < 1 - \lambda + 1/k$. Also

$$||z_k|| \leq ||z_k - \lambda x|| + ||\lambda x|| \leq 1 - \lambda + 1/k + \lambda = 1+1/k,$$

so that $||z_k|| \to 1$. Putting

$$y_k = \frac{1}{1-\lambda} z_k - \frac{\lambda}{1-\lambda} x$$

we have

$$z_k = (1 - \lambda)y_k + \lambda x$$

and

$$||y_k|| = \frac{1}{1-\lambda} \quad ||z_k - \lambda x|| \to 1.$$

Thus $||x|| = 1$, $||y_k|| \to 1$ and $||z_k|| \to 1$ which implies that $||x+y_k|| \to 2$ for z_k is a convex linear combination of x and y_k with a fixed $\lambda \neq 0,1$. It now follows from (4.2) that there is a compact set K such that the distance $d(y_k,K)$ from y_k to K tends to zero; equivalently $d(z_k,K) \to 0$. Since u is bounded above in $K + \varepsilon B$ for some $\varepsilon > 0$ this contradicts the inequality $u(z_k) > M_k$ for k large, and this contradiction proves the theorem.

The spaces ℓ^1 and c_0 are not uniformly convex and the conclusion of Theorem 4.1 does not hold for them. In fact, we have the following two examples.

EXAMPLE 4.2 The entire function

$$f(x) = \sum_2^{\infty} e^{-kx_1} x_k^k, \quad x \in \ell^1,$$

has radius of convergence $R_f(x)$ satisfying

$$R_f(x) = 1 + \mathrm{Re}\, x_1 \quad \text{when} \quad \mathrm{Re}\, x_1 \geq 0,$$

$$R_f(x) = e^{\mathrm{Re}\, x_1} \quad \text{when } \mathrm{Re}\, x_1 \leq 0.$$

Thus $|R_f(x) - R_f(y)| = ||x-y||$ when x and y are positive multiples of $e_1 = (1,0,0, \dots)$.

EXAMPLE 4.3 Put $\phi(x) = (1 - ||x||)^+$, $x \in c_0$. The continuous function

$$g(x) = \sum_2^{\infty} k\, \phi(k(x - e_k)), \quad x \in c_0,$$

where e_k as usual is the k:th unit vector, is bounded for $|x_1| \geq \varepsilon > 0$. Its radius of boundedness satisfies

$$R(te_1) = \sup (|t|, 1),$$

in particular $R(te_1) = |t|$ when $|t| \geq 1$.

It is not by chance that the function g in the above example is only continuous. For plurisubharmonic functions the conclusion of Theorem 4.1 still holds:

THEOREM 4.4 *Let* $u \in PSH(c_0(J))$ *where* J *is an infinite index set, and suppose that* u *is unbounded on some bounded set. Then*

$$|R_u(x) - R_u(y)| < ||x - y||$$

for all $x, y \in c_0(J)$ *with* $x \neq y$.

PROOF It suffices to prove that $R_u(x) > 1 - ||x||$ if $0<||x||<1$ and $R_u(0) = 1$. For a fixed x, let J_x denote the set of those indices $j \in J$ for which $|x_j| = ||x||$, and put

$$\alpha = \sup_{j \notin J_x} |x_j| < ||x||.$$

We let τ denote the (non-separated) topology of convergence of all coordinates y_j, $j \in J_x$. The τ-local radius of boundedness was introduced in [3]; the definition is given following the statement of Theorem 2.2 of the present paper. By [4, Proposition 3.5] we must have

$$R_{\tau,u}(y) \geq R_u(0) = 1 \tag{4.3}$$

for all $y \in F$ with $||y|| < 1$, where we have written F for the subspace of $c_0(J)$ consisting of all points y with $y_j = 0$ for $j \notin J_x$. Since J_x is finite, the unit ball of $c_0(J)$ is τ-quasi-compact, so Proposition 3.7 of [4] yields

$$\inf_{\substack{||y||<1 \\ y \in F}} R_{\tau,u}(x + \lambda y) \leq \lambda \tag{4.4}$$

for all $\lambda > R_u(x)$. Now $R_{\tau,u}$ is obviously Lipschitz continuous with constant at most 1 in the variables y_j, $j \notin J_x$, so we get from (4.4), writing π for the canonical projection onto F, and noting that $||\pi(x)|| = ||x||$ and $||\pi(x) - x|| = \alpha < ||x||$,

$$\inf_{\substack{||z|| \leq ||x||+\lambda \\ z\varepsilon F}} R_{\tau,u}(z) \leq \inf_{\substack{||y|| \leq 1 \\ y\varepsilon F}} R_{\tau,u}(\pi(x)+\lambda y) \leq$$

$$\leq \inf_{\substack{||y|| \leq 1 \\ y\varepsilon F}} R_{\tau,u}(x+\lambda y) + \alpha \leq \lambda + \alpha.$$

Now suppose that $R_u(x) = 1 - ||x||$. We can then let $\lambda = 1-||x||+\delta$ where δ is any positive number. Hence

$$\inf_{\substack{||z|| < 1+\delta \\ z\varepsilon F}} R_{\tau,u}(z) \leq 1 - ||x|| + \delta + \alpha, \ \delta > 0,$$

or equivalently

$$\inf_{\substack{||z|| < 1+\delta \\ z\varepsilon F}} R_{\tau,u}(z) \leq 1 - ||x|| + \alpha < 1, \ \delta > 0,$$

whereas, by (4.3),

$$\inf_{\substack{||y|| < 1 \\ y\varepsilon F}} R_{\tau,u}(y) \geq 1.$$

This means that the function

$$\phi(t) = \sup_{\substack{||y|| < 1 \\ y\varepsilon F}} (-\log R_{\tau,u}(ty))$$

which is subharmonic in $t \in \mathbb{C}$, satisfies

$$\phi(1) \leq 0, \quad \phi(1+\delta) \geq -\log\,(1-||x||+\alpha) > 0$$

for all $\delta > 0$. But ϕ is a convex function of $\log |t|$ so it must, in particular, be continuous whenever it is finite. (It is thus essential to note that $\phi(t)$ is finite for $t \neq 0$). This contradiction shows that we must have $R_u(x) > 1 - ||x||$, i.e. the desired conclusion.

5. ESTIMATES FOR THE LIPSCHITZ CONSTANT OF A RADIUS OF BOUNDEDNESS

In spite of the strict inequality $|R_u(x) - R_u(y)| < ||x-y||$,

$x \neq y$, for all $u \in PSH(E)$ (such that $R_u < +\infty$) where E is any of the spaces $\ell^p(J)$, $1 < p < \infty$, or $c_0(J)$, there is no constant $A < 1$ such that $|R_u(x) - R_u(y)| \leq A||x-y||$ holds for all plurisubharmonic functions u. This follows from the existence theorems of §4 in [4], summarized in Theorem 6.1 of the present paper. We recall here another result from [4] which gives the best Lipschitz constant for plurisubharmonic functions in $\ell^p(J)$, $1<p<\infty$, and $c_0(J)$ whose radius of boundedness decays slowly at infinity. The following theorem is slightly more precise than Propositions 3.9 and 3.11 of [4].

THEOREM 5.1 *Let* $E = \ell^p(J)$, $1 < p < \infty$, *or* $E = c_0(J)$, *where* J *is an arbitrary index set, and let* F *be a coordinate subspace of* E. *If* $u \in PSH(E)$ *and for some* $C > 0$, $\gamma \geq 0$,

$$+\infty > R_u(x) \geq C||x||^{-\gamma} \text{ for } x \in F, \ ||x|| \geq r,$$

then the Lipschitz constant of R_u *in* F *is at most*

$$A = A(p,\gamma) = \left(\frac{\gamma}{\gamma+1}\right)^{1-1/p}$$

where $E = c_0$ *corresponds to* $p = \infty$ *and* $A(\infty,\gamma) = \gamma/(1+\gamma)$.

SKETCH OF PROOF Let G be a finite-dimensional coordinate subspace contained in F. It suffices to prove that

$$|R_u(x) - R_u(y)| \leq A||x-y||$$

for $x,y \in G$. In the proof of Proposition 3.9 of [4] we used the weak topology $\sigma(E,E')$ when $1 < p < \infty$. We now use instead the topology τ of convergence of all coordinates spanning G and let

$$v(t) = \sup_{\substack{x \in G \\ ||x|| \leq |t|}} (-\log R_{\tau,u}(x)), \quad t \in \mathbb{C}.$$

The result now follows as in [4], using the logarithmic convexity of v (cf. also the proof of Theorem 3.1). We omit the details.

REMARK 5.2 The conclusion of Theorem 5.1 is trivially true for $p = 1$, and gives then no information on the problem studied in Theorem 3.1. For $1 < p < \infty$, Theorem 5.1 can be deduced from Theorem 3.1, but not for $p = \infty$, $E = c_0$.

6. PRESCRIBING THE RADIUS OF CONVERGENCE

Let $R: E \to]0, +\infty[$ be a given function on a normed space E satisfying (1.1) and (1.2). Is it then possible to find an entire function on E such that its radius of convergence is equal to R? Results in that direction were given in [4]. The exact degree of arbitrariness of a radius of convergence depending on finitely many variables in $\ell^p(J)$ or $c_0(J)$ is shown by the following theorem.

THEOREM 6.1 *For any given* $p \in [1, +\infty]$ *let* $E = \ell^p$, $1 \leq p < +\infty$, *or* $E = c_0$, $p = +\infty$, *respectively. Let* ω *be a pseudoconvex open set in* $\mathbb{C}^{n+1}$ *containing all points* $z \in \mathbb{C}^{n+1}$ *with* $z_{n+1} = 0$. *Norming* $\mathbb{C}^{n+1}$ *by*

$$||z||_p = \left(\sum_1^{n+1} |z_j|^p\right)^{1/p} \text{or} \ ||z||_\infty = \sup |z_j|$$

(i.e. as a coordinate subspace of E*) we let* $d_p(z)$ *denote the distance from* $z \in \omega$ *to* $\partial\omega$. *Then there exists an entire function* f *on* E *such that its radius of convergence* R_f *satisfies*

$$R_f(x) = d_p(x_1, x_2, \ldots, x_n, 0), \ x = (x_1, x_2, \ldots) \in E.$$

Conversely, given any $u \in PSH(E)$ *such that* $R_u(x)$ *depends only on finitely many variables* $x_1, \ldots, x_n$ *there is a pseudoconvex open set* ω *in* $\mathbb{C}^{n+1}$ *such that* $R_u(x) = d_p(x_1,\ldots,x_n,0)$.

The proof is to be found in [4], Theorems 3.8 and 4.1. (For $1 \le p < \infty$, the last part of Theorem 6.1 holds without the finiteness condition).

There are two obvious restrictions in the first part of Theorem 6.1. On the one hand R_f factors through a finite-dimensional space; on the other hand the finitely many variables on which R_f may depend are coordinates. Let us consider a case where R_f depends on just one complex variable which is not necessarily a coordinate: $E = \ell^1$, $\xi(x) = \Sigma\, \xi_j x_j$ a linear form on ℓ^1, and R_f a function of $\xi(x)$ only. We note that the unit ball is compact for the weak star topology $\sigma(\ell^1, c_0)$ and quasi-compact for the topology generated by ξ, but not for the topology generated by c_0 and ξ together if $\xi \notin c_0$. On the other hand, the inner and outer moduli as defined in [4] do not agree for the topology defined by one linear form. Thus there seems to be no topology in ℓ^1 which is suitable for the results of [4], for these require the inner and outer moduli to be the same and the unit ball to be quasi-compact. In the absence of more general results it is therefore perhaps of interest to see what can be proved in this particular case which falls outside the scope of [4].

THEOREM 6.2 *Let* ξ *be a linear form on* ℓ^1, $\xi = (\xi_j)_1^\infty \in \ell^\infty$, *and introduce*

$$\alpha = \sup |\xi_j| = ||\xi||_{\ell^\infty} \text{ and } \beta = \limsup |\xi_j| = ||\xi||_{\ell^\infty/c_0}.$$

We assume that $\alpha > 0$. *Norm* $\mathbb{C}^2$ *by*

$$||z||_{\alpha,\beta} = \sup\left(|z_2|, \tfrac{1}{\alpha}|z_1| + \left(1 - \tfrac{\beta}{\alpha}\right)|z_2|\right).$$

Let ω *be any pseudoconvex open set in* $\mathbb{C}^2$ *containing all points*

z with $z_2 = 0$, and let $d_{\alpha,\beta}(z)$ be the distance, as measured by $||\cdot||_{\alpha,\beta}$, from $z \in \omega$ to $\partial\omega$. Then there exists $f \in \mathcal{O}(\ell^1)$ whose radius of convergence R_f satisfies

$$R_f(x) = d_{\alpha,\beta}(\xi(x),0), \quad x \in \ell^1.$$

I do not know whether this theorem can be improved, i.e. $||z||_{\alpha,\beta}$ replaced by a better norm: the ideal would be

$$||z||_{\alpha,0} = \alpha^{-1}|z_1| + |z_2|,$$

corresponding to $p = 1$ in Theorem 6.1. On the other hand, Theorem 6.2 is already much better than what could be expected from the behavior of the inner and outer moduli with respect to the topology generated by ξ. In fact these satisfy

$$m(x) = 1 - ||x||_1 \text{ for } ||x||_1 < 1, \text{ and}$$

$$M(x) = 1 + ||x||_1 \text{ for } ||x||_1 < 1, \ |\xi(x)| \leq \beta,$$

so the results of [4] only yield an entire function with radius of convergence satisfying an inequality

$$d'(\xi(x),0) \leq R_f(x) \leq d''(\xi(x),0)$$

where $$d'(z_1,0) < d''(z_1,0).$$

PROOF OF THEOREM 6.2 Let $h \in \mathcal{O}(\omega)$ be a function which cannot be continued anywhere beyond the boundary of ω (i.e. for which ω is the domain of existence; see e.g. Hörmander [2, Theorem 4.2.8]) and expand h in a partial Taylor series around $(z_1,0) \in \omega$:

$$h(z) = \sum_0^\infty h_k(z_1)z_2^k, \quad z_1 \in \mathbb{C}, \ |z_2| \text{ small.}$$

Then $h_k \in \mathcal{O}(\mathbb{C})$ and

$$(6.1) \qquad (\limsup |h_k|^{1/k})^*(z_1) = \frac{1}{d(z_1,0)}, \quad z_1 \in \mathbb{C},$$

where $d(z)$ denotes the distance from z to $\partial\omega$ in the direction $(0,1)$. Define

$$(6.2) \qquad f(x) = \sum_0^\infty h_k(\xi(x))x_{n_k}^k, \quad x \in \ell^1,$$

where $(n_k)_0^\infty$ is a strictly increasing sequence of integers such that

$$|\xi_{n_k}| \to \beta \quad \text{as} \quad k \to +\infty.$$

We shall also need a sequence $(m_k)_0^\infty$ (tending to infinity or constant) such that

$$|\xi_{m_k}| \to \alpha \quad \text{as} \quad k \to +\infty,$$

and we may obviously pick (n_k) and (m_k) so that $m_k \neq n_j$ for all k and j.

We shall first prove that $R_f(a) \geq d_{\alpha,\beta}(\xi(a),0)$. Let $r < R < d_{\alpha,\beta}(\xi(a),0)$. It is then enough to show that f is bounded in $a + rB$. Now $r < R < d_{\alpha,\beta}(\xi(a),0)$ implies that

$$d(z_1,0) > R \quad \text{when} \quad |z_1 - \xi(a)| \leq \beta r, \tag{6.3}$$

and, provided $\beta < \alpha$ and r is close enough to R,

$$d(z_1,0) > \frac{\alpha R - |z_1 - \xi(a)|}{\alpha - \beta} \quad \text{when} \quad \beta r \leq |z_1 - \xi(a)| \leq \alpha r. \tag{6.4}$$

If $\beta = \alpha$, it is enough to consider (6.3) only. In view of the so called Hartogs' lemma (Hörmander [2, p.21]), (6.1) and (6.3) imply that for large k, writing $x = a + ry$, we have

$$|h_k(\xi(x))| = |h_k(\xi(a) + r\xi(y))| \leq \frac{1}{R^k} \quad \text{when} \quad |\xi(y)| \leq \beta,$$

and similarly we obtain from (6.1) and (6.4) for large k:

$$|h_k(\xi(x))| = |h_k(\xi(a) + r\xi(y))| \leq \left(\frac{\alpha - \beta}{\alpha R - r|\xi(y)|}\right)^k \text{when} \quad \beta \leq |\xi(y)| \leq \alpha.$$

To estimate f we multiply these inequalities by

$$x_{n_k}^k = (a_{n_k} + ry_{n_k})^k$$

and note that a_{n_k} tends to zero; thus for large k

$$|h_k(\xi(x))x_{n_k}^k| \leq \left(\frac{r+\varepsilon}{R}\right)^k \quad \text{when} \quad |\xi(y)| = |\xi(\tfrac{x-a}{r})| \leq \beta \tag{6.5}$$

and

$$|h_k(\xi(x))x_{n_k}^k| \leq \left[\frac{(\alpha-\beta)(r|y_{n_k}| + \varepsilon)}{\alpha R - r|\xi(y)|}\right]^k \quad \text{when} \quad \beta \leq |\xi(y)| \leq \alpha. \tag{6.6}$$

To use just $|y_{n_k}| \leq 1$ would not be enough in (6.6) so we observe that

$$(\alpha - \beta - \varepsilon)|y_n|+|\xi(y)| \leq (\alpha - \beta - \varepsilon + |\xi_n|)|y_n| + \sum_{j\neq n} |\xi_j||y_j| \leq$$

$$\leq \alpha|y_n| + \sum_{j\neq n} |\xi_j||y_j| \leq \alpha \sum |y_j| \leq \alpha$$

provided $|\xi_n| \leq \beta + \varepsilon$ (true for large n) and $||y|| \leq 1$. Thus (6.6) gives, if $\beta < \alpha$ and k is large,

$$|h_k(\xi(x))x_{n_k}^k| \leq \left(\frac{\alpha - \beta}{\alpha - \beta - \varepsilon} \cdot \frac{\alpha r - r|\xi(y)| + \alpha\varepsilon}{\alpha R - r|\xi(y)|}\right)^k \text{ when } \beta \leq |\xi(y)| \leq \alpha.$$

Now the right-hand side in this inequality assumes its maximum for $|\xi(y)| = \beta$ (if $r + \varepsilon \leq R$, which we assume), so we have

$$(6.7) \quad |h_k(\xi(x))x_{n_k}^k| \leq \left(\frac{(\alpha-\beta)(\alpha r-\beta r+\alpha\varepsilon)}{(\alpha-\beta-\varepsilon)(\alpha R-\beta r)}\right)^k = q^k, \text{ when } \beta \leq |\xi(y)| \leq \alpha,$$

where $q < 1$ for a suitable choice of $\varepsilon > 0$. Now (6.5) and (6.7) show that series defining f converges uniformly for $x \in a + rB$, hence that $R_f(a) \geq r$.

Conversely, assume that f is bounded in $a + RB$. We shall then prove that $d_{\alpha,\beta}(\xi(a),0) \geq R$, which will yield $d_{\alpha,\beta}(\xi(a),0) \geq$ $\geq R_f(a)$. For $x = a + t'R'e_{n_k} + t''R''e_{m_k}$, where e_n denotes as usual the unit vector of index n, and where $|t'| \leq 1, |t''| \leq 1$, $0 \leq R' \leq R$ and $R' + R'' = R$ we get

$$(6.8) \quad f(x) = h_k(\xi(a) + t'R'\xi_{n_k} + t''R''\xi_{m_k})(a_{n_k} + t'R')^k + \sum_{j\neq k} h_j(\xi(x))a_{n_j}^j.$$

Now $|\xi(x)| \leq |\xi(a)| + R'\alpha + R''\alpha = |\xi(a)| + R\alpha$, and for $|z_1| \leq$ $\leq |\xi(a)| + R\alpha$ we have an estimate of the form

$$|h_j(z_1)| \leq C^{j+1}$$

for some constant C. Since a_{n_j} tends to zero this shows that the sum in (6.8) is bounded. It follows that the first term to the right in (6.8) must be bounded for $x \in a + RB$, hence, for k so large that $|a_{n_k}| \leq \varepsilon$,

$$\sup_{\substack{|t'|<1 \\ |t''|\leq 1}} |h_k(\xi(a) + t'R'\xi_{n_k} + t''R''\xi_{m_k})|^{1/k}(R' - \varepsilon) \leq M^{1/k},$$

or equivalently

$$\sup_{|t|\leq 1} |h_k(\xi(a) + tR_k)|^{1/k} \leq \frac{M^{1/k}}{R' - \varepsilon}$$

where $R_k = R'|\xi_{n_k}| + R''|\xi_{m_k}|$. Let r be an arbitrary positive number smaller than $R'\beta + R''\alpha$ if the latter is positive, and let $r = R'\beta + R''\alpha = 0$ if $R'\beta + R''\alpha = 0$ (i.e. $R'' = \beta = 0$). Then $R_k \geq r$ for large k so that

$$\sup_{|t|\leq 1} |h_k(\xi(a) + tr)|^{1/k} \leq \frac{M^{1/k}}{R' - \varepsilon}$$

which in view of (6.1) gives

$$\inf_{|t|< 1} d(\xi(a) + tr, 0) \geq R' - \varepsilon.$$

Since r is arbitrarily close to $R'\beta + R''\alpha$ and $R' - \varepsilon$ is arbitrarily close to R' this means that $d_{\alpha,\beta}(\xi(a),0) \geq R' + R'' = R$. Indeed, with $R'' = 0$ we have $R' = R$ and

$$\inf_{|t|< 1} d(\xi(a) + tr,0) \geq R \tag{6.9}$$

where r is arbitrarily close to $R\beta$; this is what we need if $\beta = \alpha$. If $\beta < \alpha$ we have in addition to (6.9):

$$\inf_{|t|< 1} d(\xi(a) + tr,0) \geq R' = \frac{R\alpha - r - \delta}{\alpha - \beta} \tag{6.10}$$

where we have defined δ by the equation $r = R'\beta + R''\alpha - \delta$; now (6.9) and (6.10) together mean that $d_{\alpha,\beta}(\xi(a),0) \geq R$.

We can thus note that in this special situation at least, the technique used in proving Theorem 4.1 in [4] survives, in spite of the fact that there is no useful relation between the radius of boundedness and the local radius of boundedness. Indeed, here the boundary distance $d(\xi(a),0)$ serves the same purpose as the local radius of boundedness in [4], but we can only define it for functions of the special form (6.2).

REFERENCES

[1] GLOBEVNIK, J., On complex strict and uniform convexity. *Proc. Amer. Math. Soc.* 47, 175-178 (1975).

[2] HÖRMANDER, L., *An introduction to complex analysis in several variables*. North-Holland, 1973.

[3] KADEC, M.I., Spaces isomorphic to a locally uniformly convex space. *Izv.Vysš.Učebn.Zaved. Matematika* no. 6 (13), 51-57 (1959), and correction no.6 (25),186-187 (1961).

[4] KISELMAN, C.O., On the radius of convergence of an entire function in a normed space. Ann. Polon. Math. 33 (1976), 39 - 55.

[5] LELONG, P., Fonctions plurisousharmoniques dans les espaces vectoriels topologiques. *Lecture Notes in Mathematics* 71 (1968), 167-190.

[6] MARTINEAU, A., *Oral communication*, 1968.

[7] NACHBIN, L., *Topology on spaces of holomorphic mappings*. Springer-Verlag, 1969.

[8] NOVERRAZ, P., *Pseudo-convexité, convexité polynomiale et domaines d'holomorphie en dimension infinie*. North-Holland, 1973.

[9] TROYANSKI, S.L., On locally uniformly convex and differentiable norms in certain non-separable Banach spaces. *Studia Math.* 37, 173-180 (1970-71).

C. O. Kiselman
Dept. of Mathematics
Sysslomansgatan 8
S- 752 23 Uppsala, Sweden

Infinite Dimensional Holomorphy and Applications, Matos (ed.)

HOLOMORPHIE ET THEORIE DES DISTRIBUTIONS EN DIMENSION INFINIE*

Par *PAUL KRÉE*

Que peut être la théorie des distributions sur un e.ℓ.c.s. réel Y de dimension infinie? Peut-elle servir dans l'étude de l'holomorphie en dimension infinie?

La transposition de la théorie développée il y a une trentaine d'années par L. Schwartz n'a pas été immédiate, peut être parce que cette transposition fait apparaître simultanément de nombreux phénomènes spécifiques à la dimension infinie. Par exemple, si ϕ est une fonction ℓ fois Frechet dérivable sur Y, la dérivée $D^{\ell}\phi(y)$ ne prend pas ses valeurs dans $\underset{\ell}{\otimes}(Y'^{c})$, où Y'^{c} désigne le complexifié du dual de Y; mais dans un espace vectoriel plus grand. Il existe d'ailleurs en dimension infinie d'autres calculs différentiels que celui de Frechet, par exemple les calculs différentiels de Gateaux et de L. Gross. Ainsi, la première question qui se pose est de choisir un calcul différentiel pour définir les fonctions d'épreuve. Il est naturellement préférable de n'en choisir aucun à *priori* ce qui motive la démar-

---

* Conférence à Campinas (Août 1975).

che suivante [7]. Soit Y un espace de Banach et soit N un certain type de normes tensorielles, par exemple π, ε..... . Pour $\ell = 0, 1, 2...$ le complété de $\otimes(Y'^c)$ pour la norme N est noté E^N_ℓ. Au départ, toutes les fonctions d'épreuve considérées sont supposées cylindriques, ce qui permet d'utiliser seulement le calcul différentiel de Newton et de Leibnitz. Pour chaque choix de la famille (E^N_ℓ), une théorie des distributions est développée en utilisant le formalisme des prodistributions et des protenseurs distributions [8]. Les opérations sont définies par des arguments de transposition analogues à ceux de la théorie de L. Schwartz. Puis, en utilisant un argument de bitransposition, à chaque choix d'une famille $(E^N_\ell)_\ell$, il est associé un calcul différentiel généralisé sur Y. (Les fonctions dérivables de cette théorie sont supposées universellement Lusin mesurables, mais elles ne sont pas supposées continues). Ainsi, le calcul différentiel de Frechet est généralisé en prenant $E^N_\ell = \hat{\otimes}_\ell Y'^c$, le calcul différentiel de L. Gross est prolongé en prenant un triplet de Wiener $Y' \subset X \subset Y$ et $E^N_\ell = \hat{\hat{\otimes}}_\ell X^c$... ce qui donne déjà une application de la théorie des distributions.

Dans la conférence au colloquium, il a été tenté de donner un aperçu d'ensemble de la théorie. Ceci est assez long car le seul concept de distributions relatif à la dimension finie se diversifie en dimension infinie, en six concepts différents:

- prodistributions
- protenseurs distributions
- espaces de Sobolev scalaires
- espaces de Sobolev vectoriels
- distributions

- tenseurs distributions

Ces deux dernières théories ont été développées jusqu'ici seulement dans le cadre banachique [5] [8]. Et vue la diversité des normes tensorielles *naturelles* sur $\underset{\ell}{\otimes} Y'^c$, il apparaît encore des sous-classes. Ainsi dans le cas d'un triplet de Wiener, Y étant supposé hilbertien, il apparaît trois types d'espaces de Sobolev scalaires et neuf types de distributions scalaires sur Y !

Le texte qui suit est limité à l'exposé de résultats relatifs à deux cas particuliers qui sont reliés à l'holomorphie; il s'agit de l'étude des profonctionnelles analytiques, des fonctionnelles analytiques (avec une application à la physique); et de l'étude des distributions et du calcul différentiel sur des e.ℓ.c.s. Cette dernière étude est motivée par le résultat de Grothendieck montrant la coincidence des topologies π et ε sur $\underset{\ell}{\otimes} Y'^c$ si Y'^c est nucléaire, d'où un seul calcul différentiel raisonnable dans ce cas; alors qu'il y a beaucoup de calculs différentiels si Y est un espace de Banach non nucléaire. Il est raisonnable de penser que ceci est lié au fait que l'holomorphie pour les espaces à dual nucléaire semble se formuler plus facilement que l'holomorphie concernant les espaces de Banach de dimension infinie.

1. FONCTIONNELLES ANALYTIQUES DE TYPE EXPONENTIEL

Ce paragraphe présente une extension de la théorie topologique des mesures de Radon aux fonctions entières de type exponentiel. Cette extension est motivée par la mécanique quantique.

(1.1) NOTATIONS Si Y est un espace vectoriel et si θ est une topologie localement convexe sur Y, (Y, θ) désigne l'espace localement convexe correspondant et $(Y, \theta)'$ désigne le dual topologique de (Y, θ). Si X désigne un espace topologique complètement regulier, $B^0(X)$ désigne l'espace des fonctions continues bornées $\phi : X \to \mathbb{C}$. Soit β la boule unité de $B^0(X)$.

(1.2) $$\beta = \{\phi \in B^0(X) \; ; \; ||\phi||_\infty = \sup |\phi(x)| \leq 1\}$$

L'espace M(X) des mesures de Radon bornées sur X est l'espace des mesures complexes bornées m sur la tribu borélienne de X, telles que pour tout $\varepsilon > 0$, il existe une partie compacte K de X telle que $|m|(X \setminus K) \leq \varepsilon$. Soit t_k la topologie sur $B^0(X)$ de la convergence uniforme sur les parties compactes de X. La topologie stricte τ sur $B^0(X)$ est la topologie localement convexe la plus fine sur $B^0(X)$ qui coincide avec t_k sur β. On sait [2] [3] que $M(X) = (B^0(X), \tau)'$.

(1.3) INTRODUCTION DE POIDS Des topologies strictes seront définies sur des espaces de fonctions continues, de manière à induire des topologies sur certains espaces de fonctions entières. Comme une fonction entière bornée est constante, des poids sont introduits de manière à permettre une certaine croissance à l'infini.

Soit Z un espace de Banach réel et soit m un entier positif. L'espace $C\,Exp^m(Z)$ est l'espace des fonctions continues ϕ sur Z telles que

(1.4) $$||\phi||_m = \sup |\phi(z)| \exp(-m\,||z||) < \infty$$

Cet espace a une boule unité $\beta^m = \{\phi, ||\phi||_m \leq 1\}$, une topologie t_k, d'où une topologie stricte τ^m. Le dual $M\,Exp'^m(Z)$ de $(C\,Exp^m(Z), \tau^m)$ est l'espace des mesures de Radon μ sur Z telles que

(1.5) $$\int \exp (m \, ||z||) \; d \, |\mu| \, (z) < \infty$$

L'espace C Exp(Z) = $\bigcup_m$ (C $Exp^m(Z)$) est l'espace des fonctions continues sur Z, à croissance exponentielle. Il peut être muni de la topologie $\theta = \lim_{\to} \tau^m$. Et le dual M Exp'(Z) de (C Exp(Z),θ) est l'espace des mesures de Radon μ à décroissance exponentielle, c'est-à-dire des mesures $\mu \in M(X)$ qui vérifient (1.5) pour tout $m > 0$.

(1.6) FONCTIONNELLES ANALYTIQUES DE TYPE EXPONENTIEL *Soit* Z *un espace de Banach complexe, et soit* Exp(Z) *le sous-espace vectoriel topologique de* (C Exp(Z), θ) *formé par les fonctions entières à croissance exponentielle. Le dual* Exp'(Z) *de* Exp(Z) *est appelé l'espace des fonctionnelles analytiques de type exponentiel sur* Z.

Soit θ' la trace de θ sur Exp(Z). Une application du théorème de Hahn Banach donne immédiatement la:

(1.7) CARACTÉRISATION DE Exp'(Z) *Soit* T *une forme linéaire définie sur un sous-espace dense de* (Exp(Z), θ')*.Alors* $T \in$ Exp'(Z) *si et seulement si* T *est représentable par une mesure de Radon à décroissance exponentielle.*

(1.8) TRANSFORMATION DE FOURIER (T.F.) Pour tout $\zeta \in Z'$, la fonction $e_\zeta : z \to \exp(-\sqrt{-1}\, z\, \zeta)$ appartient à Exp ζ, où $z\,\zeta$ désigne la forme bilinéaire de dualité entre Z et Z'. Alors $\mathcal{F}$T est par définition la fonction suivante sur Z' :

$$(\mathcal{F}T)\;(\zeta) = \int_Z e^{-\sqrt{-1}\, z\, \zeta}\, dT(z)$$

où l'intégrale désigne symboliquement le résultat de l'action de T sur e_ζ.

(1.9) HYPOTHÈSE (H) *On dit que l'espace de Banach* Z *vérifie l'hypothèse* (H) *s'il existe une suite* (u_n) *d'opérateurs de rang fini de* Z, *de normes uniformément majorées, tendant simplement vers l'opérateur identique de* Z.

(1.10) LEMME DE DENSITÉ *Si l'espace de Banach* Z *vérifie l'hypothèse* (H), *alors* $\mathrm{Exp}_{cyl}(Z)$ *et* $\mathrm{Pol}_{cyl}(Z)$ *sont des sous-espaces denses de* Exp(Z).

Par définition $\mathrm{Exp}_{cyl}(Z)$ et $\mathrm{Pol}_{cyl}(Z)$ sont les sous-espaces de Exp(Z) constitués respectivement par les fonctions cylindriques et par les fonctions polynomiales cylindriques.

PREUVE a) Pour $f \in \mathrm{Exp}(Z)$, soit $f_n(z) = f(u_n z)$. Alors il vient pour tout n et pour tout $z \in Z$:

$$|f_n(z))| \leq C\, e^{m\, ||u_n(z)||} \leq C \;\mathrm{esp}(C'\, ||z||)$$

De plus d'après [12] par exemple, la suite (f_n) converge vers f uniformément sur toute partie compacte de Z. Donc (f_n) converge vers f pour θ'.

b) Vu a), il suffit de montrer la densité de Pol(Z)dans Exp(Z) si Z est un espace vectoriel de dimension finie. Il suffit d'approcher f par les sommes partielles de son développement de Taylor.

(1.11) PROPOSITION *La transformée de Fourier de toute fonctionnelle analytique de type exponentiel est entière. La transformation de Fourier est injective si l'espace de Banach* Z *vérifie* (H).

DÉMONSTRATION a) D'après (1.7), il existe $\mu \in M\,\mathrm{Exp}(Z)$ telle que

$$\hat{T}(\zeta) = (\mathcal{F}\, T)\ (\zeta) = \int e^{-iz\zeta}\, d\mu(z)$$

La fonction $\hat{T}$ est bornée sur les bornés de Z' car

$$|\hat{T}(\zeta)| \leq \int e^{||z|| \cdot ||\zeta||}\, d|\mu|(z)$$

Soit $\phi(\zeta) = -\, i \int z\, e^{-iz\zeta}\, d\mu(z)$

Il vient $A = \hat{T}(\zeta+\Delta\zeta) - \hat{T}(\zeta) - \phi(\zeta)\,\Delta\,\zeta = \int e^{-iz\zeta}(e^{-iz\zeta} - 1 + i\, z\,\Delta\,\zeta)\, d\mu(z)$

Or pour tout nombre complexe u

$$|e^{u} - 1 - u| \leq \frac{|u|^2}{2}\, e^{|u|}$$

Donc $|A| \leq \dfrac{||\Delta\,\zeta||^2}{2} \displaystyle\int e^{||z||(||\zeta|| + ||\Delta\,\zeta||)}\, d|\mu|\ (z)$

Ceci prouve que $\hat{T}$ est $\mathbb{C}$-dérivable; et $\hat{T}$ est entière et de plus $D\,\hat{T}(0) = \phi(0)$.

b) Par récurrence sur l'entier k, on peut montrer:

$$D^k\, \hat{T}(0) = (-\, i)^k \int (\underset{k}{\otimes}\, z)\ d\mu(z)$$

Si donc $\hat{T}$ est nulle, T s'annule sur les fonctions polynomiales cylindriques. Vu le lemme de densité, T est identiquement nulle.

(1.12) IMAGE PAR UNE APPLICATION LINÉAIRE Soient Z et U deux espaces de Banach complexes et soit λ une application linéaire continue de Z dans U. L'application linéaire:

$$\mathrm{Exp}(U) \xrightarrow{\ \alpha\ } \mathrm{Exp}(Z)$$

$$\phi \longrightarrow \phi \circ \lambda$$

est continue. Par transposition, une application α' de Exp'(Z) dans Exp'(U) est définie. Pour toute $T \in$ Exp'(Z), son image par λ est définie par $\lambda\, T = \alpha'(T)$, soit

$$\forall\, \psi \in \mathrm{Exp}(U) \qquad <\lambda\, T,\ \psi> \;=\; <T,\ \psi \circ \lambda>$$

Ceci entraine la relation suivante entre les transformées de Fourier de T et de λ T

$$(\mathcal{F}(\lambda\ T))(\zeta) = (\mathcal{F}\ T)(\lambda'\ \zeta)$$

où λ' désigne la transposée de λ.

(1.13) PRODUIT PAR UN ÉLÉMENT $\phi \in \mathrm{Exp}(Z)$ Comme en théorie des distributions, le produit ϕ T est défini par

$$< \phi\ T,\ \psi > = < T,\ \phi\ \psi >$$

pour toute ψ de Exp(Z)

(1.14) PRODUIT TENSORIEL Soient Z^1 et Z^2 deux expaces de Banach complexes. Alors le produit tensoriel des deux formes linéaires associées à $T_1 \in \mathrm{Exp}'(Z')$ et à $T_2 \in \mathrm{Exp}'(Z^2)$ est une forme linéaire λ sur le sous-espace $E = \mathrm{Exp}(Z^1)\otimes \mathrm{Exp}(Z^2)$ de Exp (Z) avec $Z = Z^1 \times Z^2$. Il est supposé que Z vérifie (H); il en est par exemple ainsi si Z^1 et Z^2 vérifient (H). Alors E est un sous-espace dense de Exp(Z), car si $T \in \mathrm{Exp}'(Z)$ est orthogonale à E, alors $\mathcal{F}T = 0$, ce qui entraine l'annulation de T d'après (1.1). De plus, si $\mu_j \in M\ \mathrm{Exp}(Z^j)$ représente T_j, alors $\mu_1 \otimes \mu_2 \in M\ \mathrm{Exp}(Z)$ représente λ. Donc d'après (1.7), $\lambda \in \mathrm{Exp}'(Z)$. Il est posé $\lambda = T_1 \otimes T_2$.

(1.15) FORMULE DE FUBINI L.SCHWARTZ Si Z^1 et Z^2 vérifient (H), alors

$$\iint \phi(x,y)\ d\ (T_1 \otimes T_2) = \int_{Z^1} d\ T_1(x) \int_{Z^2} \phi(x,y)\ d\ T_2(y)$$

pour toute $\phi \in \mathrm{Exp}(Z)$.

En effet:

Soient respectivement (u_k) et (v_ℓ) des suites d'opérateurs de rang fini de Z^1 et de Z^2, de norme au plus un, tendant vers l'identité uniformément sur toute partie compacte. Alors

$$\iint \phi(u_k\, x, v_\ell\, y)\, d\,(T_1 \otimes T_2) = \int_{Z^1} d\, T_1(x) \int_{Z^2} \phi(u_k\, x, v_k\, y) d\, T_2(y)$$

Il suffit alors de faire vendre k et ℓ vers $+\infty$.

(1.16) ILLUSTRATION Soit E un espace de Banach réel et soit $E^c = E + \sqrt{-1}\, E$ le complexifié de E. Soit m une mesure de Radon sur E à décroissance exponentielle et soit $m' = m(X) \otimes \delta_o(y)$ l'extension de m à E^c. Soit θ un angle quelconque, et soit $R(\theta)$ la rotation d'angle θ dans E^c

$$E^c \longrightarrow E^c$$

$$z = x + \sqrt{-1}\, y \longrightarrow z\, e^{i\theta}$$

Alors cette transformation $\mathbb{C}$ - linéaire de E^c transforme m' en une fonctionnelle analytique $R(\theta)m$ de transformée de Fourier $\hat{m}(e^{2i\theta} z)$. Ceci prolonge en dimension infinie un résultat bien connu dans le cas particulier où

$$E = \mathbb{R}\,;\ m = \pi^{-1/2} \exp(-\frac{x^2}{2})\, dx\ ;\quad \theta = \frac{\pi}{2}$$

En effet, dans le tome 1 de leur traité, Gelfand et Silov ont noté que la fonction $\exp \frac{x^2}{2}$ est la transformée de Fourier d'un élément de $Z'(\mathbb{R})$ représenté par l'extension au plan complexe d'une mesure gaussienne sur l'axe imaginaire.

(1.17) Il faut noter, même dans ce cas particulier, que $R(\theta)m$ agit non seulement sur les éléments de $Exp(E^c)$, mais sur des fonctions ϕ beaucoup plus générales. En effet, il suffit que la trace de ϕ sur la variété linéaire de E^c déduite de E par $R(\theta)$ puisse être définie; et que cette trace soit intégrable par rapport $R(\theta)m$.

2. PROFONCTIONNELLES ANALYTIQUES DE TYPE EXPONENTIEL

Le concept de promesure généralise le concept de mesure de Radon. De la même manière, la notion de fonctionnelle analytique de type exponentiel est prolongée par la notion de profonctionnelle analytique.

(2.1) NOTATION Soit Z un espace localement convexe séparé complexe et soit $F_m(Z) = (A_i)_{i \in I}$ la famille des sous-espaces fermés de codimension finie de Z, ordonnée par l'ordre inverse de celui défini par la relation d'inclusion: $i \geq j$ signifie $A_i \subset A_j$. Soit s_{ij} la surjection canonique de $Z_i = Z / A_i$ sur $Z_j = Z / A_j$, définie si $i \geq j$. Soit s_i la surjection de Z sur Z_i. Pour $i \geq j$, on a une injection de $\mathrm{Exp}(Z_j)$ dans $\mathrm{Exp}(Z_i)$, et la limite inductive $\mathrm{Exp}_{cyl}(Z)$ de ces espaces s'identifie à un espace de fonctions cylindriques sur Z.

(2.2) DÉFINITION DE $\mathrm{Exp}'_{cyl}(Z)$ *Une profonctionnelle analytique de type exponentiel* T *sur* Z *est une forme linéaire sur* $\mathrm{Exp}_{cyl}(Z)$ *dont la restriction à chaque* $\mathrm{Exp}(Z_i)$ *est représentée par une fonction réelle* $T_i \in \mathrm{Exp}'(Z_i)$. *L'ensemble de ces formes linéaires est noté* $\mathrm{Exp}'_{cyl}(Z)$.

Pour $\psi = \psi_i \circ s_i \in \mathrm{Exp}_{cyl}(Z)$, on écrit

$$(2.3) \qquad \langle T, \psi \rangle = \int_Z \psi(z)\, dT(z)$$

D'une manière équivalente, $\mathrm{Exp}'_{cyl}(Z)$ est l'espace des familles $T = (T_i)_i$ avec $T_i \in \mathrm{Exp}'(Z_i)$, ces familles étant cohérentes au sens suivant:

$$(2.4) \qquad i \geq j \qquad T_j = s_{ij}(T_i)$$

D'une manière équivalente, on peut munir pour tout i, l'espace

$Exp(Z_i)$ de la trace θ_i' de la topologie "stricte" sur $C\ Exp(Z_i)$, puis $Exp_{cyl}(Z)$ peut être muni de la topologie $\varinjlim \theta_i'$. Alors $Exp'_{cyl}(Z)$ apparaît comme le dual de l'espace $(Exp_{cyl}(Z), \varinjlim \theta_i')$.

(2.5) INJECTION DE Exp'(Z) dans $Exp'_{cyl}(Z)$ a) L'injection naturelle

$$(2.6) \qquad (Exp_{cyl}(Z),\ \lim \theta_i') \xhookrightarrow{J} (Exp(Z),\ \theta')$$

est continue. En effet, vue la propriété universelle des limites inductives (P U L I), il suffit de montrer que pour tout $i \in I$, l'injection

$$(2.7) \qquad (Exp(Z_i),\ \theta_i') \xrightarrow{J_i} (Exp(Z),\ \theta')$$

est continue. Il suffit de montrer la continuité de

$$(C\ Exp(Z_i),\ \theta_i) \hookrightarrow (C\ Exp(Z),\ \theta')$$

Vue P U L I, il suffit de montrer la continuité de

$$(C\ Exp^m(Z_i),\ Z_i^m) \xhookrightarrow{J_i^m} (C\ Exp(Z),\ \theta')$$

Vue P U L I, il suffit de montrer la continuité de la restriction de J_i^m à la boule unité de $C\ Exp^m(Z_i)$. Vu un lemme de Grothendieck il suffit de montrer la continuité à l'origine; et ceci est clair.

b) Si Z est un Banach vérifiant (H), alors J a une image dense d'après (1.10) Par conséquent, dans ce cas, la transposée de J est une injection de Exp'(Z) dans $Exp'_{cyl}(Z)$. Les éléments de $Exp'_{cyl}(Z)$ appartenant à Im J sont caractérisés par (1.7).

(2.8) OPÉRATIONS USUELLES SUR LES PROFONCTIONNELLES ANALYTIQUES DE TYPE EXPONENTIEL La transformation de Fourier, l'image par une application linéaire continue, la produit tensoriel la convolution se définissent naturellement. Par exemple la T.F. de

$T \in Exp'_{cyl}(Z)$ est la fonction suivante définie sur le dual Z' de Z

$$(2.9) \qquad \hat{T}(\zeta) = \int e^{-\sqrt{-1}\, z\, \zeta}\, dT(z)$$

Soit λ une application linéaire continue de Z dans l'espace localement convexe complexe U. L'application $\psi \longrightarrow \psi \circ \lambda$ applique $Exp_{cyl}(U)$ dans $Exp_{cyl}(Z)$. Et l'image λT de $T \in Exp'_{cyl}(Z)$ est définie par

$$(2.10) \qquad \forall\, \psi \in Exp_{cyl}(U) \qquad < \lambda\ T,\ \psi > = < T, \psi \circ \lambda >$$

Pour définir la produit tensoriel $T = T^1 \otimes T^2$ des $T^j \in Exp'_{cyl}(Z^j)$, on introduit d'abord les systèmes projectifs $(Z^1_k)_k$ et $(Z^2_\ell)_\ell$ associés respectivement à $Z^1 \times Z^2$. On veut définir

$$(2.11) \qquad < T,\ \psi > = < T^1 \otimes T^2,\ \psi >$$

pour toute $\psi \in Exp_{cyl}(Z)$ avec $Z = Z^1 \times Z^2$.Comme ψ admet une base du type $Z^1_k \times Z^2_\ell$, ψ admet la factorisation:

$$(2.12) \qquad \psi = \psi_{k\ell} \circ (s^1_k \times s^2_\ell) \qquad \psi_{k\ell} \in Exp\ (Z^1_k \times Z^2_\ell)$$

Il suffit de poser $< T\ \ \psi > = < T^1_k \otimes T^2_\ell,\ \psi_{k\ell} >$

Le produit de convolution $T * U$ de $T \in Exp'_{cyl.}(Z)$ et $U \in Exp'_{cyl}(Z)$ est défini comme étant l'image de $T \otimes U$ par l'application somme $x, y \longrightarrow x + y$ de $Z * Z$ dans Z.

(2.13) RADONIFICATION DES PROFONCTIONNELLES ANALYTIQUES L'injection canonique

$$(Exp_{cyl}(Z),\ \lim_{\rightarrow} \theta'_i) \hookrightarrow (C\ Exp_{cyl}(Z),\ \lim_{\rightarrow} \theta_i)$$

est continue. Donc toute promesure à décroissance exponentielle μ sur Z définit canoniquement une profonctionnelle analytique $\tilde{\mu}$ de type exponentiel. Soit λ une application linéaire continue de Z dans un Banach U transformant μ en une mesure de Radon à décroissance exponentielle. Alors $\lambda(\tilde{\mu})$ est une fonctionnelle analytique de type exponentiel d'après (1.7). Ainsi, les résultats de la théorie des triplets de Wiener de L. Gross, ou de la théorie des applications radonifiantes de L. Schwartz permettent de montrer que certaines profonctionnelles analytiques sont transformées par certaines applications linéaires en des fonctionnelles analytiques de type exponentiel.

(2.14) APPLICATIONS Le fait que le changement du temps t en $\sqrt{-1}$ t permet de passer de l'équation de diffusion de certains processus de Markov à des équations du type Schrödinger, entraine une analogie entre théorie des diffusions et mécanique quantique. Cette analogie conduit à rechercher l'analogue quantique de la mesure de Wiener. On rappelle que cette mesure P s'obtient par radofification de la promesure normale canonique ν de l'espace de Hilbert réel:

$$(2.15) \qquad X = \{\phi(t) = \int_0^t g(\theta)\, d\theta \quad ; \quad g \in L^2(I)\}$$

avec $I = [0, 1]$ par exemple. Le produit scalaire dans X étant $< \phi, \psi > = \int_0^1 \phi(t)\ \psi(t)\ dt$. Plus précisément, P est l'image de ν par l'injection canonique λ de X dans l'espace de Banach réel Y des fonctions continues sur I. Ceci a conduit C.B. De Witt [1] à définir la pseudo-mesure de Feynman W sur X come étant la collection des mesures W_i sur les sous-espaces X_i de dimension finie de X; où W_i admet pour transformée de Fourier la restriction à X_i de la fonction suivante définie sur X

$$\hat{W}(x) = \exp\left(-\frac{\sqrt{-1}}{2}\,||x||^2\right)$$

Signalons que la théorie des prodistributions permet d'interpré ter W comme une prodistribution à décroissance rapide sur X; et ainsi, W définit une forme linéaire sur l'espace des fonctions cylindriques à croissance très lente sur X : [8]. Le problème de la "radonification" de W consiste à trouver une application linéaire continue de X dans un certain espace de Banach E, transformant W en une forme linéaire sur un vaste espace de fonctions non cylindriques sur E. Une solution de ce problème est donnée par la théorie qui précède, en considérant W comme une profonctionnelle analytique de type exponentiel sur le complexifié X^c de X.

PROPOSITION *La complexifiée* λ^c *de l'injection* λ *de* X *dans* Y *transforme* W *en une fonctionnelle analytique de type exponentiel sur* Y^c.

La remarque (1.17) montre d'ailleurs que $\lambda^c(W)$ définit une forme linéaire sur une classe beaucoup plus vaste que $\text{Exp}(Y^c)$.

DEMONSTRATION L'application λ^c peut être considérée comme la composée de trois applications:

- La rotation d'angle $-\frac{\pi}{4}$ dans X^c, transformant W en la promesure normale canonique ν.
- l'injection λ^c transformant $\nu(x) \otimes \delta_0(y)$ en la mesure de Radon $P(x) \otimes \delta_0(y)$ sur Y^c.
- la rotation d'angle $\frac{\pi}{4}$ dans Y^c, transformant $P(x) \otimes \delta_0(y)$ en une mesure de Radon sur Y^c à décroissance exponentielle, concentrée sur un plan faisant un angle de $\frac{\pi}{4}$ avec Y.

Par conséquent, vu le théorèm de représentation (1.7), $\lambda^c(W)$ est une fonctionnelle analytique de type exponentiel sur Y^c.

Pour d'autres applications à la physique voir [10].

3. DISTRIBUTIONS ET CALCULS DIFFERENTIELS GENERALISES RELATIFS A DES ESPACES LOCALEMENT CONVEXES

Dans [7] ont été présentés la théorie des distributions et les calculs différentiels généralisés pour les espaces de Banach. Le but de ce paragraphe est d'indiquer comment s'effectue le prolongement de ces résultats aux espaces localement convexes. On se limite aux distributions bornées d'ordre borné; pour définir des distributions non bornées d'ordre quelconque, il suffit d'utiliser la technique de localisation développée au paragraphe 1 de [11].

(3.1) MESURES DE RADON VECTORIELLES La théorie des distributions de L. Schwartz est un prolongement du théorème de Riess concernant la représentation des mesures de Radon scalaires sur un compact K comme formes linéaires sur $B^0(K)$. La théorie des distributions en dimension infinie a pour point de départ la représentation comme formes linéaires des mesures de Radon vectorielles sur des espaces complètement réguliers: voir [5],[6] dans le cas de mesures à valeurs dans des Banach et [4] dans le cas de mesures à valeurs dans des espaces localement convexes.

Soit X un espace complètement régulier et E un espace localement convexe. Soit $B_K(X, E)$ l'espace des fonctions définies sur X, à valeurs dans E, dont l'image est une partie relativement compacte de E. A. Katsaras introduit une famille filtrante croissante (p_ℓ) de semi-normes définissant la topologie de E. Pour tout $\ell \in L$,. β_ℓ désigne la topologie localement convexe sur $B_K(X,E)$ telle que l'origine admette pour base de voisinages la famille des convexes disqués absorbants W tels que pour tout $r > 0$, il existe un compact K de X et $\eta > 0$ avec

$\{\phi;\ \sup\ \{p_\ell(\phi(x)),\ x \in K\} < \eta;\ \sup\ \{p_\ell(\phi(x)),\ x \in X\} < r\} \subset W$

Puis la topologie localement convexe β est définie comme la limite projective des topologies β_ℓ.

Si B est la tribu de Baire de X, $M_\ell(X,E')$ est défini comme l'espace des fonctions additives d'ensemble m: B → E' telles que $\sup \sum_i m(G_i)\, s_i$ soit fini, pour toute partition finie $(G_i)_i$ de X et pour toute famille $(s_i)_i$ de vecteurs de E tels que $p_\ell(s_i) \leq 1$. Le dual de $(B_K(X,E),\ \beta)$ est le sous-espaces $M(X,E')$ de la réunion des $M_\ell(X,E')$ formée par les mesures vectorielles tendues.

(3.4) LA SITUATION GÉOMÉTRIQUE Soit Y un e.ℓ.c.s. réel et $(Y_i,\ s_{ij})$ le système projectif d'espaces de dimension finie intervenant usuellement en théorie des probabilités cylindriques: $(Y_i)_{i \in I}$ est la famille des quotients de Y par les sous-espaces fermés A_i de codimension fine de Y; la surjection $s_{ij}: Y_i \to Y_j$ est définie si $i \geq j$, c'est-à-dire si $A_i \subset A_j$. Soit k un nombre qui est ou entier ≥ 0, ou égal à $+\infty$. Soit $B^k(Y_i)$ l'espace des fonctions $\phi_i : Y_i \to \mathbb{C}$ de classe C^k, dont les dérivées d'ordre au plus k sont bornées sur Y_i. Donc pour tout $\ell \leq k$, la dérivée $D^\ell\, \phi_i$ est une application continue bornée de Y_i dans $\underset{\ell}{\odot} Y_i'^{\,c}$. On note $B^k_{cyl}(Y)$ la limite inductive des espaces $B^k(Y_i)$, car cette limite inductive s'identifie à un espace de fonctions cylindriques sur Y. Plus précisément pour $\phi_i \in B^k(Y_i)$, posons $\phi = \phi_i \circ s_i$. La dérivée $D^\ell\, \phi$ est une fonction vectorielle cylindrique admettant la factorisation

$$\begin{array}{ccc} Y & \xrightarrow{\quad D^\ell \phi \quad} & \underset{\ell}{\odot} Y'^{\,c} \\ \big\downarrow s_i & & \big\uparrow \odot s_i'^{\,c} \\ Y_i & \xrightarrow{\quad D^\ell \phi_i \quad} & \odot Y_i'^{\,c} \end{array} \tag{3.5}$$

Posant $B^o_{cyl}(Y, \underset{\ell}{\odot} Y'^c) = \underset{\rightarrow}{\lim}\ B^o(Y_i, \underset{\ell}{\odot} Y_i'^c)$,

on a $D^\ell \phi \in B^o_{cyl}(Y, \odot Y'^c)$

(3.6) DÉFINITION DES T-DISTRIBUTIONS Supposons donnée pour tout $\ell \leq k$ une complétion E^T_ℓ de $\underset{\ell}{\odot} Y'^c$ relativement à une certaine topologie localement convexe. On peut prendre par exemple les traces des topologies π ou ε sur les produits tensoriels $\otimes Y'^c$. Soit T la famille des espaces $E^T_o, E^T_1, E^T_2 \ldots$. Noter que $E^T_o = \mathbb{C}$, mais que E^T_1 peut différer de Y'^c. On introduit le plongement canonique

$$B^k_{cyl}(Y) \longmapsto \prod_{\ell=o}^{k} B^o_{cyl}(Y, \underset{\ell}{\odot} Y'^c)$$

$$\phi \longmapsto (\phi, D\phi, \ldots, D^k \phi)$$

Chaque facteur est muni de la topologie induite par la topologie stricte de $B_k(Y, E^T_\ell)$. Et $B^k_{cyl}(Y)$ est muni de la topologie θ^k induite par la topologie produit. L'espace $B'^k_T(Y)$ des T-distributions bornées d'ordre au plus k est le dual de $(B^k_{cyl}(Y), \theta^k)$.

(3.7) PLONGEMENT DE $B'^k_T(Y)$ DANS LES PRODISTRIBUTIONS Pour tout i, on a un plongement canonique

$$B^k(Y_i) \longmapsto \prod_{\ell=o}^{k} B^o(Y_i, \underset{\ell}{\odot} Y_i'^c)$$

Si chaque facteur est muni de la topologie stricte et si $B^k(Y_i)$ est muni de la trace θ^k_i de la topologie produit, alors le dual $B'^k(Y_i)$ de $(B^k(Y_i), \theta^k_i)$ s'identifie à l'espace des distributions intégrables au sens de L. Schwartz, d'ordre au plus k. Par transposition de la bijection continue

$$(B^k_{cyl}(Y), \underset{\rightarrow}{\lim}\ \theta^k_i) \longmapsto (B^k_{cyl}(Y), \theta^k)$$

on obtient une injection canonique de $B'^k_T(Y)$ dans $B'^k_{cyl}(Y)$. En particulier, pour k = 0, on obtient une injection de l'espace

$B'^{o}(X) = M(X)$ des mesures de Radon bornées sur X, dans l'espace $B'^{o}_{cyl}(Y)$ des promesures bornées. Il faut noter que pour $k \neq 0$ et pour Y de dimension infinie, on obtient un phénomène analogue à celui obsevée par L. Nachibin [13] en théorie de l'holomorphie en dimension infinie. Et pour divers choix de T on obtient divers types de distributions.

(3.8) STRUCTURE DES DISTRIBUTIONS *Toute T-distributions est une somme de divergences de mesures de Radon vectorielles.*

La démonstration de ce théorème résulte du théorème de Hahn Banach et de la théorie des protenseurs distributions. En effet, pour toute $U \in B'^{k}_{T}(Y)$, il existe k' fini $\leq k$ et des mesures $\mu_j \in M(Y, (E^T_\ell)')$ telles que

$$\forall \phi \in B^k_{cyl}(Y) \qquad < U, \phi > = \sum_{j=o}^{k'} <\mu_j, D^j \phi >$$

$$\text{Or} \qquad < \mu_j, D^j \phi > = (-1)^j < \text{div}_j \mu_j, \phi >$$

D'où le théorème.

(3.9) THÉORIE DES DISTRIBUTIONS ET CALCUL DIFFÉRENTIEL A chaque famille T peut être associé un calcul différentiel généralisé sur Y. La construction de ce calcul différentiel est très différente de l'extension du calcul différentiel déduite de la théorie des distributions en dimension finie. Le principe consiste à partir d'une opération *triviale* de dérivation

$$B^1_{cyl}(Y) \longmapsto B^o_{cyl}(Y, Y'^c)$$

$$\phi \longmapsto D\,\phi$$

Mais comme cette application linéaire α est continue pour des topologies du type stricte, α est prolongée par sa bitransposée α''. Cette bitransposée donne l'extension voulue puisque α'' peut être restreinte à cetaines fonctions boréliennes. Voir [4].

BIBLIOGRAPHIE

[1] CECILE B. DE WITT, Feynman's path integral.Definition without limiting procedure. Comm.Math.Phys. Berlin. t 28. 1972 p. 47-67.

[2] D.H. FREMLIN, D.J.H. GARLING et R.G. HAYDON, Bounded measues on topological spaces. Proc.Lond. Math.Soc. 3º série. t 25. 1972. p. 115-136.

[3] D.J.H. GARLING, A generalized form of inductive limit topology for vector space. Proc.London Math. Soc. 3º série. t 14. 1964. p. 1-28.

[4] A.K. KATSARAS, Locally convex topologies on spaces of continuous vector functions. A paraître. Math. Nachrichten.

[5] P. KRÉE, Distributions sur les espaces de Banach. Comptes Rendus. 280. série A. 1975.

[6] P. KRÉE, Mesures de Radon vectorielles définies sur des espaces complètement réguliers. Comptes Rendus. 20 octobre 1975.

[7] P. KRÉE, Théories des distributions et calculs différentiels sur des espaces de Banach. A paraître au séminaire P.Lelong 1974-1975. Lectures Notes in mathematics. Springer.

[8] P. KRÉE, Equations aux dérivées partielles en dimension infinie. Séminaire P.K.- 1e année. Publié le secré

tairat mathématique de l'Institut H. Poincaré. 1975 - (PARIS).

[9] P. KRÉE, Courants et courants cylindriques sur les variétés de dimension infinie. "Linear operators and approximation: Proceed of the conf. held at the Math. Institute at Obercvolfach 1971" p.159-174. Baseland Stuttgart. Birkhauser Verlag 1972.International. Series of numerical mathematies, 20.

[10] P. KRÉE, et R. RACZKA, Kernel and symbols of operators in quantum field theory. (en préparation).

[11] P. KRÉE, Solutions prodistributions d'équations aux dérivées fonctionnelles (à paraître).

[12] J. LESMES, On the approximation of continuously differentiable functions on Hilbert spaces. Revista Colombiana de Mat.Vol. VIII. (1974) pp. 217-223.

[13] L. NACHBIN, Topology on spaces of holomorphic mappings. Collection jaune. Springer Verlag. Berlin (1968).

P. Krée

Département de mathématiques

Université de Paris VI

Place Jussieu - Paris 5ème

Infinite Dimensional Holomorphy and Applications, Matos (ed.)

SUR L'APPLICATION EXPONENTIELLE DANS L'ESPACE DES FONCTIONS ENTIÈRES

Par *PIERRE LELONG*

1. INTRODUCTION

Soit $E = A(C^n)$ l'algèbre des fonctions entières $f(z) = f(z_1,\ldots,z_n)$ de n variables complexes. Muni de la topologie de la convergence uniforme sur les compacts de C^n, E est un espace de Fréchet-Montel; on notera

(1) $\quad p_K(f) = \sup |f(z)| \;,\; z \in K,\; f \in E$

la semi-norme relative au compact K de C^n.

On se propose ici de préciser l'image de l'application exponentielle; celle-ci notée

(2) $\quad \exp \;:\; f \longrightarrow e^f$

sera considérée comme une application de E dans E. L'image $\eta = \text{Exp.}(E)$ est évidemment le cône de sommet O formé des fonctions entières sur C^n qui ne s'annulent jamais. Nous avons montré dans [3,a] que η est un ensemble fermé et polaire (au sens des fonctions plurisousharmoniques). Plus précisément, si N est le sous-espace fermé défini dans E par $f(O) = O$, il existe une fonction plurisousharmonique $U_o(f)$ <u>définie sur</u> E et telle qu'on ait

(3) $$\eta = \text{Exp.}(E) = [f \in E-N;\ U_o(f) = -\infty].$$

Dans la suite nous dirons qu'une fonction plurisousharmonique U est continue si e^U l'est, ou encore si U est continue en tout point où sa valeur est finie. Avec cette définition $U_o(f)$ est continue sur E-N; on peut encore énoncer en posant $V_o(f) = \exp. U_o(f)$: l'image de Exp. dans $E = A(C^n)$ est un cône fermé dans E-N et est l'ensemble des zéros d'une fonction plurisousharmonique $V_o(f)$ positive et continue sur E-N.

Ce résultat laissait ouvert un problème essentiel: l'ensemble η est-il un sous-ensemble analytique dans l'espace

$$E_* = E - \{0\}$$

où $\{0\}$ désigne l'ensemble constitué par la fonction identiquement nulle de E. On donnera une réponse négative à cette question. Pour cela on utilisera un résultat récent [1] de H. Alexander: si e est dans le plan complexe un ensemble polaire fermé, il existe une fonction entière $g(x,y) \in A(C^2)$ telle que pour tout $y \in e$, $g_y(x) = g(x,y)$ ne s'annule pas. H. Alexander établit ainsi une réciproque d'un résultat que nous avions donné en 1942 (cf. [3b] et [3c]): les y tels que $g_y(x)$ ne s'annule pas forment un ensemble fermé de capacité nulle dans le plan;ce résultat a été retrouvé indépendamment par M. Tsuji [5]. La réciproque donnée dans [1] par H. Alexander caractérise l'ensemble des y pour lesquels une relation entière $g(x,y) = 0$ n'a aucune solution en x; elle répond ainsi à un problème posé par G. Julia [2]. On utilisera ici ce résultat de la dimension finie pour montrer l'énoncé suivant qui résout le problème posé plus haut.

THÉORÈME 1. *Soit* $E = A(C^n)$ *l'espace des fonctions entières de* C^n*; soit* N *le sous-espace fermé défini par* $f(O) = O$ *et soit* $E_* = E - \{O\}$. *L'image* η *de l'application exponentielle* (2) *dans* E *est un cône de sommet l'origine, épointé à l'origine, et fermé dans* E_*. *Pour* $f \in E_*$, *soit* $d(f) \geq 0$ *la distance de l'origine de* C^n *à l'ensemble* $f = 0$*; la fonction*

$$(4) \qquad U_o(f) = -\log d(f) + \log |f(O)|$$

est plurisousharmonique et continue de f *dans* E - N *et se prolonge sur tout* E *en une fonction plurisousharmonique continue sur* E. *L'image* η *de l'exponentielle est le cône polaire défini dans* E-N *par* $U_o(f) = -\infty$.

2. *L'image* η *est dans* E_* *le complémentaire de la projection selon* C^n *d'un ensemble analytique* M *défini dans* $E \times C^n$*;* η *n'est pas, même localement, un sous-ensemble analytique dans* E_*. *Plus précisément, pour toute fonction* $f_o \in \eta$, *pour tout voisinage* W_o *de* f_o *dans* E_* *il n'existe pas d'application analytique* H *de* W_o *dans un espace vectoriel* F *localement convexe séparé telle que* $\eta \cap W_o$ *soit l'ensemble* $H^{-1}(O) \cap W$.

L'énoncé obtenu est à rapprocher d'un résultat plus précis obtenu pour les formes linéaires (c'est-à-dire les fonctionelles analytiques linéaires): si une telle forme $\lambda(f)$ s'annule sur η, elle est la constante nulle. En effet soit μ une mesure à support compact dans C^n représentant une telle fonctionelle; si $\mu(f) = 0$ pour tout $f \in \eta$, on a

$$\int d\mu(a)\, e^{g(a)} = 0$$ pour toute $g \in E$; en particulier pour $g = \langle a,z \rangle = \sum_1^n a_k z_k$, on aura

$$L_\mu(z) = \int e^{\langle a,z \rangle} d\mu(a) = 0$$

pour tout $z \in C^n$ ce qui entraîne $\mu = 0$ et la nullité de la fonctionnelle analytique linéaire. On notera que la méthode suivie dans le cas non linéaire est très différente et correspond au fait que η est (cf. le corollaire 1) le complémentaire de la projection sur E_* d'un ensemble analytique défini dans $E_* \times C^n$.

2. PROPRIÉTÉS DE $U_o(f)$

Précisons les résultats donnés dans [3,a] en rattachant la plurisousharmonicité de $-\log d(f)$ sur E-N à une propriété classique.

PROPOSITION 1. *Soit* G *un domaine de* C^n. *Pour* $f \in A(G)$ *et* $z \in G$, *l'application*

$$(5) \qquad \Phi : (f,z) \longrightarrow f(z)$$

est une fonction analytique dans $G \times E$.

En effet elle est continue de (f,z) en (f_o,z_o): soit B une boule de centre z_o, compacte dans G, et $r > 0$ son rayon; soit $p_B(f) = \sup|f(z)|$ pour $z \in B$.

Pour $||z - z_o|| \leq r' < \frac{r}{2}$, on a par la formule de Cauchy

$$|f_o(z) - f_o(z_o)| \leq C\,||z - z_o||$$

où $C = 2r^{-1}p_B(f_o)$. D'autre part soit $f \in E$; on a aussi pour $||z - z_o|| < r'$

$$|f(z) - f_o(z)| \leq p_B(f - f_o).$$

D'où

$$(6) \qquad |f(z) - f_o(z_o)| \leq p_B(f - f_o) + C\,||z - z_o||$$

qui établit la continuité de ϕ sur E. De plus Φ est G-analytique car on a pour $u \in C$; $h \in C^n$

$$(7) \qquad \Phi(f_1 + uf_2, z_o + uh) = f_1(z_o + uh) + uf_2(z_o + uh)$$
$$= \phi(u).$$

Il suffit de montrer que l'application $C \longrightarrow C$ donnée en (7) par $\phi(u)$ est une fonction analytique pour $u = 0$. Or $u \longrightarrow f_1(z_o + uh)$ et $u \longrightarrow f_2(z_o + uh)$ sont, pour f_1, f_2 fixés dans E, pour h fixé dans C^n et pour z_o pris dans G, des fonctions analytiques de u pour $|u|\ ||h|| < r$, donc au voisinage de $u = 0$.

COROLLAIRE 1 *Soit* $M = \bigcup_f m(f)$, $f \in E_*$, *où* $m(f)$ *est l'ensemble des zéros de* f *dans* C^n : M *est un ensemble analytique dans* $\Delta = E_* \times C^n$, *et l'image* η *de Exp. est le complémentaire dans* E_* *de la projection de l'ensemble analytique* M *sur* E_* *selon* C^n.

En effet $\Phi(f,z) = 0$ définit un ensemble analytique M' dans $E \times C^n$; on a

$$M' = M_1 \cup M_2$$

où $M_2 = \{0\} \times C^n$ est obtenu dans $E \times C^n$ comme ensemble des zéros de la fonction identiquement nulle (origine de E). On élimine M_2 en considérant l'ensemble analytique $M' \cap (E_* \times C^n) = M_1$.

PROPOSITION 2 *Pour* $f \in A(C^n) = E$, *la distance* d(f) *de l'origine de* C^n *à l'ensemble* m(f) *des zéros de* f *est une fonction semi-continue supérieurement de* f *sur* E_*, *à valeurs dans* $[0, +\infty]$.

On a seulement à établir la propriété au voisinage de f_o pour lequel on suppose $f_o \not\equiv 0$ et $d(f_o) = d_o$ où $d_o \geq 0$ est fini. Il existe alors $z_o \in C^n, ||z_o|| = d_o \geq 0$ tel qu'on ait $f_o(z_o)=0$.

D'autre part $f_o \not\equiv 0$ entraîne pour tout $\delta > 0$ donné, l'existence de $h \in C^n$, $||h|| < 1$, tel que $f_o(z_o + uh) = \psi(u)$ ait $u= 0$ comme seul zéro pour $|u| < 1$. Il existe alors ρ, $0 < \rho < \varepsilon$ tel que $|\psi(\rho e^{i\theta})|$ ait un minimum non nul, soit $m > 0$. Dans ces conditions si K est un compact de C^n contenant le disque δ défini par $z = z_o + uh$, $|u| < \rho$, les f appartenant dans E au voisinage

W de f_o défini par $p_K(f - f_o) < \frac{m}{2}$ s'annulent sur δ donc dans la boule $||z|| < ||z_o|| + \varepsilon$ ce qui établit

$$d(f) < d(f_o) + \varepsilon \quad \text{pour} \quad f \in W$$

et la semi-continuité supérieure de d(f); le raisonnement est valable si $d(f_o) = 0$.

COROLLAIRE 2 *L'image* η *de l'exponentielle est un ensemble fermé sans point isolé dans* E_*.

En effet $E_c = [f \in E_*,\ d(f) < c]$ est pour tout $c > o$ un ouvert dans E_* d'après la proposition 2; il en est de même du complémentaire de η dans E_* qui est la réunion des E_c pour $0<c<$ $<\infty$.

L'ensemble η est de plus parfait (c'est-à-dire sans point isolé) dans E_*. En effet soit $f_o \in \eta$, $f_o \not\equiv 0$; les fonctions $f = uf_o$ pour $u \in C$, $u \neq 0$, appartiennent à η; soit W un voisinage de f_o déterminé par $p_K(f - f_o) < \varepsilon$; W contient les fonctions $f = uf_o$ pour $u \in C$ vérifiant $|u| > 0$, $|u - 1| < \varepsilon\, p_K^{-1}(f_o)$.

PROPOSITION 3 *La fonction*

$$U_1(f) = - \log d(f)$$

est plurisousharmonique au voisinage de toute fonction $f_o \in E$ *pour laquelle on a* $f_o(0) \neq 0$.

Soit $|f_o(0)| = b$, $b > 0$; soit K la boule compacte $||z|| \leq 2$, et $W(f_o)$ le voisinage de f_o défini par

$$W(f_o) = [f \in E,\ p_K(f-f_o) < \tfrac{b}{2}] \ . \tag{8}$$

On a

$$p_K(f) < p_K(f_o) + \frac{b}{2} = m \quad \text{pour} \quad f \in W(f_o). \tag{9}$$

Soit h un vecteur unitaire de C^n, et $D_h f(z)$ la dérivée de f en z dans la direction h, c'est-à-dire $D_h\, f(z) = \left[\frac{\partial}{\partial u} f(z+uh)\right]_{u=o}$, $u \in C$. La formule de Cauchy et (9) donnent

$$|D_h\, f(z)| \leq m \quad \text{pour} \quad ||z|| \leq 1, \text{ et } f \in W(f_o).$$

Si z_o est dans C^n le zéro de f de plus petite distance à l'origine, on a alors si $||z_o|| \leq 1$:

$$|f(z_o)-f(0)| = |f(0)| \leq m\, ||z_o|| .$$

Mais $||z_o|| = d(f)$; on a donc si $d(f) \leq 1$

$$(10) \qquad d(f) \geq m^{-1}\, |f(0)| \geq m^{-1}\, \frac{b}{2} \text{ pour } f \in W(f_o)$$

et

$$U_o(f) = -\log d(f) + \log |f(0)| \leq \log m \text{ pour } f \in W(f_o),\ d(f) \leq 1.$$

Soit $d(f,k)$ la distance de l'origine de C^n à l'ensemble $m(f)$ des zéros de f parallèlement au vecteur unitaire k de C^n:

$$d(f) = \inf_k\, d(f,k).$$

Interprétons $d(f,k)$ comme la distance dans $E \times C^n$ et parallèlement à k du point $(f, z = 0)$ à la frontière du domaine Δ qui est le domaine d'holomorphie de la fonction $[\phi\,(f,z)]^{-1}$: Δ est $E_* \times C^n$ dont on retranche l'ensemble M_1 des zéros de Φ. D'après un résultat classique (cf. [5]), $-\log d(f,k)$ est une fonction plurisousharmonique de $f \in E^*$. De plus près (10) et (11), pour $f \in W(f_o)$ on a

$$-\log d(f,k) \leq -\log d(f) \leq \log_+ \left[m|f^{-1}(0)|\right]$$

où $\log_+ a = \sup(\log a, 0)$. D'après (8) on a aussi $|f(0)| \geq \frac{b}{2}$, donc

$$-\log d(f,k) \leq \log_+ 2m\, b^{-1} \text{ pour } f \in W(f_o).$$

Les fonctions $-\log d(f,k)$ sont ainsi majorées uniformément (par rapport à k) dans $W(f_o)$; leur enveloppe supérieure régularisée est donc plurisousharmonique, (cf. [3,d]). D'autre part $d(f)$ est semi-continu inférieurement de f dans $W(f_o)$ car $d(f)$

est la distance de $(f,0)$ à l'ensemble fermé défini par $\Phi(f,z)=0$ dans $E_* \times C^n$, distance prise parallèlement à C^n et $d(f)$ est bornée inféricurement dans $W(f_o)$ d'après (10). Ainsi $U_1(f) =$ $= -\log d(f)$ est l'enveloppe supérieure régularisée de $-\log d(f,k)$ pour $||k|| = 1$, et est semi-continue supérieurement; $U_1(f)$ est donc plurisousharmonique pour $f \in W(f_o)$, c'est-à-dire, finalement au voisinage de tout point de E-N.

On peut alors énoncer d'après les propriétés 2 et 3:

COROLLAIRE 3 *La fonction* $U_1(f) = -\log d(f)$ *est plurisousharmonique et continue sur* E-N; *elle l'est en particulier sur un voisinage de image* η *de l'exponentielle dans l'espace* $E_* = E - \{0\}$.
D'autre part on a :

COROLLAIRE 4 *L'image* η *de* Exp. *est un cône sur* $\mathbb{C}$ *épointé en son sommet* 0 *d'intérieur vide et fermé dans* E_*.

Le corollaire 4 a déjà été donné dans [3,a]; il est immédiat que $f \in \eta$ entraîne $\lambda f \in \eta$, pour tout $\lambda \in C$, $\lambda \neq 0$, donc η est un cône épointé à son sommet qui est l'origine. De plus, η est exactement l'ensemble $V_1(f) = 0$ dans $E - N$ où $V_1(f)=d^{-1}(f)$ $= \exp [U_1(f)]$; $V_1(f)$ est plurisousharmonique continue sur l'ouvert $E - N$ et tend vers $+\infty$ quand f tend vers $f_o \in N - \{0\}$; ainsi η est un cône fermé dans E_*. Enfin son intérieur est vide puisqu'il est polaire dans $E - N$.

3. L'IMAGE η DE Exp. N'EST PAS UN ENSEMBLE ANALYTIQUE DANS E_*

La demonstration comprend trois parties:

a) On établira d'abord

PROPOSITION 4 *Soit $g(z_1,\ldots,z_n,y) = g(z,y)$ une fonction entière sur $C^{n+1}(z,y)$: l'application $G : y \longrightarrow g(z,y) = g_y(z)$ est une application analytique.*

En effet G est continue: posons $M(R,R') = \sup |g(z,y)|$ pour $|x| \leq R$, $|y| \leq R'$. Pour $||z|| < R$ et $|y| < \frac{R'}{2}$, $|\frac{\partial g}{\partial y}|$ est majoré par la formule de Cauchy et l'on a, pour $|y| \leq \frac{R}{2}$, $|y_o| \leq \frac{R}{2}$:

$$p_K \left[G(y) - G(y_o)\right] \leq |y-y_o| \left[2R'^{-1} M(R,R')\right]$$

valable en semi-norme p_K, si K est un compact contenu dans $||z|| < R$. De même on majore le module de la derivée $\frac{\partial^q g}{\partial y^q}$ pour $|y_o| \leq \frac{R'}{2}$, $||z|| < R$.

Si h est un vecteur unité de C^n, $u \longrightarrow G(y_o+uh)$ est analytique de u au voisinage de u = O car on a

$$G(y_o+uh) = G(y_o)+ \sum_1^\infty \frac{u^q}{q!} \frac{\partial^q g(z,y_o)}{\partial y^q}$$

et l'on obtient une série majorante en norme p_K qui converge pour $|u| < \frac{R'}{2}$ d'après

$$p_K \left[\frac{u^q}{q!} \frac{\partial^q g}{\partial y^q} (z,y_o)\right] \leq M(R,R') \left(\frac{2|u|}{R'}\right)^q .$$

Ainsi G(y) est une application holomorphe de $\mathbb{C}$ dans E.

b) Rappelons alors l'énoncé suivant [1] de H. Alexander: étant donné un ensemble fermé e de capacité nulle dans le plan complexe C(y), il existe une fonction entière $g(z,y) \in A(C^2)$ telle que e soit l'ensemble des y pour lesquels $g_y(z) = g(z,y)$ ne s'annule pour aucune valeur $z \in \mathbb{C}$.

Nous prendrons pour e un ensemble fermé sans point isolé de capacité nulle, construit à partir d'un segment fermé [a,b] de l'axe réel de C(y) en utilisant une suite d'entiers $S=\{p_1,p_2\ldots\}$ fortement croissante. Si ℓ_1 est la longueur de [a,b] l'opération (p_1) consiste à retirer de [a,b] l'intervalle ouvert médian de

longueur $\ell_1(1 - p_1^{-1})$ de manière qu'il ne reste que l'ensemble $e(p_1)$ formé de deux segments fermés
$[a, a + \ell_1(2p_1)^{-1}]$, $[b - \ell_1(2p_1)^{-1}, b]$ de longueur $\ell_1(2p_1)^{-1}$. L'opération (p_2) est appliquée à chacun de ces deux segments et donne $e(p_1,p_2)$, etc. Ce procédé, imité de la construction de l'ensemble de Cantor fournit une suite d'ensembles fermés $E(p_1,\ldots,p_n)$; leur intersection $e(S)$ est un ensemble fermé sans point isolé; si la série de terme général $2^{-q} \log p_q$ diverge, $e(S)$ est de capacité nulle (cf. [4]).

Dans ces conditions on utilisera l'énoncé suivant

PROPOSITION 5 *Il existe une fonction entière* $g(z,y) \in A(C^2)$ *telle que les* y *pour lesquels* $g_y(z) = g(z,y)$ *appartient à l'image* η *de* Exp. *forment un ensemble* e *non vide, fermé et sans point isolé.*

c) On a alors

PROPOSITION 6 *Soit* W_o *un voisinage donné dans* E_* de $f_o \in \eta$: *il n'existe pas d'application holomorphe* $H : W_o \longrightarrow F$ *où* F *est un espace vectoriel localement convexe séparé telle que* $\eta \cap W_o$ *soit défini dans* W_o par $H(f) = 0$. *En d'autres termes dans* $E_* = E - \{0\}$ *l'ensemble* η *est un ensemble fermé mais n'est un ensemble analytique au voisinage d'aucun de ses points.*

Montrons en effet que si f_o et W_o sont donnés dans E et $f_o \in \eta$, il n'existe pas d'espace F localement convexe séparé et d'application holomorphe $H : W_o \longrightarrow F$ telle que

$$W_o \cap \eta = W_o \cap H^{-1}(0). \tag{11}$$

Dans ce but nous utiliserons l'ensemble $e(S)$ construit sur les réels de C et une fonction entière $g(z,y) \in A(C^2)$ telle que ensemble des $y \in \mathbb{C}$ où l'on a $g_y(z) = g(z,y) \in \eta$ soit $e(S)$.

Prenons un point y_o dans $e(S)$; $g(z_1,y_o) = e$ s'annule pas et on posera

$$g'(z_1,z_2,\dots,z_n,y) = g(z_1,y)\ [g(z_1,y_o)]^{-1}\ f_o(z_1,\dots,z_n).$$

L'application $G : y \longrightarrow g'_y = g'(z_1,\dots,z_n,y) \in E$ est holomorphe (proposition 4) et l'on a $G(y_o) = f_o$. De plus on a $G(y) \in \eta$ si et seulement si y appartient à $e(S)$. L'application G' étant continue, il existe $\alpha > 0$ de manière qu'on ait

$$G(y) \in W_o \quad \text{pour} \quad |y - y_o| < \alpha.$$

On a alors

$$G(y) \in W_o \cap \eta \quad \text{pour} \quad |y - y_o| < \alpha,\ y \in e(S).$$

Supposons que, dans W_o, η soit défini par $H(f) = 0$ où H est une application $W_o \longrightarrow F$, holomorphe, F étant un espace localement convexe séparé. Alors l'application composée

$$y \longrightarrow H \circ G(y)$$

est holomorphe de $\mathbb{C}$ dans F. Soit $\phi \in F'$ une forme linéaire continue sur F: $h_\phi = \phi \circ H \circ G(y)$ est une fonction holomorphe $\mathbb{C} \rightarrow \mathbb{C}$. Elle s'annule pour $|y - y_o| < \alpha$, $y \in e(S)$, c'est-à-dire sur un ensemble fermé non vide n'ayant aucun point isolé; elle est donc la constante nulle; on a alors $h_\phi \equiv 0$ pour tout $\phi \in F'$, ce qui entraîne

$$h(y) = H \circ G(y) = 0 \quad \text{pour} \quad |y - y_o| < \alpha.$$

On a alors

$$G(y) \in H^{-1}(0) \quad \text{pour} \quad |y - y_o| < \alpha.$$

On a d'autre part $G(y) \in W_o$ pour $|y - y_o| < \alpha$, donc

$$G(y) \in W_o \cap H^{-1}(0) \quad \text{pour} \quad |y - y_o| < \alpha.$$

Si donc (11) était réalisé, on aurait

$$G(y) \in W_o \cap \eta \quad \text{pour} \quad |y - y_o| < \alpha$$

c'est-à-dire $g'_y(z_1 \ldots z_n) \in \eta$ pour tout y vérifiant $|y - y_o| < \alpha$, contrairement à l'hypothèse faite, ce qui établit la proposition 6, et la seconde partie du théorème(1).

4. PROLONGEMENT DE $U_o(f)$

On a montré dans [3a] que la fonction $U_o(f) = -\log\ d(f)$ demeurait bornée au voisinage du sous-espace fermé N défini par $f(O) = O$. Elle se prolonge donc à travers N. Montrons que ce prolongement est continu sur N, c'est-à-dire que $U_o(f)$ ainsi prolongée est une fonction plurisousharmonique continue sur E.

PROPOSITION 7 *On a*

(12) $$U_o(f) = -\log d(f) + \log |f(O)| \leq \log p_K(f)$$

où K *est la boule compacte* $||z|| \leq 2$ *de* C^n.

En effet si $d(f) < 1$, on a en opérant comme en (10)

$$d(f) \geq |f(O)|\ p_K^{-1}(f)$$

d'où résulte (12); dans le cas $d(f) \geq 1$, on a $U_o(f) \leq \log|f(O)| \leq \log p_K(f)$, c'est-à-dire encore (12).

Soit alors $f \in N$, $f(O) = O$ et un voisinage W de f défini dans E par $p_K(f' - f) < b$; on a $p_K(f') \leq p_K(f) + b$ pour $f \in W$ et $U_o(f)$ demeure borné supérieurement dans W d'après (12). D'après un résultat connu (cf. [5]), $U_o(f)$ se prolonge alors d'une manière unique en une fonction plurisousharmonique à travers N, définie par

$$U_o(f) = \limsup\ U(f_m),\ f_m \longrightarrow f \in N,\ f_m \in E - N.$$

Montrons que ce prolongement est encore continu sur N.

Posons $z = \alpha\rho$, $\rho > O$, $||\alpha|| = 1$, et

$$f(z) = f(\alpha\rho) = f(O) + \rho P_1(\alpha) + \rho^2 P_2(\alpha) + \ldots$$

où $P_k(\alpha)$ est un polynome homogène de degré k des coordonnées de . Supposons $P_1(\alpha) \neq 0$. Alors on a

$$f_m(\alpha\rho) = f_m(0) + \rho P_{1,m} + \rho^2 P_{2,m}(\alpha) + \ldots$$

où $f_m(0) \longrightarrow 0$, $P_{1,m}(\alpha) \longrightarrow P_1(\alpha)$, la convergence $f_m \longrightarrow f$ étant uniforme sur tout compact selon la topologie de E.

Soit $M_m = \max |P_{1,m}(\alpha)|$ pour $||\alpha|| = 1$. On a

$d(f_m) = M_m^{-1} |f_m(0)| (1 + \varepsilon_m)$, $\varepsilon_m \longrightarrow 0$ quand $f_m \longrightarrow f$.

D'où

$$\lim_{m=\infty} U_o(f_m) = \lim_m \log M_m.$$

On a donc, pour valeur du prolongement

(15) $$U_o(f) = \log \max_{||\alpha||=1} Df_\alpha(0) \quad , f \in N$$

qui montre que $U_o(f)$ est continu en $f \in N$ si la differentielle de f à l'origine n'est pas nulle.

Si l'on a $f(0) = 0$ et $D_\alpha f(0) = 0$ pour tout α, mais $f \not\equiv 0$, c'est-à-dire $f \in N - \{0\}$ la résultat est encore valable: le prolongement de $U_o(f)$ demeure continu avec valeur $-\infty$. En effet, au besoin après un changement de coordonnée unitaire, f equivant à un pseudo-polynome π en z et on a

$$f(z) = p(z) \quad \pi(z)$$

valable dans un domaine $D = [z' \in d, |z_n| < \rho]$, choisi tel que f(z) ne s'annule pas sur le compact $z' \in \bar{d}$, $|z_n| = \rho$; $\pi(z)$ est un pseudo-polynome de degré $q \geq 2$

$$\pi(z) = z_n^q + A_1(z') z_n^{q-1} + \ldots + A_q(z')$$

où z_n est la variable distinguée et $z' = (z_1, \ldots, z_{n-1}) \in d$; $p(z) \neq 0$ dans D; les racines $\zeta_i(z')$ de π s'annulent à l'origine $z' = 0$ ainsi que les coefficients $A_k(z')$ de π.

Alors soit $f_m \longrightarrow f$ est une suite de fonctions entières qui converge uniformément vers f dans un voisinage de l'origine contenant $\bar{D}$. Il existe m_o tel que pour $m > m_o$, f_m ait dans D une décomposition holomorphe

$$f_m = p_m \cdot \pi_m \quad , \; p_m(z) \neq 0 \text{ pour } z \in D$$

$p_m \longrightarrow p$, $p_m(0) \neq 0$, $\pi_m \longrightarrow \pi$ uniformément sur $\bar{D}$, et a q racines $\zeta_{i,m}(z')$ de module inférieur à ρ pour $z' \in d_1 \subset\subset d$. On a alors

$$d(f_m) \geq \inf_i |\zeta_{i,m}(z')|$$

$$f_m(0) = |p_m(0)| \prod_1^q |\zeta_{i,m}(z')| .$$

(16) $$f_m(0)\,[d(f_m)]^{-1} \leq |p_m(0)| \prod |\zeta_{i',m}(z')|$$

où le produit au second membre concerne les racines de π_m dont on a enlevé celle de plus petit module: comme on a $q \geq 2$, le produit tend vers 0 quand $f_m \longrightarrow f$, cependant qu'on a $p_m(0) \longrightarrow p(0) \neq 0$. Le premier membre de (16) tend vers 0 quand $f_m \longrightarrow f$ et la valeur de prolongement de $U_o(f)$ est $-\infty$, égale à celle donnée par (15), ce qui établit la continuité de U_o au voisinage d'une fonction $f \in N - \{0\}$.

Enfin $U_o(f)$ est encore continu à l'origine de $E(f \equiv 0)$, avec valeur $U_o(0) = -\infty$. En effet si K est un compact contenant la boule $||z|| \leq 2$, on a d'après (12): $U_o(f) \leq \log p_K(f)$. Donc si $f_m \longrightarrow f \equiv 0$, $p_K(f_m) \longrightarrow 0$ et $U_o(f_m) \longrightarrow -\infty$.

Finalement $U_o(f)$, compte-tenu du Corollaire 3, est une fonction plurisousharmonique continue sur tout E, ce qui achève la démonstration du théorème.

BIBLIOGRAPHIE

[1] ALEXANDER (H.), On a problem of G. Julia. Duke Math. J., t. 42, n° 2, p. 326-332, 1975.

|2| JULIA (G.), Sur le domaine d'existence d'une fonction implicite définie par une relation entière $G(x,y) = 0$. Bull. Soc. Math. France, t.54, p. 26-37, 1926.

|3| LELONG (P.), a) Fonctions plurisousharmonique dans les espaces vectoriels topologiques. Colloque International du C.N.R.S., 1972, publié dans Agora Mathematica, vol. 1. 1974, p. 95-116 (Gauthier-Villars).

b) Sur certaines fonctions multiformes, C.R.Ac. Sci. Paris, t. 214, 1942, p. 53.

c) Sur les valeurs lacunaires d'une relation à deux variables complexes. Bull.des Sciences Math., t. 56, p. 103-112, 1942.

d) Séminaire d'Analyse, Lecture-Notes Springer, N°71, p. 167-190 et N° 116, p. 1-20.

|4| NEVANLINNA (R.), Eindeutige analytische Funktionen, Springer, cf. p. 149, 1936.

|5| NOVERRAZ (Ph.), Pseudo-convexité et convexité polynomiale.... North. Holland, 1973.

|6| TSUJI (M.), Theory of meromorphic functions in a neighboorhood of a closed set of capacity zero, Jap. J. Math., t. 19, p. 139-154, 1944.

(1) - Récemment (Mai 1977), M. ZRAIBI, puis J.SICIAK ont donné une démonstration de la proposition 6 sans utiliser le résultat de H.ALEXANDER: la nullité de $H[\exp(f + tf)]$ pour $f \in A(C^n)$ au voisinage de $t = 0$ entraîne celle du développement de Taylor de H en f_o.

DÉPARTEMENT DE MATHÉMATIQUES,
UNIVERSITÉ DE PARIS VI,
4 PLACE JUSSIEU,
PARIS 5, FRANCE.

Infinite Dimensional Holomorphy and Applications, Matos (ed.)

HOLOMORPHIC GERMS ON INFINITE DIMENSIONAL SPACES.

By *JORGE MUJICA*.

1. INTRODUCTION

$\mathcal{H}(U)$ denotes the vector space of all complex-valued holomorphic functions on an open subset U of a complex Banach space E. τ_ω denotes the Nachbin topology on $\mathcal{H}(U)$. We recall the definition of τ_ω; see [11] . A seminorm p on $\mathcal{H}(U)$ is said to be ported by a compact subset K of U if for each open set V, with $K \subset V \subset U$, there exists $c(V) > 0$ such that $p(f) \leq c(V) \sup_{x \in V} |f(x)|$ for all $f \in \mathcal{H}(U)$. The locally convex topology τ_ω is defined for all such seminorms.

$\mathcal{H}(K)$ denotes the locally convex space of all complex-valued holomorphic germs on the compact set K, endowed with the inductive topology coming from

$$\mathcal{H}(K) = \operatorname*{ind\,lim}_{\varepsilon > 0} \mathcal{H}^\infty(K_\varepsilon)$$

where $K_\varepsilon = \{x \in E : \text{dist}(x,K) < \varepsilon\}$ and $\mathcal{H}^\infty(K_\varepsilon)$ denotes the Banach space of all bounded holomorphic functions on K_ε, with the sup norm. In this survey we show how the spaces $\mathcal{H}(K)$ play an important role in the study of the locally convex space

$(\mathcal{H}(U), \tau_\omega)$. Most of the results presented here are taken from [9]. Some of them were announced without proof in [10].

2. LOCAL MULTIPLICATIVE CONVEXITY OF $\mathcal{H}(K)$ AND $(\mathcal{H}(U), \tau_\omega)$.

We begin with

THEOREM 1. *$\mathcal{H}(K)$ is a locally multiplicatively convex algebra, i.e. its topology is defined by the continuous seminorms p such that $p(fg) \leq p(f)\, p(g)$ for all $f, g \in \mathcal{H}(K)$.*

For the proof we write $V_n = K_{\varepsilon_n}$, where (ε_n) is a sequence of positive numbers decreasing to zero. For each n and $0 < \delta_n < 1$ we define

$$\mathcal{U}_{n,\delta_n} = \{f \in \mathcal{H}^\infty(V_n) : \sup_{x \in V_n} |f(x)| < \delta_n\}$$

Since $\mathcal{H}(K) = \text{ind lim } \mathcal{H}^\infty(V_n)$ the sets

$$\mathcal{U}_\delta = \text{convex hull of } \bigcup_{n=1}^{\infty} \mathcal{U}_{n,\delta_n}$$

form a base of convex, balanced neighborhoods of 0 in $\mathcal{H}(K)$, as the sequence $\delta = (\delta_n)$ varies. Since

$$\mathcal{U}_\delta = \{\text{finite sums } \Sigma\lambda_n f_n : \lambda_n > 0,\ \Sigma\lambda_n = 1,\ f_n \in \mathcal{U}_{n,\delta_n}\},$$

it is easy to see that $\mathcal{U}_\delta \cdot \mathcal{U}_\delta \subset \mathcal{U}_\delta$, ending the proof.

THEOREM 2. *$(\mathcal{H}(U), \tau_\omega)$ is a locally multiplicatively convex algebra.*

Theorem 2 answers a question raised by Matos [8]. We sketch the proof of Theorem 2. For each compact $K \subset U$, we let M^K denote the image of the canonical mapping $\mathcal{H}(U) \to \mathcal{H}(K)$. For each $\varepsilon > 0$ we define $M_\varepsilon^K = M^K \cap \mathcal{H}^\infty(K_\varepsilon)$, with the norm induced by $\mathcal{H}^\infty(K_\varepsilon)$, and endow M^K with the inductive topology coming from

$$M^K = \underset{\varepsilon > 0}{\text{ind}} \lim M_\varepsilon^K$$

We have the following diagram

$$\begin{array}{ccc} M^K & \hookrightarrow & \mathcal{H}(K) \\ \uparrow & & \uparrow \\ M^K_\varepsilon & \hookrightarrow & \mathcal{H}^\infty(K_\varepsilon) \end{array}$$

Theorem 2 follows from Lemma 1 and Lemma 2 below.

LEMMA 1. M^K *is a locally multiplicatively convex algebra.*

LEMMA 2. $(\mathcal{H}(U), \tau_\omega) = \text{proj}\lim_{K \subset U} M^K$

The proof of Lemma 1 is similar to that of Theorem 1. the proof of Lemma 2 is straightforward.

3. BOUNDED SUBSETS OF $\mathcal{H}(K)$.

The characterization of the bounded subsets of $\mathcal{H}(K)$ is an important tool for obtaining further results about $\mathcal{H}(K)$. The main result in this section is the following.

THEOREM 3. *Given a bounded subset* X *of* $\mathcal{H}(K)$, *there exists* $\varepsilon > 0$ *such that*

(a) X *is contained and bounded in* $\mathcal{H}^\infty(K_\varepsilon)$.

(b) *Every net* $(f_\alpha) \subset X$ *which is Cauchy in* $\mathcal{H}(K)$ *is also Cauchy is* $\mathcal{H}^\infty(K_\varepsilon)$.

We say then that $\mathcal{H}(K) = \text{ind}\lim \mathcal{H}^\infty(K_\varepsilon)$ *is a Cauchy regular inductive limit.*

Theorem 3 was first given by Chae [2] and Hirschowitz [6], but the original proofs were incomplete. We give a different proof, based on Theorem A and Lemma 3 below.

THEOREM A (Grothendieck [4]). *If* $X = \text{ind}\lim X_n$ *is the inductive limit of an increasing sequence of normed spaces* X_n *whose union is* X, *then* X *is a* (DF)*-space and every bounded subset of* X

is contained in the X*-closure of a bounded subset of some* X_n.

LEMMA 3. *Let* X *be a bounded subset of* $\mathcal{H}^\infty(K_\varepsilon)$ *and let* $0 < \delta < \varepsilon$. *Then:*

(a) *Every net* $(f_\alpha) \subset X$ *which is Cauchy in* $\mathcal{H}(K)$ *is also Cauchy in* $\mathcal{H}^\infty(K_\delta)$.

(b) *The closure of* X *in* $\mathcal{H}(K)$ *is contained and bounded in* $\mathcal{H}^\infty(K_\delta)$.

The proof of Lemma 3 is straightforward. From theorem 3 we see at once that every closed, bounded subset of $\mathcal{H}(K)$ is complete, and hence that $\mathcal{H}(K)$ itself is complete, being a (DF)--space; see [4].

4. COMPLETENESS OF $(\mathcal{H}(U), \tau_\omega)$.

The following theorem answers a question raised by Nachbin [12].

THEOREM 4. $(\mathcal{H}(U), \tau_\omega)$ *is always complete.*

Earlier partial results had been given by Dineen [3], Chae [2] and Aron [1] for certain open sets U. For the proof of Theorem 4 we use the notation of Section 2, and define

$$\tilde{M}^K_\varepsilon = \text{closure of } M^K_\varepsilon \text{ in } \mathcal{H}^\infty(K_\varepsilon)$$

$$\tilde{M}^K = \bigcup_{\varepsilon > 0} \tilde{M}^K_\varepsilon$$

$\tilde{M}^K_\varepsilon$ is the completion of the normed space M^K_ε, and $\tilde{M}^K$ is endowed with the inductive topology coming from

$$\tilde{M}^K = \operatorname{ind\,lim}_{\varepsilon > 0} \tilde{M}^K_\varepsilon$$

We have the following diagram.

$$\begin{array}{ccccc} M^K & \hookrightarrow & \tilde{M}^K & \hookrightarrow & \mathscr{H}(K) \\ \uparrow & & \uparrow & & \uparrow \\ M^K_\varepsilon & \hookrightarrow & \tilde{M}^K_\varepsilon & \hookrightarrow & \mathscr{H}^\infty(K_\varepsilon) \end{array}$$

Theorem 4 follows from from Lemma 4 and Lemma 5 below.

LEMMA 4. $\tilde{M}^K$ *is the completion of* M^K.

LEMMA 5. $(\mathscr{H}(U), \tau_\omega) = \operatorname{proj\,lim}_{K \subset U} \tilde{M}^K$

It is clear that M^K is sequentially dense in $\tilde{M}^K$, and that every continuous seminorm on M^K extends uniquely to a continuous seminorm on $\tilde{M}^K$. To prove that $\tilde{M}^K$ is complete we use Lemma 3 and Theorem A to show that $\tilde{M}^K = \operatorname{ind\,lim} \tilde{M}^K_\varepsilon$ is a Cauchy regular inductive limit. This shows Lemma 4. Lemma 5 follows from Lemma 2 and Lemma 4.

5. OPEN SETS WITH THE RUNGE PROPERTY.

We continue using the notation of Section 2 and Section 4. A compact set $K \subset U$ is said to be U-Runge if M^K is sequentially dense in $\mathscr{H}(K)$. U is said to have the Runge property if every subset of U is contained in another one which is U-Runge. When U is an open set with the Runge property then we can improve Lemma 2 and Lemma 5 as follows.

THEOREM 5. *If* U *is an open set with the Runge property then*

$$(\mathscr{H}(U), \tau_\omega) = \operatorname{proj\,lim}_{K \subset U} \mathscr{H}(K)$$

Theorem 5 follows from Lemma 5 and Lemma 6 below.

LEMMA 6. *For* $K \subset U$ *we have:*

(a) $\tilde{M}^K$ = sequential closure of M^K *in* $\mathscr{H}(K)$.

(b) *If* K *is* U-*Runge then for each* $\varepsilon > 0$ *there exists* δ, *with* $0 < \delta < \varepsilon$, *such that* $\mathcal{H}^\infty(K_\varepsilon) \subset \tilde{M}^K_\delta$. *In particular*, $\tilde{M}^K = \mathcal{H}(K)$ *as topological vector spaces.*

The proof of (a) is an immediate application of Theorem 3. To prove (b) we use (a) and Theorem B below.

THEOREM B (Grothendieck [5]). *Let* X *be a Hausdorff locally convex space which is the union of an increasing sequence of Fréchet spaces* X_n *and assume that each inclusion mapping* $X_n \to X$ *is continuous. Then any continuous linear mapping* $T : Y \to X$ *from a Fréchet space* Y *into* X *can be factored continuously through some* X_n, *i.e. there exists* n *and a continuous linear mapping* $T_n : Y \to X_n$ *such that the following diagram is commutative.*

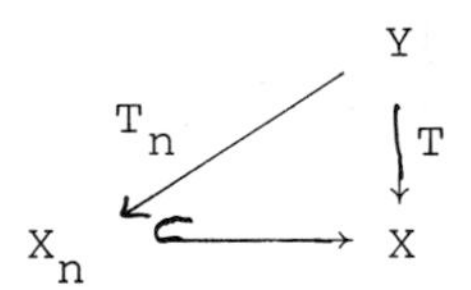

To apply Theorem B in the proof of Lemma 6 (b) we take $X = \mathcal{H}(K)$, $X_n = \tilde{M}^K_{\varepsilon_n}$, $Y = \mathcal{H}^\infty(K_\varepsilon)$, where (ε_n) is any sequence of positive numbers decreasing to zero.

The following theorem gives several characterizations of the compact subsets of U which are U-Runge.

THEOREM 6. *For any compact set* K U *the following conditions are pairwise equivalent.*

(a) K *is* U-*Runge, i.e.* M^K *is sequentially dense in* $\mathcal{H}(K)$.

(b) *Every bounded subset of* $\mathcal{H}(K)$ *is contained in the* $\mathcal{H}(K)$-*closure of a bounded subset of* M^K.

(c) *Given* $X \subset \mathcal{H}(K)$ *bounded, there exists* $\delta > 0$ *with* $K \subset K_\delta \subset U$ *such that* $X \subset \mathcal{H}^\infty(K_\delta)$ *and for each* $f \in X$ *there exists a sequence* $(f_j) \subset M^K_\delta$ *with* $f_j \to f$ *in* $\mathcal{H}^\infty(K_\delta)$.

(d) *Given* $\varepsilon > 0$ *with* $K \subset K_\varepsilon \subset U$ *there exists* δ *with* $0<\delta<\varepsilon$ *such that for each* $f \in \mathcal{H}^\infty(K_\varepsilon)$ *there exists a sequence* $(f_j) \subset M_\delta^K$ *with* $f_j \to f$ *in* $\mathcal{H}^\infty(K_\delta)$.

The implication (a) $\Rightarrow$ (d) follows from Lemma 6. Each of the implications (d) $\Rightarrow$ (b) and (b) $\Rightarrow$ (c) follows from Theorem 3. Finally the implication (c) $\Rightarrow$ (a) is obvious.

6. OPEN PROBLEMS.

PROBLEM 1. *Can we construct the envelope of holomorphy of a domain* $U \subset E$ *as a subset of the spectrum of* $(\mathcal{H}(U), \tau_\omega)$, *at least when* E *is separable? For a attempt see* [8] . *By a counterexample of Josefson* [7], *with* $E = c_0(A)$, A *uncountable, this cannot be done in general.*

PROBLEM 2. *Theorem 1 and Theorem 2 remain true for any metrizable locally convex space* E. *On the other hand, we have been able to entend Theorems 3 through 6 only to a certain class of metrizable locally convex spaces; see* [9] . *Do Theorems 3 through 6 remain true for any metrizable locally convex space? For this purpose it certainly suffices to extend Lemma 3.*

PROBLEM 3. *With the notation of Section 2, for* $K \subset U$, *does* M^K *have the induced topology of* $\mathcal{H}(K)$? *This is certainly the case when* K *is* U-*Runge, by Lemma 4 and Lemma 6. Does this remain true in general?*

PROBLEM 4. *We defined* K *to be* U-*Runge if* M^K *is sequentially dense in* $\mathcal{H}(K)$. *Is this equivalent to saying that* M^K *is dense in* $\mathcal{H}(K)$?

PROBLEM 5. *Very few examples of open sets with the Runge property are known; see* [9]. *Can we find new examples? Can we improve the results in* [13] *to show, for instance, that every pseu doconvex open subset of a Banach space with basis has the Runge property ? This question was raised by Schottenloher; see* [14].

REFERENCES

[1] R. ARON; Holomorphy types for open subsets of a Banach space, Studia Math. 45 (1973), 273-289.

[2] S. B. CHAE; Holomorphic germs on Banach spaces, Ann. Inst. Fourier 21 (1971), 107-141.

[3] S. DINEEN; Holomorphy types on a Banach space, Studia Math. 39 (1971), 241-288.

[4] A. GROTHENDIECK; Sur les especes (F) et (DF), Summa Brasil, Math. 3 (1954), 57-122.

[5] A. GROTHENDIECK; Produits tensoriels topologiques et espaces nucléaires, Memoirs Am. Math. Soc. 16 (1955).

[6] A. HIRSCHOWITZ; Bornologie des especes de fonctions analytiques en dimension infinie, Séminaire Lelong 1970, Lecture Notes in Math. 275, Springer-Verlag (1971), 21-33.

[7] B. JOSEFSON; A counterexample in the Levi problem, Proceedings on infinite dimensional holomorphy, University of Kentucky 1973, Lecture Notes in Math. 364, Springer-Verlag (1974), 168-177.

[8] M. MATOS; Holomorphic mappings and domains of holomorphy, Centro Brasileiro de Pesquisas Físicas, Rio de Janeiro (1970).

[9] J. MUJICA; Spaces of germs of holomorphic functions, Thesis, University of Rochester (1974), to appear in Advances in Mathematics.

[10] J. MUJICA; On the Nachbin topology in spaces of holomorphic functions, Bull. Am. Math. Soc. 81 (1975), to appear.

[11] L. NACHBIN; Topology on spaces of holomorphic mappings, Ergebnisse der Mathematik und ihrer Grenzgebiete 47, Springer-Verlag (1969).

[12] L. NACHBIN; Concerning spaces of holomorphic mappings, Rutgers University (1970).

[13] PH. NOVERRAZ; Approximations of holomorphic or plurisubharmonio functions in certain Banach spaces, Proceedings on infinite dimensional holomorphy University of Kentuchy 1973, Lecture Notes in Math. 364, Springer-Verlag (1974), 178-185.

[14] M. SCHOTTENLOHER; The Levi problem and Oka-Weil approximation, this conference.

INSTITUTO DE MATEMÁTICA,
UNIVERSIDAD CATOLICA DE CHILE,
CASILLA 114-D, SANTIAGO,
CHILE.

Current Address
INSTITUTO DE MATEMÁTICA, ESTATÍSTICA E CIÊNCIA DA COMPUTAÇÃO-IMECC
UNIVERSIDADE ESTADUAL DE CAMPINAS
CAMPINAS-SP - BRASIL.

Infinite Dimensional Holomorphy and Applications, Matos (ed.)

ON A PARTICULAR CASE OF SURJECTIVE LIMIT

By *PHILIPPE NOVERRAZ*

1. We shall consider the following situation: Let E be a non separable complex Banach space and let $(E_i)_{i \in I}$ be a directed (by inclusion) family of closed subspaces of E such that:

(a) $E = \bigcup_{i \in I} E_i$,

(b) for any i in I, there is a projection u_i from E onto E_i ,

(c) for any i and j in I, $i \leq j$, there is a projection u_{ij} from E_j onto E_i such that the following diagramm commutes:

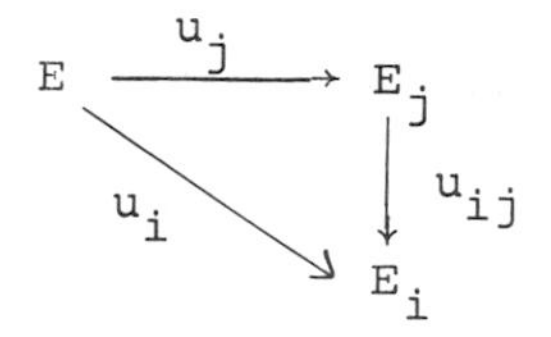

We consider the surjective limit (in the sense of (4)) of $(E, u_i, u_{ij})_{i,j \in I}$ and we denote by $E_{\leftarrow}$ this surjective limit. As a set, we have $E = E_{\leftarrow}$, but the projective topology of $E_{\leftarrow}$ is always strictly weaker than the initial norm topology. It is obvious, via the open mapping theorem, that E is an open surjective limit and, so, any holomorphic function on $E_{\leftarrow}$ factorizes trough an E_i (i.e. for any f in $H(U)$ there is i in I

and $\tilde{f}$ in $H(U_i)$ such that $f = \tilde{f} \circ u_i$).

Let us consider a few examples in which this situation arises in a natural way:

1) (E_i) is the set of all finite dimensional subspaces of a Banach space E and the projective topology is the weak topology $\sigma(E,E')$.

2) $E = C_o(A)$ for A uncountable, endowed with the sup norm, and the spaces (E_i) are the spaces $C_o(A')$, A' countable in A, considered as subspaces of $C_o(A)$ by identifying $x = (x_\alpha)_{\alpha \in A'}$, with $y = (y_\alpha)_{\alpha \in A}$ where $y_\alpha = x_\alpha$ if $\alpha \in A'$ and $y_\alpha = 0$ if $\alpha \in A \setminus A'$.

3) $E = \ell^p(A)$, A uncountable, $1 \leq p \leq +\infty$, and the spaces E_i are the spaces $\ell^p(A')$, A' countable in A.

4) K is a non metrizable compact set and E is the space of all continuous functions with metrizable support. The spaces E_i are the spaces $\mathcal{C}(K')$ for $K' \subset K$, compact and metrizable, the mappings u_i are the restriction mappings which are onto by the Tietze extension theorem.

5) E is a non separable Hilbert space and (E_i) the family of all separable closed subspaces and u_i the orthogonal projections.

We shall also suppose that the following condition is satisfied: (*) for any countable subset I' of I, there is an index i' in I such that

$$E_{i'} \supset \bigcup_{i \in I'} E_i .$$

This condition is satisfied in the examples 2,3, 4 and 5 (but not in example 1).

PROPOSITION 1. *Suppose* E *and* $E_{\leftarrow}$ *as considered above and assume that condition* (*) *is satisfied, then the compact subsets in* E *and* $E_{\leftarrow}$ *are the same.*

PROOF: If K is a compact subset of $E_{\leftarrow}$ then, for any i, $u_i(K)$ is compact in E_i. If (x_n) is a sequence in K, condition (*) implies that (x_n) is contained in a subspace E_i. As u_i is a projection, it follows that (x_n) is contained in the compact $u_i(K)$ and contains a subsequence converging in $u_i(K)$. Since K is closed in E and since E is metrizable, the proposition is proved.

On E and $E_{\leftarrow}$ we have the same hypoanalytic functions (i.e. G - analytic with continuous restrictions to any compact subsets) and, as E is normed, any hypoanalytic function is analytic (i.e. G - analytic and continuous). In the case $E = c_o(A)$, we have $H(E) = H(E_{\leftarrow})$ as pointed out by Josefson (5). This is not the case in general, for instance consider $1 \leq p < +\infty$ and $E = \ell^p(A)$. The function f, defined by $f(x) = \sum_{\alpha \in A} x_\alpha^{[p]}$, where $[p]$ means the integral part of p, is polynomial and continuous for the norm topology. Since it cannot factorizes through any subspace $\ell^p(A')$, A' countable, the function f is not continuous for the projective topology. This also shows that, in an open surjective limit, hypoanalytic functions do not factorize.

Let us recall that a subset B of E is said to be bounding if every holomorphic function on E is bounded on B. If E

is a weakly compactly generated (WCG) Banach space, it is known (8) that the bounding subsets of E are exactly the compact subsets. The converse is not always true: the space $E = \ell^1(A)$, A uncountable, is not WCG (6) but for that space the bounding sets are compact: if B is bounding in $H(E)$, it is bounding in $H(E_{\leftarrow})$; for any i, $u_i(B)$ is bounding in E_i and then compact since E_i is a separable Banach space. The set B is then compact for the projective topology and also, according to the proposition compact for the norm topology.

2. We shall consider now a slightly more general situation:

(a) Let I be a directed set such that for any sequence (i_n) in I there is an index i such that $i_n \leq i$ for all n.

(b) Let E be a locally convex space, for each i of I, let E_i be a metrizable locally convex space and $u_i : E \to E_i$ a linear, surjective, continuous and open mapping such that a basis for the topology of E is given by the sets $u_i^{-1}(V)$ for any i in I and any open set V of E_i. Assume also that any compact subset of E_i is contained in the image of compact subset of E.

(c) For any couple $i \leq j$, let u_{ij} be a linear, continuous and surjective mapping from E_j to E_i such that the following diagram commutes

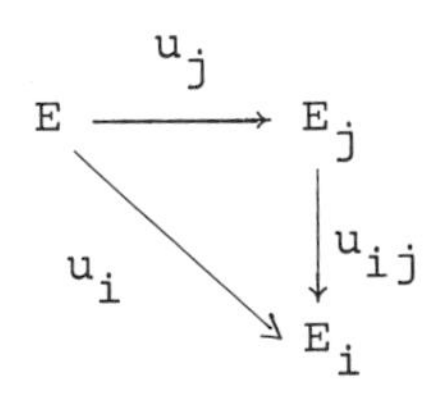

In other words, we consider an open and compact surjective limit (4) of matrizable spaces with an extra condition of countable stability on the indexing set. This condition excludes the case of a lcs with the weak topology.

If E is a locally convex space, let us denote on $H(E)$ as usual by T_o the compact open topology, T_ω the topology of Nachbin generated by all the semi-norms ported by compact subsets, T_δ the topology generated by the semi-norms p such that for any increasing countable and open covering (U_n) of E there is a constant C and an index n_o such that $p(f) \leq C|f|_{U_{n_o}}$ for all f in $H(E)$.

Let us index by b the bornological topology T_b associated with T, for instance $T_{o,b}$ or $T_{\omega,b}$.

Since holomorphic functions on E factorizes, we can identify each $H(E_i)$ with a subspace of $H(E)$ by the relation $f \to \tilde{f}$ where $f = \tilde{f} \circ u_i$, and it is natural to ask about the inductive limit of the spaces $H(E_i)$ for different topologies on $H(E_i)$. For each i in I, we have the following diagram:

$$\begin{array}{ccccc} H(E),T_\delta & \longrightarrow & H(E),T_\omega & \longrightarrow & H(E),T_o \\ \uparrow & & \uparrow & & \uparrow \\ H(E_i),T_{\delta,i} & \longrightarrow & \varinjlim H(E_i),T_{\omega,i} & \longrightarrow & \varinjlim H(E),T_{o,i} \end{array}$$

In the following diagram, an arrow indicates continuity of the identity mapping:

$$\begin{array}{ccccc} H(E),T_\delta & \longrightarrow & H(E),T_\omega & \longrightarrow & H(E),T_o \\ \uparrow & & \uparrow & & \uparrow \\ \varinjlim H(F_i),T_{\delta,i} & \longrightarrow & \varinjlim H(E_i),T_{\omega,i} & \longrightarrow & \varinjlim H(E_i),T_{o,i} \end{array}$$

THEOREM *If* E *is a surjective limit as considered above, all the topologies in the diagram have the same bounded sets.*

PROOF: It is well known that two topologies have the same bounded sets if and only if they have the same bounded sequences. If (f_n) is a bounded sequence in $H(E),T_o$ each f_n factorizes through a space E_{i_n} and then, by hypothesis, there is an index i such that the sequence (f_n) factorizes through E_i. If we identify f_n with $\tilde{f}_n$ such that $f_n = \tilde{f}_n \circ u_i$, the set $(\tilde{f}_n)$ is then bounded in $H(E_i),T_{o,i}$: let K be a compact subset of E_i, there is a compact subset K' of E such that $K \subset u_i(K')$, hence $\sup_n |\tilde{f}_n|_K \leq \sup_n |f_n|_{K'} < +\infty$. As E_i is metrizable, the topologies $T_{o,i}$, $T_{\omega,i}$ and $T_{\delta,i}$ have the same bounded sets (3). The sequence $(\tilde{f}_n)$ is bounded in $H(E_i),T_{\delta,i}$ and then also in $\varinjlim H(E_i),T_{\delta,i}$, which proves the theorem.

REMARK: *In general, a bounded subset of* $H(E)$ *does not factorize through a space* E_i *as the following example shows: Let* $E = \ell^p(A)$, *A uncountable, be considered as a projective limit of the spaces* $\ell^p(A')$, *A' countable in A, and let* $B = (f_\alpha)_{\alpha \in A}$ *where each* f_α *is defined by* $f_\alpha(x) = x_\alpha$, *the set B is bounded in* $H(E)$ *for the compact open topology but does not factorize.*

The bornological topology associated with the six topologies considered in the theorem coincides with the following topologies: T_δ, $\varinjlim H(E_i),T_{\delta,i}$, $\varinjlim H(E_i),T_{\omega,i,b}$ and $\varinjlim H(E_i),T_{o,i,b}$.

PROPOSITION 2. *If* E *is a surjective limit as considered above, we have:*

(a) $H(E), T_{\omega,b} = \varinjlim H(E_i), T_{\omega,i,b}$,

(b) T_δ *is the bornological topology associated with* T_ω *or* T_o ,

(c) *if* $H(E_i), T_{\omega,i}$ *is bornological for any* i , *then* $H(E), T_\delta = \varinjlim H(E_i), T_{\omega,i}$.

The hypothesis (c) is verified for instance if $E = C_o(A)$ or $\ell^p(A)$, $1 \leq p < +\infty$, A uncountable (see (2)).

REMARK: *It is known (1) that any bounded set of* $H(\mathbb{C}^I)$ *is contained and bounded in a subspace* $H(\mathbb{C}^J)$, J *finite in* I, *and so, on* $H(\mathbb{C}^I)$ *we have* $H(\mathbb{C}^I), T_\delta = \varinjlim_{\substack{J \text{ finite} \\ J \subset I}} H(\mathbb{C}^J)$ *although the theorem does not apply in that case.*

3. We shall now consider the special case $E = C_o(A)$, A uncountable, where the holomorphic functions for the norm and for the projective topologies are the same (5). Let us index by 1 the topologies on $H(E)$ associated with the norm topology (for instance $T_{o,1}$, $T_{\omega,1}$ and $T_{\delta,1}$) and by 2 the topologies associated with the projective topology ($T_{o,2}$, $T_{\omega,2}$ and $T_{\delta,2}$). We have the following diagram:

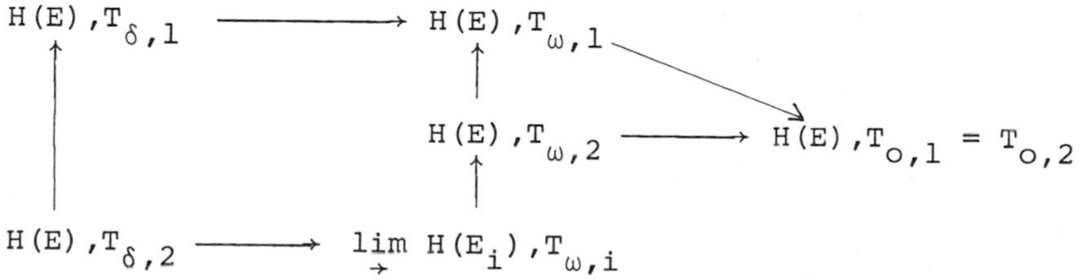

$$\begin{array}{ccccc}
H(E), T_{\delta,1} & \longrightarrow & H(E), T_{\omega,1} & \searrow & \\
\uparrow & & \uparrow & & \\
 & & H(E), T_{\omega,2} & \longrightarrow & H(E), T_{o,1} = T_{o,2} \\
\uparrow & & \uparrow & & \\
H(E), T_{\delta,2} & \longrightarrow & \varinjlim H(E_i), T_{\omega,i} & &
\end{array}$$

From the theorem, we know that all these topologies have

the same bounded sets and so $T_{\delta,1} = T_{\delta,2}$. In fact, it is possible to prove that the two topologies $T_{\omega,1}$ and $T_{\omega,2}$ coincide: let p be a semi-norm on H(E) continuous for $T_{\omega,2}$. If p is not $T_{\omega,1}$ continuous, for any compact K of E there is as T_1-open set U, $U \supset K$ such that $p(f_{n,K}) \geq n|f_{n,K}|_U$. For each K, the sequence $(f_{n,K})$ factorizes through a space $C_o(A_K)$, A_K countable. Is u is the restriction mapping from $C_o(A)$ onto $C_o(A_K)$, the set $U_K = u^{1} \circ u(U)$ is open for the projective topology and we have $|f_{n,K}|_{U_K} = |f_{n,K}|_U$ for each n. Since the compact sets are the same for the norm and for the projective topologies, we can deduce that the semi-norm p is not $T_{\omega,2}$ continuous. Contradiction.

Finally, on the space $C_o(A)$ there are two distinct topologies defining exactly the same holomorphic functions and on H(E) exactly the same topology T_δ (resp. T_ω, T_o). So, from the holomorphic point of view, it seems that there is no reason to distinguish between these two topologies. Nevertheless, for the norm topology the Levi problem has in general no solution while for the projective topology the Levi problem has always a solution.

BIBLIOGRAPHY

[1] J. A. BARROSO AND L. NACHBIN, - Sur certaines propriétés bornologiques des espaces d'applications holomorphes, Colloque de Liège (1970).

[2] S. DINEEN, - Holomorphic functions on (C_o, X_1) - modules, Math. Ann. 1972, t. 196, p. 106 - 116.

[3] S. DINEEN, - Holomorphic functions on lcs, Ann. Fourier, 1973, t. 23, p. 19 - 54.

[4] S. DINEEN, - On surjective limits, to appear in the Bull. SMF.

[5] B. JOSEFSON, - A counterexample to the Levi problem, Lexington conference, Springer Lecture notes nº 364.

[6] J. LINDENSTRAUSS, - Weakly compact sets, Symposium on Infinite Dimensional Topology, Annals of Math. Studies nº 69, 1972.

[7] Ph. NOVERRAZ, - Sur la topologie tonnelée et la topologie bornologique associée à la topologie de Nachbin, CR. Acad. Cs. t. 279, 1974, p. 459 - 463.

[8] M. SCHOTTENLOHER, - Über analytische Fortsetzung in Banachräumen, Math. Ann. (1972), t. 199, p. 313-336.

Université de Nancy I
Mathématiques
Case Officielle 140
54037 - NANCY CEDEX
FRANCE

Infinite Dimensional Holomorphy and Applications, Matos (ed.)

THE CONNECTED FINITE DIMENSIONAL LIE SUB-GROUPS OF THE GROUP Gh(n,C)

BY *DOMINGOS PISANELLI*

With this work we continue our researches about the infinite Lie group of germs of holomorphic invertible transformations that preserve the origin in C^n, which in [1] we called Gh(n,C). A natural problem is the research of it's sub-groups. In [2] we found the one parameter sub-groups and we showed that these are given by $\alpha \in C \to \exp \alpha\, a$, where exp x is the exponential

$$\sum_{m\geq 0} \frac{1}{m!} X_x^m(\mathrm{id}) \in Gh(n,C),$$

$x \in gh(n,C)$ the Lie algebra of Gh(n,C), of germs of holomorphic transformations around the origin that preserve the origin, id the unity of Gh(n,C) and $X_x = x_1 \frac{\partial}{\partial t_1} + \dots + x_n \frac{\partial}{\partial t_n}$. Though exp is not invertible around the origin of gh(n,C) and the identity of Gh(n,C) ([2]), we show here that it is locally injective when restricted to a finite dimensional vector sub-space. This is possible because a finite dimensional vector sub-space of a Hausdorff locally convex vector space has a topological supplement. We can then construct M the connected finite dimensional sub-group associated to a finite dimensional sub-algebra of gh(n,C) (theorem 1), and we give the inverse of a chart around the identity of M. This is equivalent to the construction of the local finite dimensional transformation's group. This is the content of §1. In order to obtain a self-contained paper we give concisely in §2 the construction of a Lie group theory in an open set of a locally convex vector space, including the construction of the local canonical Lie group in a Banach space and some properties of the group Gh(n,C). This result will appear in [1] and [2].

§1

THEOREM 1: *There exists a 1-1 mapping from the finite dimensional connected Lie sub-groups of* Gh(n,C) *onto the finite dimensional Lie sub-algebras of* gh(n,C). *This mapping associates to each connected finite dimensional Lie sub-group the Lie algebra of the tangent space at the unity.*

I) If M is an r-dimensional connected Lie sub-group of Gh(n,C), there exists a chart ϕ_e around $e=id(\phi_e(e)=0)$ with values in C^r. $\psi_e = (\phi_e^{-1})'(0)$ is 1-1 (linear) map from C^r onto T_e (the tangent space at M in e). We can define a local Lie group G_e in T_e. $\phi_e^{-1} \circ \psi_e^{-1}$ will be a local homomorphism (definition (§2,II)) from G_e in Gh(n,C), and its differential at the origin = identity, will be an homomorphism of the Lie algebra in gh(n,C) (§2,II). *Then* T_e *is a Lie sub-algebra of* gh(n,C) *and its dimension is* r.

The local Lie group G_e is locally isomorphic to a local canonical Lie group C_e in T_e, with operation ϕ_1, infinitesimal transformation $L_1(x_1)$, the inverse of this $\mathcal{L}_1(x_1)$ and same algebra of G_e (§2,II). The isomorphism (local) between C_e and G_e is given by an exponential mapping from T_e in T_e : $\exp_\phi$ (§2,II). $\theta = \phi_e^{-1} \circ \psi_e^{-1} \circ \exp_\phi$ is a local isomorphism between C_e and M, i.e. $\theta(\phi_1(x_1,y_1)) = \phi(\theta(x_1), \theta(y_1))$ (x_1,y_1 around the origin in T_e). Then

$$\theta'(\phi_1(x_1,y_1))(\phi_1)'(x_1,o)h_1 = \phi_y'(\theta(x_1),\theta(o))\,\theta'(o)h_1.$$

But $\theta'(o)h_1 = h_1$ ($\forall\ h_1 \in T_e$) then:

$$(1)\quad \begin{cases} \theta'(x_1)k_1 = J(\theta)\,\mathcal{L}_1(x_1)k_1 \\ \theta(o) = e \end{cases}$$

(see (§2,II) for the mean of $J(\theta)$).

We will see later (§1,II) that $\theta = \exp$.

M is connected, then $M = \bigcup_{n\geq 1} U^n = \bigcup_{n\geq 1} [\theta(U_1)]^n$, where U is a neighbourhood of e in M, $U = U^{-1}$ and U_1 is a neighbourhood of zero in T_e. We have then *the uniqueness of the sub-group M as a set.*

If in M we have another atlas $\tilde{\phi}$, and $(M,\tilde{\phi})$ is a Lie subgroup with the same tangent space T_e at the identity e, we will have obviously:

$$\tilde{\phi}_e^{-1} \circ \tilde{\psi}_e^{-1} \circ \exp_{\tilde{\phi}} = \exp = \phi_e^{-1} \circ \psi_e^{-1} \circ \exp_\phi\ ,$$

then

$$\phi_e \circ \tilde{\phi}_e^{-1} = (\psi_e \circ \exp_\phi) \circ (\tilde{\psi}_e^{-1} \circ \exp_{\tilde{\phi}})^{-1}$$

and $\tilde{\phi}_e \circ \phi_e^{-1}$ holomorphic. The identity mapping i_M of M is holomorphic at e with holomorphic inverse . But i_M is the composition of

$$x \in (M,\phi) \to a^{-1}x \in (M,\phi) \to a^{-1}x \in (M,\tilde{\phi}) \to a(a^{-1}x) = x \in (M,\tilde{\phi})\ ,$$

$$x \in (M,\tilde{\phi}) \to a^{-1}x \in (M,\tilde{\phi}) \to a^{-1}x \in (M,\phi) \to a(a^{-1}x) = x \in (M,\phi)\ .$$

This gives the holomorphy of i_M and of its inverse at any $a \in M$. (M,ϕ) *is then isomorphic to* $(M,\tilde{\phi})$.

II) Conversely let T_e be a finite dimensional sub-algebra of $gh(n;C)$. Let $G_e = (\phi_1, U_1, V_1, W_1)$ be the local canonical group in T_e whose algebra is T_e (§2,II). Let $L_1(x_1)$ be its infinitesimal transformation and $\mathcal{L}_1(x_1)$ its inverse. Let be the system:

$$(2) \begin{cases} \theta'(x_1)h_1 = J(\theta)\, \mathcal{L}_1(x_1)h_1 \\ \theta(o) = e \end{cases}$$

In order to solve the last system we must solve the ordinary system:

$$(3) \begin{cases} \frac{d}{dt}g = J(g)\, \mathcal{L}_1(tx_1)x_1 = J(g)x_1 \\ g(o) = e \end{cases}$$

whose solution is obviously $g(t,x_1) = \exp tx_1$ (§2,II).

A) The right member of (2) satisfies the integrability condition:

$(J(\theta)\,\mathcal{L}_1(x_1)h_1)'_{x_1}\, k_1 + (J(\theta)\,\mathcal{L}_1(x_1)h_1)'_{\theta}(J(\theta)\,\mathcal{L}_1(x_1)\, k_1)$

is symmetric in $(h_1,k_1) \in T_e \times T_e$ $(\forall\, \theta \in Gh(n,C), x_1 \in U_1)$.

Let us put $L(\theta) = J(\theta)$, $\mathcal{L}(\theta) = J^{-1}(\theta)$. When $h, k \in T_e$ we have:

$$\mathcal{L}(\theta)\, [Lh, Lk]_{\theta} = [h,k] = \mathcal{L}_1(x_1)\, [L_1h, L_1k]_{x_1} \quad \text{i.e.}$$

$$(4) \begin{cases} \mathcal{L}(\theta)\, [L'(\theta)(L(\theta)k)h - L'(\theta)(L(\theta)h)k] = \\ = \mathcal{L}_1(x_1)\, [L_1'(x_1)(L_1(x_1)k)h - L_1'(x_1)(L_1(x_1)h)k]\,. \end{cases}$$ But

$$\mathcal{L}_1(x_1)L_1(x_1)h = h \qquad (\forall\, h \in T_e),$$

$$\mathcal{L}_1'(x_1)k(L_1(x_1)h) + \mathcal{L}_1(x_1)\cdot L_1'(x_1)kh = 0 \qquad \text{and}$$

$$\mathcal{L}_1'(x_1)(L_1(x_1)k)(L_1(x_1)h) + \mathcal{L}_1(x_1)(L_1'(x_1)(L_1(x_1)k)h) = 0.$$

(4) gives then:

$$(5)\begin{cases} \mathcal{L}(\theta)\ [L'(\theta)(L(\theta)k)h - L'(\theta)(L(\theta)h)k] = \\ \\ = -\mathcal{L}_1'(x_1)(L_1(x_1)k)(L_1(x_1)h) + \mathcal{L}_1'(x_1)(L_1(x_1)h)(L_1(x_1)k). \end{cases}$$

Computing $L(\theta)$ in both members of (5) and by substitution of h and k by $\mathcal{L}_1(x_1)h$ and $\mathcal{L}_1(x_1)k$ we obtain:

$$L'(\theta)(L(\theta)(\mathcal{L}_1(x_1)k)\mathcal{L}_1(x_1)h) - L'(\theta)(L(\theta)(\mathcal{L}_1(x_1)h)(\mathcal{L}_1(x_1)k)$$

$$= -L(\theta)\mathcal{L}_1'(x_1)kh + L(\theta)\mathcal{L}_1'(x_1)hk \quad , \text{ i.e.}$$

$$(L(\theta)\mathcal{L}_1(x_1)h)_\theta'(L(\theta)\mathcal{L}_1(x_1)k) + L(\theta)(\mathcal{L}_1'(x_1)h)k =$$

$$= (L(\theta)\mathcal{L}_1(x_1)k)_\theta'(L(\theta)\mathcal{L}_1(x_1)h) + L(\theta)(\mathcal{L}_1'(x_1)k)h \quad \text{i.e. A).}$$

From (§2,IV) we obtain that $g(1,x_1)$ the solution of (3) at $t = 1$, is the solution of (2).

B) θ gives a local homomorphism from G_e into $Gh(n,C)$.

By using the Lie equation (§2,II):

$$(\theta(\phi_1(x_1,y_1)))_{y_1}' h_1 = \theta'(\phi_1(x_1,y_1))(\phi_1)_{y_1}'(x_1,y_1)h_1$$

$$= L(\theta(\phi_1(x_1,y_1)))\mathcal{L}_1(x_1)h_1 \quad .$$

$$(\phi(\theta(x_1),\theta(y_1)))_{y_1}' h_1 = \phi_y'(\theta(x_1),\theta(y_1))\theta'(y_1)h_1$$

$$= L(\phi(\theta(x_1),\theta(y_1)))\mathcal{L}_1(x_1)h_1 \ .$$

But $\theta(\phi_1(x_1,o)) = \theta(x_1)$

$$\phi(\theta(x_1),\theta(o)) = \phi(\theta(x_1),e) = \theta(x_1).$$

$\phi(\theta(x_1,\theta(y_1))$ and $\theta(\phi_1(x_1,y_1))$ satisfy the same differential equation with the same initial conditions, then

$$\theta(\phi_1(x_1,y_1)) = \phi(\theta(x_1),\theta(y_1))$$

when x_1 and y_1 are around $(0,0) \in T_e \times T_e$. This proves B).

C) θ is locally injective.

T_e has a supplementary topological sub-space S_e.

Let p_1 and p_2 the projections from $gh(n,C)$ onto T_e and S_e and $\theta_1 = p_1 \circ \theta$, $\theta_2 = p_2 \circ \theta$. We obtain $\theta = \theta_1 + \theta_2$.

Let $\tilde{\theta} : x_1 + y_1 \in T_e \oplus S_e \to \theta(x_1) + y_1 \in T_e \oplus S_e$

$$\tilde{\theta}'_{x_1}(o)h_1 = \theta'(o)h_1 = h_1 \qquad (\forall\, h_1 \in T_e),$$

$$\tilde{\theta}'_{x_1}(o)h_1 = \theta'_1(o)h_1 + \theta'_2(0)h_1 \qquad (\forall\, h_1 \in T_e),$$

then $\theta'_1(o) =$ identity.

θ_1 an analytic mapping from T_e in T_e, is locally invertible and $\tilde{\theta}$ too:

$$\tilde{\theta}^{-1} \begin{cases} x_1 = \theta_1^{-1}(x) \\ y_1 = -\theta_2 \circ \theta_1^{-1}(x) + y. \end{cases}$$

θ the restriction of $\tilde{\theta}$ on $y_1 = 0$ is locally injective, this proves C.

By restriction of the open sets U_1, V_1 and W_1, we can suppose that θ is injective in W_1. Then by B) and C)

$$(o, \theta(U_1), \theta(V_1), \theta(W_1))$$

is a local topological group in $Gh(n,C)$. $G_U = \bigcup_{n \geq 1} U^n$ is a topological group $(U = \theta(U_1))$ and the aU $(a \in U^{n-1}, n \geq 1, U^o = \{e\})$ are open in G_U(§2,I).

We will define in G_U a Lie sub-group structure (§2,III):

a) $G_U = \bigcup_{n \geq 1} \bigcup_{a \in U^{n-1}} a\,U$

b) $\theta_a : x \in a\,U \to \theta^{-1}(a^{-1}x) \in U_1$

is 1-1 surjective. θ_a is an homeomorphism too.

c) $\theta_a(a\,U \cap b\,U) = \theta^{-1}(a^{-1}(b\,U))$ is open because $a^{-1}(b\,U)$ is open G_U.

d) $x_1 \in \theta_a(a\,U \cap b\,U) \to \theta_b \circ \theta_a^{-1}(x_1) = \theta^{-1}(b^{-1} \cdot \theta(ax_1))$

$$= \tilde{\theta}^{-1}(b^{-1} \cdot \theta(ax_1))$$

is holomorphic.

e) The identity mapping from G_U in Gh(n,C) is holomorphic because so is $\theta_a^{-1} = a\,\theta$ $(\forall\ a \in M)$.

f) $h_1 \in T_e \to (\theta_a^{-1})'(x_1)h_1 = J(a)\theta'(x_1)h_1$ is injective.

g) $\theta_{ab}(\theta_a^{-1}(x_1)\cdot\theta_b^{-1}(y_1)) = \theta^{-1}((ab)^{-1}\cdot a\,\theta(x_1)\cdot b\,\theta(y_1))$

$= \theta^{-1}(b^{-1}\cdot\theta(x_1)\cdot b\cdot\theta(y_1)) = \tilde{\theta}^{-1}(b^{-1}\cdot\theta(x_1)\cdot b\cdot\theta(y_1))$

is holomorphic in $(x_1,y_1) \in U_1 \times U_1$ $(\forall\ a,b \in G_U)$.

This gives the holomorphy of the product in G_U.

h) $\theta_{a^{-1}}([\theta_a^{-1}(x_1)]^{-1}) = \theta^{-1}(a[a\cdot\theta(x_1)]^{-1})$

$= \theta^{-1}(a[\theta(x_1)]^{-1}\,a^{-1}) = \tilde{\theta}^{-1}(a(\theta(x_1))^{-1}\,a^{-1})$ is holomorphic.

This gives the holomorphy of the inverse in G_U.

The tangent space of G_U at e is

$$\theta'(o)T_e = T_e.$$

By restriction we can suppose that U_1 is connected and $U = \theta(U_1)$ too. We then have the connection of U^n and of G_U too.

§2

I) LOCAL TOPOLOGICAL GROUPS

Let W be a topological space, $U \subset V \subset W$ open and the mapping:

$$\phi:\ (x,y) \in V \times V \to xy \in W$$

$$U\cdot U \subset V$$

continuous such that

$$(xy)z = x(yz) \qquad \forall\ x,y,x \in U.$$

There exists $e \in U$ such that

$$xe = ex = x \qquad \forall\ x \in U.$$

To each $x \in U$ corresponds $x^{-1} \in U$ such that

$$xx^{-1} = x^{-1}x = e$$

and let the mapping $x \in U \to x^{-1} \in U$ be continuous.

As a consequence we have the uniqueness of the inverse element x^{-1} of $x \in U$ and $(x^{-1})^{-1} = x$ $(\forall\, x \in U)$. Then $U = U^{-1}$. We will use the symbol (ϕ,U,V,W) in order to indicate a *local topological group*.

When $U = V = W$ we will have a *topological group* and we will use the symbol (ϕ,U).

When G is a group (algebraic) with operation ϕ and (ϕ,U,V,W) is a local topological group $(U,V,W \subset G)$, we can define a topological group in $G_U = \bigcup_{n\geq 1} U^n$, where aU is open

$$(\forall\, a \in U^{n-1},\ n \geq 1,\ U^0 = \{e\}).$$

A *local homomorphism* of the topological local group (ϕ_1,U_1,V_1,W_1) into the topological local group (ϕ,U,V,W), is a continuous mapping f defined in an open set of W_1 with values in W, such that

$$f(\phi_1(x_1,y_1)) = \phi(f(x_1),f(y_1))$$

when (x_1,y_1) is around (e_1,e_1).

A *local isomorphism* of the topological local group (ϕ,U_1,V_1,W_1) into the topological local group (ϕ,U,V,W) is an homomorphism f that is invertible in its image that we suppose open and f^{-1} continuous.

An *homomorphism* from the topological group (ϕ_1,U_1) into (ϕ,U) is a continuous mapping that preserves the operations ϕ_1 and ϕ.

An *isomorphism* from the topological group (ϕ_1,U_1) onto the topological group (ϕ,U) is an homomorphism that has a continuous inverse on U .

II) LIE LOCAL GROUP

Let T be a complex locally convex space. Hausdorff and sequentially complete. A *local Lie group* is a local topological group (ϕ,U,V,W) where U,V,W are open in T, ϕ and $x \in U \to x^{-1} \in U$ are G-analytic ([5]).

The mapping, linear in h, $(x,h) \in V \times T \to L(x)h$, where $L(x)h = \{\frac{d}{d\alpha}\phi(x,e+\alpha h)\}_{\alpha=0}$, is called the *infinitesimal transformation* of the Lie local group.

There exists $\mathcal{L}(x)$ the inverse of $L(x) \in \mathcal{L}(T)$ (the set of the linear mappings of T) for any $x \in U$. When $U = V = W$ we will have a *Lie group* that we will indicate (ϕ,U).

Let H(U) be the complex vector space of G-analytic mappings in $U \subset T$ with values in T.

H(U) is a Lie algebra when we define the bracket

$$[f,g] = f'g - g'f \qquad f,g \in H(U).$$

The mappings $x \in U \to L(x)h \in T$ are a Lie sub-algebra in H(U) and there exists $[h,k] \in T$ such that

$$[Lh,Lk] = L[h,k]$$

We have then a Lie algebra isomorphic to T when we define in T the bracket

$$[h,k] = L'(e)k\,h - L'(e)h\,k\,.$$

The operation ϕ satisfies the *Lie equation*

$$\begin{cases} \phi'_y(x,y)h = L(\phi)\,\mathcal{L}(y)h \\ \\ \phi(x,e) = x \end{cases} \qquad x,y \in U$$

EXAMPLE: Let gh(n,C) be the complex locally convex space of germs of holomorphic transformations that preserve the origin of C^n ([1]).

Let $Gh(n,C)$ be the open sub-set of invertible germs, with the group operation $x \circ y$ $(x,y \in Gh(n,C))$. $Gh(n,C)$ *is a Lie group with unity id (the germ of the identity mapping), and the inverse of* $x \in Gh(n,C)$ *is the germ of the inverse transformation of a representative of* x.

The infinitesimal transformation is $L(x)h = J(x)h$, *where* $J(x)$ *is the Jacobian of* $x \in Gh(n,C)$. *The Lie algebra of* $gh(n,C)$ *is the classical:*

$$[h,k] = J(k)h - J(h)k$$

The system

$$\begin{cases} \dfrac{dg}{dt} = J(g)\ x \\ g(o) = id \end{cases}$$

has the solution $g(t,x)$ *holomorphic in* $C \times gh(n,C)$ *and*

$$\exp x = g(1,x)$$

is an exponential mapping, i.e.

$$\exp \alpha x \circ \exp \beta x = \exp(\alpha + \beta)x$$

$$(\forall\ \alpha, \beta \in C,\ x \in gh(n,C)).$$

We have too:

$$\exp x = \sum_{m \geq 0} \frac{1}{m!} X_x^m(id)$$

where $X_x = x_1 \dfrac{\partial}{\partial t_1} + \ldots + x_n \dfrac{\partial}{\partial t_n}$ ([2]).

A *local homomorphism* of the local Lie group (ϕ,U,V,W) into is an holomorphic (G-analytic and continuous) mapping f defined is an open set contained in U_1 containing $e_1 \in U_1$, with values in W such that:

$$f(\phi_1(x_1,y_1)) = \phi(f(x_1),f(y_1))$$

around (e_1,e_1).

A *local isomorphism* from the local Lie group (ϕ_1, U_1,V_1,W_1) into the local Lie group (ϕ,U,V,W), is a local homomorphism of the first into the second that is invertible, it's image is open in W and f^{-1} is holomorphic.

When f *is a local homomorphism of the local Lie group* (ϕ_1, U_1, V_1, W_1) *into the local Lie group* (ϕ, U, V, W), $f'(e_1)$ *is an homomorphism of the correspondent Lie algebras.*

A *local canonical Lie group* is a local Lie group (ϕ_1, U_1, V_1, W_1) where the unity is the origin and $\mathcal{L}_1(x_1)x_1 = x_1 \quad (x_1 \in U_1)$.

Every local group in a Banach space T *is locally isomorphic to a local canonical Lie group in* T . *This local isomorphism is given by the exponential* $\exp x = g(1,x)$, *where* $g(t,x)$ *is the solution of the system*

$$\begin{cases} \dfrac{dg}{dt} = L(g)x \\ \\ g(o) = e \end{cases}$$

we have too $(\exp'(o) = \text{identity}$.

To each continuous Lie algebra in a Banach space, there exists only a canonical Lie group whose algebra is the given, ([6] pg. 251).

III) LIE SUB-GROUPS

Let (ϕ, U) be a Lie group in T a locally convex, complex vector space, Hausdorff, sequentially complete.

A *Lie sub-group* of (ϕ, U) is a part $M \subset U$ such that:

α) M is an analytic manyfold: A family of parts $0_i \subset M$ $(i \in I)$ is given:

a) $M = \bigcup_{i \in I} 0_i$

b) Mappings ϕ_i $(i \in I)$ are given, 1-1, on an open set of a locally convex, complex, Hausdorff sequentially complete X $(\forall\ i \in I)$.

c) $\phi_i(0_1 \cap 0_j)$ is open in X $(\forall\ i,j \in I)$.

d) $\phi_j \circ \phi_i^{-1}$ is holomorphic in $\phi_i(0_i \cap 0_j)$ $(\forall\ i,j \in I)$.

e) The canonical mapping id from M in U is holomorphic.

f) The differential of id is injective, i.e.

$h_1 \in X \to (\phi_i^{-1})'(x_1)h_1 \in T$ is injective $(x_1 = \phi_i(x)$, $\forall\ x \in 0_i$, $i \in I)$.

β) M is a sub-group of U (in the algebraic sense).

g) The product of U when restricted to M is holomorphic.

h) The inverse of U when restricted to M is holomorphic.

IV) Let X and Y be locally convex space, complex, Hausdorff and sequentially complete. A connected and open in X, B open in Y, f LF-analytic in $A \times B \times X$ ([5]), $h \in X \to f(x,y)h \in Y$ linear $(\forall (x,y) \in A \times B)$. Let us suppose the integrability condition:

$$f'_x(x,y)h \cdot k + f'_y(x,y) \cdot f(x,y)h \cdot k$$

symmetric in $(h,k) \in X \times X$ $\quad (\forall (x,y) \in A \times B)$.

THEOREM: *The system*

$$(5) \quad \begin{cases} y'(x)h = f(x,y)h \\ \\ y(x_o) = y_o \qquad (x_o,y_o) \in A \times B \end{cases}$$

has only one solution LF-*analytic in* A, *provided that the ordinary differential equation*

$$(6) \quad \begin{cases} \dfrac{dg}{dt} = f(x_o + t(x - x_o),g)x \\ \\ g(o) = y_o \end{cases}$$

has a solution LF-*analytic in the connected set*

$$D = \{(t,x) \in C \times X \mid x_o + t(x - x_o) \in A\}.$$

PROOF: The uniqueness is a consequence of the Taylor expansion and the connectedness of A .

Let us prove the existence. We may suppose without loss of generality that $x_o = 0$, $y_o = 0$. Let us differentiate the solution $g(t,x)$ with respect to x with increment h :

$$(7) \quad \frac{d}{dt} g'_x(t,x)h = f'_x(tx,g(t,x))th\cdot x + f'_y(tx,g(t,x))g'_x(t,x)h\cdot x + \\ + f(tx,g(t,x))h.$$

$$(8) \quad \frac{d}{dt} (tf(tx,g(t.x))h) = f(tx.g(t.x))h + f'_x(tx.g(t.x))tx \cdot h + \\ + t f'_y (tx.g(t.x))f(tx.g(t.x))x \quad h.$$

Subtracting (7) and (8) and using the integrability condition we have:

$$\frac{d}{dt} (g'_x(t,x)h - t f (tx,g(t,x)h) =$$

$$= f'_y(tx,g(t,x))(g'_x(t,x)h - t f (tx,g(t,x))h)x.$$

The ordinary differential equation:

$$\begin{cases} \frac{d}{dt} a(t) = f'_y(tx,g(t,x))a(t) x \\ \\ a(o) = 0 \end{cases}$$

has the unique analytic solution $a(t) = 0$. But $g(o,x) = 0$ then

$$g'_x(o,x)h = 0 \qquad (\forall\ x \in A,\ h \in X),$$

$$g'_x(t,x)h - t f (tx,g(t,x))h = 0 \qquad (\forall (t,x) \in D),$$

and

$$g'_x(1,x)h = f(x,g(1,x))h \qquad (\forall\ x \in A).$$

From (6) we have $\frac{d}{dt} g(t,o) = 0 \quad (\forall\ t \in C)$ and

$$(10) \quad g(1,o) = g(o,o) = 0.$$

(9) and (10) show that $g(1,x)$ is the solution of (5).

In an other work ([3]) we made the assumption that X and Y are Banach scale and we gave sufficient conditions on f in order that (6) has a solution.

BIBLIOGRAPHY

[1] PISANELLI D., An example of infinite Lie group. 1975. To appear in "Proceedings of the American Mathematical Society".

[2] PISANELLI D., An extension of the exponential of a matrix and a counter-example of the inversion theorem of an holomorphic mapping in a space H(K). To appear in "Rendiconti di Matematica", of 'Istituto di Alta Matematica". 1975.

[3] PISANELLI D., Théorèmes d'Ovcyannicov, Frobenius, d'inversion et groupes de Lie locaux dans un'échelle d'espaces de Banach. C. R. Academ. Sci. Paris, Sér. A-B 277 (1973). A943-A946.

[4] PISANELLI D., Solutions of a non linear abstract Cauchy-Kovalewsky system as a local Banach analytic manyfold. To appear in "Proceedings of Colloquium of functional Analysis", Campinas, Brazil, 1974

[5] PISANELLI D., Applications analytiques en dimension Infinie. Bull. Sc. Math. 96 (1972), 181-191.

[6] MAISSEN B., Lie Gruppen mit Banachraüm als Parameterraüm, Acta Math. 108 (1962), 229-269.

INSTITUTO DE MATEMÁTICA E ESTATÍSTICA
DA UNIVERSIDADE DE SÃO PAULO

Infinite Dimensional Holomorphy and Applications, Matos (ed.)

MAXIMAL ANALYTIC EXTENSIONS OF RIEMANN DOMAINS OVER TOPOLOGICAL VECTOR SPACES

By K. RUSEK AND J. SICIAK (KRAKÓW) (*)

1. INTRODUCTION

Let E be a Hausdorff topological vector space (t.v.s.) over $\mathbb{K}$ ($\mathbb{K} = \mathbb{C}$ or $\mathbb{K} = \mathbb{R}$). Given a Riemann domain (X,p) over E let A denote a linear subspace of $\mathcal{O}(X)$ - the space of all complex-valued $\mathbb{K}$-analytic functions on X. Given a locally convex topology τ on A let A* denote the topological dual of (A,τ) endowed with the topology of pointwise convergence. Let $\Phi : X \longrightarrow A^*$ denote the evaluation mapping defined by

$$\Phi(x)(f) = f(x),\ x \in X,\ f \in A.$$

Under suitable assumptions on E, A and τ we prove that Φ is $\mathbb{K}$-analytic and, if $(X_\Phi, j_\Phi, \hat{\Phi})$ denotes the canonical maximal analytic extension of X with respect to Φ, the triple (X_Φ, j_Φ, A_Φ) is a maximal analytic extension of X with respect to A. The sym

---

(*) The present version of this paper was partially done by the second author during his stay (February-March, 1976) at the Uppsala University as its guest.

bol A_Φ denotes the set of all functions $\hat{f} \in \mathcal{O}(X_\Phi)$ defined by

$$\hat{f}(x) := \hat{\Phi}(x)(f),\ x \in X_\Phi,\ f \in A.$$

Let us mention here two examples where this construction of the maximal analytic extension of X with respect to A may be applied.

EXAMPLE 1 Let E be an arbitrary Hausdorff t.v.s. over C, A - an arbitrary linear subspace of $\mathcal{O}(X)$, and τ - the strongest locally convex topology on A. Then (X_Φ, j_Φ, A_Φ) is a m.a.e. of the pair (X,A) (see 5).

EXAMPLE 2 Let X be an open connected subset of $E = \mathbb{R}^n$ and let A be a linear space of complex-valued $\mathbb{R}$-analytic functions such that (A,τ_c) is a Frechet space, τ_c denoting the topology of uniform convergence on compact subsets of X. Then X_Φ is a Riemann domain over $\mathbb{C}^n$ and (X_Φ, j_Φ, A_Φ) is a m.a.e. of the pair (X,A) (see 9).

In particular we may take A = H(X) - the Frechet space of all complex-valued harmonic functions on X. Then the "harmonic envelope of holomorphy" of X is identical with the domain X_Φ (over $\mathbb{C}^n$) - the canonical domain of existence of the evaluation mapping $\Phi: X \longrightarrow H(X)^*$ (compare with [4], [5], [6]).

Our results presented in 6-7 improve some of the results due to Coeuré [2] and Schottenloher ([10], [11]). In particular, if E is a complex Baire or metrizable t.v.s. and $A = \mathcal{O}(X)$ is endowed with the bornological topology τ_{cb} associated with the topology τ_c, then a comparison of our construction with the Bishop construction of the envelope of holomorphy $\hat{X}$ of X leads to the conclusion that $\hat{X} = \hat{\Phi}(X_\Phi)$, i.e. each point of $\hat{X}$ may be identified with a linear multiplicative functional $h : \mathcal{O}(X) \longrightarrow \mathbf{C}$ that is τ_{cb}-continuous.

Moreover, if $(Y, j, \mathcal{O}(Y))$ is any analytic extension of $(X, \mathcal{O}(X))$ then $j^* : (\mathcal{O}(Y), \tau_{cb}) \longrightarrow (\mathcal{O}(X), \tau_{cb})$ is a topological isomorphism.

The corresponding results by Coeuré and Schottenloher were obtained under the assumption that E is metrizable.

In 10 we prove a theorem on natural Frechet spaces A of analytic functions in X which is an improvement of Theorem 2.3 of [2]; namely we do not assume that A separates the points of X.

All the notions of the theory of topological vector spaces, used in this paper, may be found [9].

Concerning the notion of an analytic function defined on open subset of a Hausdorff t.v.s. E with values in a locally convex sequentially complete t.v.s. F, along with the corresponding notation and theorems we follow [1]. All the vector spaces considered in 2 - 7 are complex.

CONTENTS

2. CATEGORY OF RIEMANN DOMAINS OVER A TOPOLOGICAL VECTOR SPACE

Let E be a Hausdorff topological vector space (t.v.s.) over $\mathbb{C}$.

2.1 DEFINITION *A Riemann domain over* E *is a pair* (X,p) *where* X *is a Hausdorff connected topological space and* $p: X \longrightarrow E$ *is a local homeomorphism.*

2.2 DEFINITION *The family of all Riemann domains over* E *will be denoted by* $\mathcal{R}(E)$ *and called the category of Riemann domains over* E. *If* (X,p) *and* (Y,q) *are elements (objects) of* $\mathcal{R}(E)$, *then any continuous mapping* $j: X \longrightarrow Y$ *such that* $p = q \circ j$ *will be called a morphism of the object* (X,p) *into the object* (Y,q).

2.3 DEFINITION *We say that a subset* $U \subset X$, *where* $(X,p) \in \mathcal{R}(E)$, *is schlicht, if* $p|U$ *is a homeomorphism of* U *onto* p(U).

One can easily check that any morphism j is a local homeomorphism.

2.4 LEMMA *Given* $(X,p) \in \mathcal{R}(E)$ *and a fixed point* $x \in X$, *there exist an open balanced neighbourhood* B(x,X) *of* $0 \in E$ *and a schlicht open neighbourhood* $\hat{B}(x,X)$ *of* x *such that*

(i) $p(\hat{B}(x,X)) = p(x) + B(x,X)$;

(ii) *If* $\tilde{U}$ *is a schlicht open neighbourhood of* x *and* U *is an open balanced neighbourhood of* $0 \in E$ *with* $p(\tilde{U}) = p(x) + U$, *then* $\tilde{U} \subset \hat{B}(x,X)$ *and* $U \subset B(x,X)$.

PROOF Let $\mathcal{U}_x$ denote the family of all pairs $(\tilde{U},U)$ satisfying the condition of (ii). Since p is a local homeomorphism, $\mathcal{U}_x \neq \emptyset$.

Put $\hat{B}(x,X) := \bigcup \tilde{U}$, $B(x,X) := \bigcup U$, where the unions are taken over all the pairs $(\tilde{U},U) \in \mathcal{U}_x$. It is obvious that $p(\hat{B}(x,X)) = p(x) + B(x,X)$. So it remains to show that $p|\hat{B}(x,X)$ is injective. This will be done if we show that $p|\tilde{U} \cup \tilde{V}$ is injective, if $(\tilde{U},U)$, $(\tilde{V},V) \in \mathcal{U}_x$. Put

$$t_U := (p|\tilde{U})^{-1} \mid p(\tilde{U}) \cap p(\tilde{V}), \quad t_V := (p|\tilde{V})^{-1} | p(\tilde{U}) \cap p(\tilde{V}).$$

Since $\tilde{U} \cap \tilde{V} \neq \emptyset$ and the set $p(\tilde{U}) \cap p(\tilde{V}) = p(x) + U \cap V$ is connected, the set

$$Z := \{z \in p(\tilde{U}) \cap p(\tilde{V}) : t_U(z) = t_V(z)\}$$

is identical with $p(\tilde{U}) \cap p(\tilde{V})$. So $p|\tilde{U} \cap \tilde{V}$ is injective, and consequently $p|\tilde{U} \cup \tilde{V}$ is injective (see Lemma 1.7 in [12]).

2.5 DEFINITION *The set* $\hat{B}(x,X)$ *will be called the maximal balanced neighbourhood of* x. *We define* $p_x := p|\hat{B}(x,X)$.

Observe that if $j: (X,p) \longrightarrow (Y,q)$ is a morphism, then $j(\hat{B}(x,X)) \subset \hat{B}(j(x),Y))$, $j_x := j|\hat{B}(x,X)$ is injective and $p_x = q_{j(x)} \circ j_x$. In particular $q_{j(x)}^{-1} = j_x^{-1} \circ p_x^{-1}$ in $p(x) + B(x,X)$.

The following statements 2.6 - 2.10 are direct consequences of the corresponding definitions or may be found in the first chapter of [2].

2.6 *If* $(X,p) \in \mathcal{R}(E)$, *then* p *determines (uniquely) the structure of an* E-*analytic manifold such that* p *is a local biholomorphism.*

2.7 *Each morphism is a holomorphic (and locally biholomorphic) mapping. Isomorphisms of the category* $\mathcal{R}(E)$ *are biholomorphisms.*

2.8 *(Identity principle). Two morphisms having the same value at one point are identical. If two analytic mappings of* (X,p) *into* (Y,q) *are identical in an open non-empty subset of* X *then they are identical in* X.

2.9 *If* (X,p), (Y,q), $(Z,r) \in \mathcal{R}(E)$ *and there are isomorphisms* $\phi : X \longrightarrow Y$, $\psi : X \longrightarrow Z$ *and morphisms* $u : Y \longrightarrow Z$, $v : Z \longrightarrow Y$ *such that* $\phi = u \circ \psi$ *and* $\psi = v \circ \phi$, *then* u *and* v *are isomorphisms and* $u = v^{-1}$.

3. ANALYTIC EXTENSIONS OF A RIEMANN DOMAIN (X,p) WITH RESPECT TO A FAMILY A OF ANALYTIC FUNCTIONS ON X

Given $(X,p) \in \mathcal{R}(E)$ and a locally convex sequentially complete t.v.s. F, we denote by $\mathcal{O}(X,F)$ the vector space of all analytic functions $f: X \longrightarrow F$. If $F = \mathbb{C}$, we shall write (X) instead of $\mathcal{O}(X,\mathbb{C})$.

Let A be a subset of $\mathcal{O}(X,F)$.

3.1 DEFINITION *We say that a triple* (Y,j,B) *is an analytic extension* (a.e.) *of the pair* (X,A) *if* $(Y,q) \in \mathcal{R}(E)$, $j: X \longrightarrow Y$ *is a morphism,* $B \subset \mathcal{O}(Y,F)$, $g \circ j \in A$ *for every* $g \in B$ *and the mapping* $j^*: B \ni g \longrightarrow g \circ j \in A$ *is surjective (then* j^* *- by the identity principle - is bijective).*

3.2 DEFINITION *An analytic extension* $(\tilde{X},\tilde{j},\tilde{A})$ *of* (X,A) *is called maximal (m.a.e.), if for evey a.e.* (Y,j,B) *of* (X,A) *there exists a morphism* $u: (Y,q) \longrightarrow (\tilde{X},\tilde{p})$ *such that* $\tilde{j} = u \circ j$.

The domain $(\tilde{X},\tilde{p})$ will be called the A(X,F)-envelope of holomorphy of (X,p). If $F = \mathbb{C}$ and $A = \mathcal{O}(X)$, then the $\mathcal{O}(X)$-envelope of holomorphy of (X,p) is called shortly the envelope of holomorphy of (X,p) and is denoted by $(\hat{X},\hat{p})$.

The following lemma and the Propositions 3.4 - 3.11 are simply consequences of the definitions or may be found in the quoted references.

3.3 LEMMA (*Coeuré* [2], *Narasimhan* [8], *Schottenloher* [11]). *There exists a maximal analytic extension for every* $(X,p) \in \mathcal{R}(E)$ *and for every subset* $A \subset \mathcal{O}(X,F)$.

3.3a REMARK *This lemma also follows from Remarks* 3.9 - 3.10 *and Theorem* 5.1 *and in this way we obtain its new proof.*

3.4 *Any two* $A(X,F)$*-envelopes of holomorphy of a domain* $(X,p) \in \mathcal{R}(E)$ *are isomorphic in the category* $\mathcal{R}(E)$.

3.5 *If* $(Y,q) \in \mathcal{R}(E)$ *is isomorphic with an* $A(X,F)$*-envelope of holomorphy of* $(X,p) \in \mathcal{R}(E)$, *then* (Y,q) *is also an* $A(X,F)$*-envelope of holomorphy of* (X,p).

3.6 *If* (Y,j,B) *is an a.e. of* (X,A) *and if* $(\tilde{X},\tilde{j},\tilde{A})$ *is a m.a.e. of* (X,A) *such that there exists a morphism* $u: \tilde{X} \longrightarrow Y$ *with* $j = u \circ \tilde{j}$, *then* (Y,j,B) *is also a m.a.e. of* (X,A).

3.7 *If* (Y,j,B) *is an a.e. of* (X,A), *where* A *is a vector space (an algebra), then* B *is also a vector space (an algebra) and* $j^* : B \longrightarrow A$ *is an algebraic isomorphism.*

3.8 $\mathcal{O}(\hat{X})$ *separates the points of* $\hat{X}$.

3.9 *Let* $\mathcal{O}_{E,F}$ *denote the covering space of the sheaf of germs of analytic functions from open subsets of* E *to* F. *Let* $\pi : \mathcal{O}_{E,F} \longrightarrow E$ *denote the canonical projection.*

Given $(X,p) \in \mathcal{R}(E)$ *and a fixed function* $f \in \mathcal{O}(X,F)$, *the mapping*

$$j_f : X \ni x \longrightarrow (f \circ p_x^{-1})_{p(x)} \in \mathcal{O}_{E,F}$$

is continuous ($(f)_a$ *denoting the germ of a function* f *at a point* $a \in E$).

Let X_f *denote the connected component of the space* $_{E,F}$ *containing the set* $j_f(X)$. *Finally define*

$$\hat{f}(x) := x(\pi(x)), \ x \in X_f,$$

where $x(\pi(x))$ *denotes the value of the germ* **x** *at the point* $\pi(x)$.

Then $(X_f,p_f) \in \mathcal{R}(E)$ $(p_f := \pi | X_f)$, $\hat{f} \in \mathcal{O}(X_f,F)$ *and the triple* $(X_f,j_f,\hat{f})$ *is a m.a.e. of the pair* (X,f).

The domain (X_f,p_f) is called the natural domain of existence of f and the triple $(X_f,j_f,\hat{f})$ is called the canonical m. a.e. of the pair (X,f).

3.10 *When looking for an a.e. of* (X,A), *without loss of generality, one may assume that* $A \subset \mathcal{O}(X,F)$ *is a vector subspace.*

Indeed, given any subset $S \subset \mathcal{O}(X,F)$, let A denote the linear span of S; then any analytic extension of (X,p) with respect to S is an analytic extension of (X,p) with respect to A and reversaly.

3.11 *If* (X,id_X,A) *is a m.a.e. of* (X,A) *and if* $x,y \in X$, $x \neq y$, *then there exists* $f \in A$ *such that* $j_f(x) \neq j_f(y)$.

4. ADMISSIBLE TOPOLOGIES ON A LINEAR SUBSPACE A OF $\mathcal{O}(X)$ AND THE ANALYTICITY OF THE EVALUATION FUNCTION $\Phi: X \longrightarrow A^*$

4.1 *Given a domain* $(X,p) \in \mathcal{R}(E)$ *let* A *be a fixed linear subspace of* $\mathcal{O}(X)$. *Denote by* (A,τ_c) *the locally convex space obtained from* A *by endowing it with the topology* τ_c *of uniform convergence on compact subsets of* X. *Let* τ_{cb} *denote the bornological topology in* A *associated with the topology* τ_c. *Let* τ_s *denote the topology of pointwise convergence on* A.

4.2 DEFINITION *We say that a locally convex Hausdorff topology* τ *on* A *is admissible (and write* $\tau \in T_A$*) if* $\tau_s \subset \tau$ *and* $(A,\tau)^*$ *is* s-*complete,* $(A,\tau)^*$ *denoting the topological dual endowed with the topology of pointwise convergence.*

Observe that the family T_A of a admissible topologies on A is not empty, because the strongest locally convex topology on A is admissible.

4.3 PROPOSITION (a) $\tau_n := \bigcap_{\tau \in T_A} \tau$ *is admissible.*

(b) *If* $\dim E < \infty$ *and* (A,τ_c) *is a closed subspace of* $(\mathcal{O}(X),\tau_c)$ *then* $\tau_c \in T_A$.

(c) *If* E *is t.v.s. metrizable or Baire and* (A,τ_c) *is a closed subspace of* $(\mathcal{O}(X),\tau_c)$, *then* τ_{cb} *is barrelled and consequently by Banach-Steinhaus theorem* $\tau_{cb} \in T_A$.

PROOF The statements (a) and (b) are direct consequences of the Definition 4.2.

Ad (c). Since the topology τ_{cb} of A is (by its definition) bornological and $\tau_s \subset \tau_c \subset \tau_{cb}$, it remains to show that τ_{cb} is barrelled.

In order to prove that τ_{cb} is barrelled it is enough to show that the space (A,τ_c) is complete. In order to show that (A,τ_c) is complete we may assume, without any loss of generality, that X is an open connected subset of E.

Let $(f_\lambda)_{\lambda\in\Lambda}$ be a generalized Cauchy sequence in the space (A,τ_c). Then in view of the completeness of the field $\mathbb{C}$, there exists a function $f: X \longrightarrow \mathbb{C}$ such that for every $x \in X$

$$\lim_{\lambda\in\Lambda} f_\lambda(x) = f(x).$$

Therefore $f_\lambda \longrightarrow f$ uniformly on all compact subsets of X. This implies that f is G-analytic in X. Now, if E is Baire, then there exists an open non-empty subset of X on which the function f is bounded.

On the other hand, if E is metrizable, then f is locally bounded in X. So in both cases we may conclude that f is analytic in X (see Th. 6.1 in [1]).

The following lemma and its Corollary are basic for the considerations of this paper.

4.4 LEMMA *Let* U *be an open subset of* E *and* (G,τ) *- a l.c. t.v.s. such that* $(G,\tau)^*$ *is s-complete. Assume that* $\psi: U \longrightarrow (G,\tau)^*$ *is a function such that for each* $f \in G$ *the function*

$$\psi_f: U \ni x \longrightarrow \psi(x)(f) \in \mathbb{C}$$

is analytic.

Then ψ *is analytic.*

PROOF The continuity of ψ follows from the equation

$$|\psi(x)(f)-\psi(x_o)(f)|=|\psi_f(x)-\psi_f(x_o)|, \quad f \in G,\ x,x_o \in U.$$

So it is sufficient to show that ψ is G-analytic, this follows from the Morera theorem (see Th. 3.1 in [1]).

4.5 COROLLARY *If* τ *is any admissible topology on* A, *then the evaluation mapping*

$$\Phi : X \longrightarrow (A,\tau)^*$$

(where $\Phi(x)(f) = f(x)$ *for* $x \in X$ *and* $f \in A$*) is analytic.*

5. MAXIMAL ANALYTIC EXTENSION AS THE CANONICAL DOMAIN OF EXISTENCE OF THE EVALUATION MAPPING Φ

In this section (X,p) is a fixed Riemann domain over E, A is any fixed linear subspace of $\mathcal{O}(X)$, τ is any admissible topology on A and $(A,\tau)^*$ denotes its topological dual with the pointwise convergence topology. By Corollary 4.5 the evaluation mapping $\Phi : X \longrightarrow (A,\tau)^*$ is analytic.

Let $(X_\Phi, j_\Phi, \hat{\Phi})$ be the canonical m.a.e. of the pair (X,Φ). The function $f: X_\Phi \longrightarrow \mathbb{C}$ defined by

(1) $\hat{f}(x) := \hat{\Phi}(x)f$, $x \in X_\Phi$, $f \in A$,

is analytic, because $\hat{f} = \delta_f \circ \hat{\Phi}$ and the linear mapping

$\delta_f : (A,\tau)^* \ni u \longrightarrow u(f) \in \mathbb{C}$ is continuous. Define

$A_\Phi := \{\hat{f}: f \in A\}$.

Then we have the following

5.1 THEOREM *The triple* (X_Φ, j_Φ, A_Φ) *is a m.a.e. of the pair* (X,A).

PROOF Since $j_\Phi : X \longrightarrow X_\Phi$ is a morphism and the mapping $j_\Phi^* : A_\Phi \rightarrow A$ is surjective, the triple (X_Φ, j_Φ, A_Φ) is an analytic extension of the pair (X,A).

Given an arbitrary a.e. (Y,j,B) of (X,A) define $\tilde{\Phi}: Y \longrightarrow (A,\tau)^*$ by the formula

$$\tilde{\Phi}(y)(f) := j^{*-1}(f)(y), \quad y \in Y, \ f \in A.$$

We claim that $\tilde{\Phi}$ is well defined and analytic.

Indeed, put $\tilde{f} := j^{*-1}(f)$, $f \in A$. In view of Lemma 4.4 the only thing we have to do is to show that every $y \in Y$ the functional

$$(*) \qquad A \ni f \longrightarrow \tilde{f}(y) \in \mathbb{C}$$

is continuous. With this aim in mind observe that the set

$Z :=$ {$b \in Y$: the functional (*) is continuous for every y belonging to a neighbourhood of b}

is open and non-empty (because $j(X) \subset Z$).

Now given any fixed $b \in Z$ we have

$$h(f;\eta) := (\tilde{f} \circ q_b^{-1})(q(b)+\eta) = \sum_{n \geq 0} \tilde{f}_n(\eta), \ \eta \in B(b,Y), \ f \in A,$$

where

$$n!\ \tilde{f}_n(\eta) = \lim_{t \to 0} t^{-n} \sum_{j=0}^{n} \binom{n}{j} (-1)^{n-j} h(f; \tfrac{j}{n} t\eta), \quad \eta \in E.$$

Since $(A,\tau)^*$ is s-complete the functional $A \ni f \longrightarrow \tilde{f}_n(\eta) \in C$ is continuous for every $\eta \in E$ and $n \geq 0$, because $A \ni f \to h(f; \frac{j}{n} t\eta) \in C$ is continuous for every t sufficiently small. This implies, again by s-completeness of $(A,\tau)^*$, that the functional $A \ni f \longrightarrow h(f;\eta)$ is continuous for every $\eta \in \hat{B}(b,Y)$. Therefore, by putting $\eta = q(y) - q(b)$, the functional (*) is continuous for every $y \in \hat{B}(b,Y)$. Hence $Z = Y$.

It is obvious that $\tilde{\Phi} \circ j = \Phi$. Hence $(Y,j,\tilde{\Phi})$ is an a.e. of (X,Φ). Thus by 3.2 there exists a morphism $u: Y \longrightarrow X_\Phi$ such that $j = u \circ j_\Phi$.

The proof is concluded.

6. ANALYTIC EXTENSIONS OF RIEMANN DOMAINS OVER METRIZABLE OR BAIRE TOPOLOGICAL VECTOR SPACES

In this section E is a metrizable or Baire t.v.s. $(X,p) \in \mathcal{R}(E)$, and (A,τ_c) is a closed linear subspace of $(\mathcal{O}(X),\tau_c)$. We already know that the topology τ_{cb} is admissible. Therefore, the evaluation mapping $\Phi : X \longrightarrow (A,\tau_{cb})^*$ is analytic.

Let $(X_\Phi, j_\Phi, \hat{\Phi})$ denote the canonical m.a.e. of the pair (X,Φ). Define

$$A_\Phi := \{\hat{f} : f \in A\} ,$$

where $\hat{f}(x) := \hat{\Phi}(x)(f)$, $x \in X_\Phi$, $f \in A$. By Theorem 5.1 $A_\Phi \subset \mathcal{O}(X_\Phi)$ and (X_Φ, j_Φ, A_Φ) is a m.a.e. of (X,A).

6.1 LEMMA *The mapping*

$$j_\Phi^* : (A_\Phi,\tau_{cb}) \longrightarrow (A,\tau_{cb})$$

is a topological isomorphism.

PROOF We shall write j instead of j_Φ. We already know that j^* is an algebraic isomorphism. Since the topology τ_{cb} is bornological it remains to show that j^* and j^{*-1} map bounded sets onto bounded sets.

It is obvious that j^* is bounded.

In order to show that j^{*-1} is bounded let B be a bounded subset of (A,τ_{cb}). Let K be a compact subset of X_Φ. Then for any function $f \in A$ we have $||\hat{f}||_k < \infty$. It follows that the set

$$U := \{f \in A : ||\hat{f}||_K \leq 1\}$$

is absorbing. It is obvious that U is absolutely convex and closed.

Therefore U is a barrel. Thus there exists a positive con

stant r such that $B \subset rU$, i.e.

$$||j^{*-1}(f)||_K = ||\hat{f}||_K \leq r, \quad f \in B.$$

This means that $j^{*-1}(B)$ is bounded.

6.2 THEOREM *If* (X,p) *is a Riemann domain over a metrizable or Baire space* E *and if* (Y,j,B) *is an a.e. of the pair* (X,A), *then*

$$j^* : (B,\tau_{cb}) \longrightarrow (A,\tau_{cb})$$

is a topological isomorphism.

PROOF First observe that (B,τ_c) is a closed subspace of $\mathcal{O}(Y)$. Indeed, if $(g_\lambda)_{\lambda\in\Lambda}$ is a generalized sequence of elements of B converging to $g \in \mathcal{O}(Y)$ uniformly on compact subsets of Y, then $g_\lambda \circ j \longrightarrow g \circ j$ uniformly on compact subsets of X. Therefore $g \circ j \in A$, and consequently $g \in B$.

Let $\Phi: X \longrightarrow (A,\tau_{cb})^*$ and $\Psi: Y \longrightarrow (B,\tau_{cb})^*$ denote the evaluation mappings. Let (X_Φ, j_Φ, A_Φ) and (Y_Ψ, j_Ψ, B_Ψ) be the m.a.e. of (X,A) and (Y,B), respectively. Then there exists a morphism $u: Y \longrightarrow X_\Phi$ such that $j_\Phi = u \circ j$. Since (X_Φ, u, A_Φ) is an a.e. of (Y,B), so there exists a morphism $v: X_\Phi \longrightarrow Y_\Psi$ such that $j_\Psi = v \circ u$. Finally, since $(Y_\Psi, j_\Psi \circ j, B_\Psi)$ is an a.e. of (X,A) there exists a morphism $v': Y_\Psi \longrightarrow X_\Phi$ such that $j_\Phi = v' \circ j_\Psi \circ j$.

Hence, in view of 2.8, $v: X_\Phi \longrightarrow Y_\Psi$ is an isomorphism and moreover we have the equation

$$v \circ j_\Phi = j_\Psi \circ j$$

which implies that $j_\Phi^* \circ v^* = j^* \circ j_\Psi^*$.

By Lemma 6.1 the mappings j_Φ^* and j_Ψ^* are topological isomorphisms. Since the mapping $v^*: (B_\Psi,\tau_{cb}) \ni f \longrightarrow f \circ v \in (A_\Phi,\tau_{cb})$ is a topological isomorphism, so is the mapping

$$j^* = j_\Phi^* \circ v^* \circ (j_\Psi^*)^{-1}.$$

7. COMPARISON WITH THE BISHOP CONSTRUCTION

First let us recall the following well known

7.1 PROPOSITION *For any* $f \in \mathcal{O}(X)$, *where* $(X,p) \in \mathcal{R}(E)$, *define*

$$f_n(x,y) := \frac{1}{n!} \left(\frac{d}{dt}\right)^n \left[(f \circ p_x^{-1})(p(x)+ty)\right]_{t=0}, x \in X, y \in E, n \geq 0.$$

Then:

1º $f_n(\cdot,y) \in \mathcal{O}(X)$, $y \in E$, $n \geq 0$.

2º *For every* $x \in X$ *the function* $E \ni y \longrightarrow f_n(x,y) \in \mathbb{C}$ *is a continuous homogeneous polynomial of degree* n.

3º $(f \circ p_x^{-1})(p(x)+y) = \sum_{n \geq 0} f_n(x,y)$, $x \in X$, $y \in B(x,X)$, $f \in \mathcal{O}(X)$.

7.2 DEFINITION *We say that a subset* $A \subset \mathcal{O}(X)$ *is* d-*stable, if for every* $y \in E$, *for every* $f \in A$ *and for every integer* $n \geq 1$ $f_n(\cdot,y) \in A$. *We say that* A *contains coordinates, if* $E' \circ p := \{\xi \circ p : \xi \in E'\} \subset A$, *where* E' *is the topological dual of* E.

Following [2], p. 45, we shall now formulate the following

7.3 DEFINITION *Let* A *be a subalgebra of* $\mathcal{O}(X)$ d-*stable and containing coordinates. Let* S *denote the set of all* $h \in \text{Spec } A$ *– the set of all non-zero linear-multiplicative functionals on* A *– such that:*

1º *There exists a balanced open neighbourhood* V_h *of* $0 \in E$ *such that the series*

$$\sum_{n \geq 0} h_y^n(f), \text{ where } h_y^n(f) := h(f_n(\cdot,y))$$

is absolutely convergent for every $y \in V_h$ *and for all* $f \in A$.

2º *For every* $f \in A$ *the function* $V_h \ni y \longrightarrow h_y(f) := \sum_{n \geq 0} h_y^n(f) \in \mathbb{C}$ *is analytic.*

3º *There exists an element* $\pi(h) \in E$ *(this element is unique by the Hahn-Banach theorem) such that*

$$h(\xi \circ p) = \xi(\pi(h)) \text{ for every } \xi \in E'.$$

7.3 *Let now* $\Phi: X \longrightarrow (A,\tau)^*$ *denote the evaluation mapping,* τ *being any admissible topology on* A. *Then one may prove the following facts (see* [2], *pp. 46-47):*

(i) *If* $h \in S$ *then* $h_y \in S$ *for all* $y \in V_h$;

(ii) $(h_a)_b = h_{a+b}$ *for all* $a,b \in V_h$ *such that*

$$za + z'b \in V_h \quad \text{if} \quad |z| + |z'| \leq 1,\ z,z' \in \mathbb{C} ;$$

(iii) $\Phi(X) \subset S$ *and* $\pi \circ \Phi = p$;

(iv) *The family of sets* $\{h_y : y \in V_h\}_{h \in S}$ *is a basis for a Hausdorff topology on* S *(the Bishop topology). Moreover,* π *is a local homeomorphism in this topology and* $\Phi: X \longrightarrow S$ *is continuous;*

(v) *Let* S_Φ *denote the connected component of* S *containing the set* $\Phi(X)$. *Let* $\tilde{A} := \{\tilde{f} : f \in A\}$, *where* $\tilde{f}: S_\Phi \longrightarrow \mathbb{C}$ *is defined by* $\tilde{f}(h) = h(f)$, $h \in S_\Phi$. *Then* $((S_\Phi,\rho),\Phi,\tilde{A})$, *where* $\rho := \pi|S_\Phi$, *is a m.a.e. of* (X,A) ;

(vi) $\tilde{A}$ *separates point of* S_Φ.

7.4 REMARK *If a* d-*stable algebra* A *does not contain coordinates and* $(\tilde{X},j,\tilde{A})$ *is a m.a.e. of* (X,A), *then* $\tilde{A}$ *may not separate the points of* $\tilde{X}$. *For instance, let* $X = \mathbb{C}$ *and let* A *be the algebra generated by one element* e^z. *Then* $\tilde{A} = A$ *does not separate*

$\tilde{X} = \mathbb{C}$.

We shall now state the main theorem of this section.

7.5 THEOREM *If* A *is a d-stable subalgebra of* $\mathcal{O}(X)$ *containing coordinates then* $\hat{\Phi}(X_\Phi) = S_\Phi$ *and the mapping* $\hat{\Phi}: X_\Phi \longrightarrow S_\Phi$ *is an isomorphism in the category* $\mathcal{R}(E)$.

PROOF 1º First we shall that given a fixed $z \in X_\Phi$, the linear-multiplicative functional $h := \hat{\Phi}(z)$ belongs to S. We have to check that the conditions 1º - 3º of Definition 7.3 are satisfied by h. To this aim observe that

(*) $\hat{\Phi}(z)(f_n(.,y)) = \hat{f}_n(z,y)$, $f \in A$, $z \in X_\Phi$, $y \in E$.

Indeed, the functions on both sides are analytic with respect to $z \in X_\Phi$, and for every $y \in E$ they are identical for z belonging to the open non-empty subset $j_\Phi(X)$ of X_Φ. Hence by the identity principle we obtain (*).

The equation (*) implies that

$$h_y^n(f) = \hat{f}_n(z,y),\ y \in E,\ f \in A,\ n \geq 0.$$

Then for every $f \in A$ the series $\sum_{n \geq 0} h_y^n(f)$ is absolutely convergent at every point $y \in B(z,X_\Phi)$ to the sum

$$h_y(f) := \sum_{n \geq 0} \hat{f}_n(z,y) = (\hat{f} \circ q_z^{-1})(q(z)+ y),\ (q := p_\Phi),$$

that is analytic in $B(z,X_\Phi)$.

Therefore $h := \hat{\Phi}(z)$ satisfies the conditions 1º and 2º of Def. 7.3, if V_h is any open balanced neighbourhood of $0 \in E$ contained in $B(z,X_\Phi)$.

In order to show that 3º of Def. 7.3 is satisfied by h, observe that

$$\hat{\Phi}(z)(\xi \circ p) = \xi(p_\Phi(z)),\ z \in X_\Phi,\ \xi \in E'$$

(because, given $\xi \in E'$, the functions on both sides of the equation are analytic in X_Φ and identical on the open subset $j_\Phi(X)$). Therefore it is sufficient to put

(**) $\pi(h) = \pi(\hat{\Phi}(z)) := p_\Phi(z)$.

2º We claim that the mapping $\hat{\Phi}: X_\Phi \longrightarrow S$ is continuous. Fix $z \in X_\Phi$ and let $\mathcal{U} = \{h_y : y \in V_h\}$ be a basic neighbourhood in S of the point $h := \hat{\Phi}(z)$. We may assume that $V_h \subset B(z, X_\Phi)$.

Now observe that

$$h_y(f) = \sum_{n \geq 0} h_y^n(f) = \sum_{n \geq 0} \hat{\Phi}(z)(f_n(\cdot, y)) = \sum_{n \geq 0} \hat{f}_n(z,y) =$$

$$= \hat{f}(q_z^{-1}(q(z) + y)) = \hat{\Phi}(q(z) + y))(f), \quad f \in A, \ y \in V_h ,$$

where we have put $q = p_\Phi$. Therefore $\hat{\Phi}^{-1}(\mathcal{U}) \supset q_z^{-1}(q(z) + V_h))$, so that $\hat{\Phi}^{-1}(\mathcal{U})$ is a neighbourhood of z.

Thus we have obtained the required result.

3º $\hat{\Phi}(X_\Phi) \subset S_\Phi$ and $\hat{\Phi}: X_\Phi \longrightarrow S_\Phi$ is a morphism in the category $\mathcal{R}(E)$. Indeed, it follows from the equation $\Phi = \hat{\Phi} \circ j_\Phi$ that $\hat{\Phi}(X_\Phi) \cap S_\Phi \neq \emptyset$. But $\hat{\Phi}(X_\Phi)$ is connected because $\hat{\Phi}$ is continuous, therefore $\hat{\Phi}(X_\Phi) \subset S_\Phi$. Now the equation (**) implies that $\hat{\Phi}: X_\Phi \longrightarrow S_\Phi$ is a morphism.

4º In order to show that $\hat{\Phi}$ is an isomorphism it is enough to observe that by the maximality of the extension (X_Φ, j_Φ, A_Φ) there exists a morphism $u: S_\Phi \longrightarrow X_\Phi$ such that $j_\Phi = u \circ \Phi$. By 2.3 we get the result.

7.6 COROLLARY *All the linear-multiplicative functionals belonging to* S_Φ *are continuous with respect to any admissible topology on* $\mathcal{O}(X)$.

In particular, if E *is metrizable or Baire, then all the functionals of* S_Φ *are* τ_{cb}*-continuous. This gives a generalization of some results due to Coeuré [2] and Schottenloher [11].*

8. HOLOMORPHIC CONTINUATIONS OF $\mathbb{R}$-ANALYTIC FUNCTIONS DEFINED ON OPEN SUBSETS OF REAL BANACH SPACES

In this section E stands for a Banach space over $\mathbb{R}$, Ω is an open connected subset of E, F is a topological vector space over $\mathbb{C}$ and F^* denotes the space of all continuous linear mappings from F to $\mathbb{C}$ with the topology of pointwise convergence, and finally $\tilde{E} = E + iE$ denotes the complexification of E.

8.1 THEOREM *If* F *is Baire and* $\Psi: \Omega \longrightarrow F^*$ *is a function such that* $\forall\ f \in F$ *the function*

$$f^* : \Omega \ni x \longrightarrow \Psi(x)(f) \in \mathbb{C}$$

is $\mathbb{R}$*-analytic, then*

(i) $\forall\ a \in \Omega\ \exists r > 0\ \forall\ f \in F \quad \rho(T_a f^*) \geq r,$

where $\rho(T_a f^*)$ *denotes the radius of the pointwise convergence of the Taylor series of* f^* *at* a;

(ii) Ψ *is* $\mathbb{R}$*-analytic.*

Moreover, the conditions (i) and (ii) are always equivalent (F being Baire or not).

PROOF (i) Observe that $\forall\ a \in \Omega$, $\forall\ x \in E$ and $\forall\ n \geq 0$ the $\mathbb{C}$-linear mappings

$$(1) \quad F \ni f \longrightarrow f_n^*(x) := \frac{1}{n!} (\delta_a^n f^*)(x) \in \mathbb{C}$$

are continuous. Indeed

$$n!\ f_n^*(x) = \lim_{t \downarrow 0} t^{-n} \sum_{j=0}^{n} \binom{n}{j} (-1)^{n-j} f^*(a + \frac{j}{n} tx), \ f \in F,$$

and the components of the sum are continuous linear mappings of F into $\mathbb{C}$; by the Banach-Steinhaus theorem the functionals (1) are continuous.

Given a fixed $a \in \Omega$, the set

$$F_k := \{f \in F: |f_n^*(x)| \leq k^n,\ n \geq 0,\ x \in S\},\quad k = 1,2,\ldots,$$

where $S = \{x \in E: ||x|| = 1\}$, is closed and $F = \bigcup_{k \geq 1} F_k$. Indeed, if $f \in F$, then there exists $\rho > 0$ such that

$$f^*(a + x) = \sum_{n>0} f_n^*(x),\quad ||x|| < \rho,$$

whence, E having the Baire property, there exists $\rho = \rho_f > 0$ such that $\sum_{n>\nu} ||f_n^*||\ \rho^n < \infty$. Therefore $f \in F_k$, if $\frac{1}{k} < \rho$. Now, using the Baire property of F, one can find k so that int $F_n \neq \emptyset$. Hence, there exists an open neighbourhood U of $0 \in F$ such that

$$||f_n^*|| \leq 2\, k^n,\ f \in U,\ n \geq 0:$$

So finally $\forall_f \in F\ \exists M_f > 0$ such that

$$||f_n^*|| \leq M_f\, k^n,\ n \geq 0.$$

Therefore

$$(2)\qquad f^*(a + x) = \sum_{n\geq 0} f_n^*(x),\quad ||x|| < r = \frac{1}{k},\ f \in F,$$

and moreover $\sum_{n\geq 0} ||f_n^*||\ r_1^n < \infty$, if $0 < r_1 < r$.

(ii) Fix $a \in \Omega$ and put

$$\Psi_n(x)(f) := f_n^*(x),\ x \in E,\ f \in F.$$

Then $\Psi_n: E \longrightarrow F^*$ is a continuous polynomial of degree n and in view of (2)

$$(3)\qquad \Psi(a + x) = \sum_{\nu\geq 0} \Psi_n(x),\quad ||x|| < r.$$

Thus Ψ is $\mathbb{R}$-analytic in Ω.

From Theorem 8.1 and from 7 of [1] one can derive the following two Propositions.

8.2 PROPOSITION *Under the assumptions of* Th. 8.1 *there exists an open set* $\tilde{\Omega} \subset \tilde{E}$ *such that for every* $f \in F$ *one can find* $\tilde{f} \in \mathcal{O}(\tilde{\Omega})$ *so that* $\tilde{f}|\Omega = f$. *Moreover, there exists a holomorphic function* $\tilde{\Psi}: \tilde{\Omega} \longrightarrow F^*$ *such that* $\tilde{\Psi}|\Omega = \Psi$ *and* $\tilde{\Psi}(x)(f) = \tilde{f}(x)$ *for* $x \in \tilde{\Omega}$, $f \in F$.

8.3 PROPOSITION *Let* $\mathcal{A}(\Omega)$ *denote the space of all complex-valued* $\mathbb{R}$*-analytic function in* Ω. *Let* $F \subset \mathcal{A}(\Omega)$ *be a Baire t.v.s. over* $\mathbb{C}$ *such that for each* $a \in \Omega$ *the* $\mathbb{C}$*-linear mapping*

$$F \ni f \longrightarrow f(a) \in \mathbb{C}$$

is continuous.

Then the following statements are true

(*i*) $\forall\, a \in \Omega\ \exists\, r > 0$ *such that* $\forall\, f \in F$ *the Taylor series* $T_a f$ *of* f *at* a *is normally convergent in the ball* $||x-a|| < r$;

(*ii*) *The evaluation function* $\Phi: \Omega \longrightarrow F^*$ *is* $\mathbb{R}$*-analytic;*

(*iii*) *There exists an open* $\tilde{\Omega} \subset \tilde{E}$ *such that* $\Omega \subset \tilde{\Omega}$, $\forall\, f \in F\ \exists\, \tilde{f} \in (\tilde{\Omega})$ *with* $f = \tilde{f}|\Omega$, *and there exists* $\tilde{\Phi} \in \mathcal{O}(\tilde{\Omega}, F^*)$ *with* $\Phi = \tilde{\Phi}|\Omega$.

8.4 EXAMPLE Let $E = \mathbb{R}^n$ and $F = H(\Omega)$ - the space of all complex-valued harmonic functions in Ω. By the Harnack theorem $H(\Omega)$ is a Frechet space in the topology of uniform convergence on compact subsets of Ω.

Since each $f \in H(\Omega)$ is $\mathbb{R}$-analytic, so the evaluation function $\Phi: \Omega \longrightarrow F^*$ $(F = H(\Omega))$ is also $\mathbb{R}$-analytic. Moreover, there exists an open subset $\tilde{\Omega} \subset \mathbb{C}^n$ such that $\Omega \subset \tilde{\Omega}$ and $\forall\, f \in H(\Omega)\ \exists\, \tilde{f} \in \mathcal{O}(\tilde{\Omega})$ such that $f = \tilde{f}|\Omega$.

9. ANALYTIC EXTENSIONS OF OPEN CONNECTED SUBSETS OF REAL BANACH SPACES

In this section E stands for a real Banach space, Ω is an open connected subset of E, $\mathcal{A}(\Omega)$ is the space of all complex-valued $\mathbb{R}$-analytic functions on Ω and A is a subset of $\mathcal{A}(\Omega)$.

9.1 DEFINITION *The triple* $((X,p),j,\underline{A})$ *will be called an analytic extension of the pair* (X,A) , *if*

1º $(X,p) \in \mathcal{R}(\tilde{E})$;

2º $j: \Omega \longrightarrow X$ *is a continuous mapping such that* $id_\Omega = p \circ j$;

3º $\underline{A} \subset \mathcal{O}(X)$, $g \circ j \in A$ *for every* $g \in \underline{A}$ *and the mapping* $j^*\ \underline{A} \ni g \longrightarrow g \circ j \in A$ *is bijective.*

9.2 REMARK *The mapping* j^* *is bijective if and only if it is surjective.*

PROOF Let j^* be a surjection. Suppose $g_1 \circ j = g_2 \circ j$ and fix $a \in \Omega$. It follows from the equation $id_\Omega = p \circ j$ that there exists an open subset U of E such that $a \in U \subset \Omega$ and $j(U) \subset \hat{B}(j(a),r)$ where $\hat{B}(j(a),r)$ denotes the "ball with center $j(a)$ and radius r" in X. Then $p_a := p|\hat{B}(j(a),r)$ is a homeomorphism of $\hat{B}(j(a),r)$ onto the ball $B(a,r) \subset \tilde{E}$. In particular $p \circ j(U) = U \subset B(a,r)$. Since the functions $g_1 \circ p_a^{-1}$ and $g_2 \circ p_a^{-1}$ are holomorphic in $B(a,r)$ and identical on U, so by the identity principle (Prop. 6.6. II of [1]) they are identical in $B(a,r)$. Therefore g_1 and g_2 are identical in $\hat{B}(j(a),r)$ and, consequently, in X.

9.3 DEFINITION *We say that an analytic extension* $((X,p),j,\underline{A})$, *of the pair* (Ω,A) *is maximal, if for every analytic extension* $((X',p'),j',\underline{A}')$ *of* (Ω,A) *there exists a morphism* $u: X' \longrightarrow X$ *such that* $j = u \circ j'$.

Observe that $(X,u,\underline{A})$ *is a m.a.e. of* $(X',\underline{A}')$.

9.4 Let A be a vector subspace of $\mathcal{A}(\Omega)$ such that there exists an open connected set $\tilde{\Omega} \subset \tilde{E}$ and a vector subspace $\tilde{A} \subset \mathcal{O}(\tilde{\Omega})$ so that the restriction mapping

$$r: \tilde{A} \ni g \longrightarrow g|\Omega \in A$$

is surjective (and therefore an algebraic isomorphism by the principle of analytic continuation).

Consider (A,τ) as a t.v.s. endowed with the maximal locally convex topology $\tau = \tau_{max}$. The function $\tilde{\Phi}: \tilde{\Omega} \longrightarrow A^* := (A,\tau)^*$ defined by

$$\tilde{\Phi}(z)(f) := r^{-1}(f)(z), \quad f \in A, \ z \in \tilde{\Omega}$$

is $\mathbb{C}$-analytic and its restriction $\Phi := \tilde{\Phi}|\Omega$ is $\mathbb{R}$-analytic in Ω.

Let $(\mathcal{O}_{\tilde{E},A^*},\pi)$ denote the sheaf of germs of holomorphic functions defined on open subsets of $\tilde{E}$ with values in A^*. The mapping

$$(*) \qquad j_\Phi : \Omega \ni x \longrightarrow (\tilde{\Phi})_x \in \mathcal{O}_{\tilde{E},A^*}$$

is continuous. Let Ω_Φ denote the connected component of $\mathcal{O}_{\tilde{E},A^*}$ containing $j_\Phi(\Omega)$. One can easily check that if $\hat{\Phi}$, $\hat{f}$ and $\hat{A}$ are defined by

$$\hat{\Phi}: \Omega_\Phi \ni x \longrightarrow x(\pi(x)) \in A^* ;$$

$$(**) \qquad \hat{f}(x) := \hat{\Phi}(x)(f), \ x \in \Omega_\Phi, \ f \in A ;$$

$$\hat{A} := \{\hat{f}: F \in A\} ,$$

then $\hat{\Phi} \in \mathcal{O}(\Omega_\Phi, A^*)$, $\hat{A} \subset \mathcal{O}(\Omega_\Phi)$, $\Phi = \hat{\Phi} \circ j_\Phi$ and $j^*_\Phi: \hat{A} \ni g \longrightarrow g \circ j_\Phi \in A$

is surjective.

9.5 THEOREM *Under the assumptions of 9.4 the following statements are true:*

(i) $(\Omega_\Phi, j_\Phi, \hat{\Phi})$ *is a m.a.e. of* (Ω, Φ) ;

(ii) $(\Omega_\Phi, j_\Phi, \hat{A})$ *is a m.a.e. of* (Ω, A).

PROOF (i) Let $((Y,q),k,\Psi)$ $((Y,q) \in \mathcal{R}(\tilde{E}))$ be any a.e. of (Ω,Φ). By a standard reasoning one may check that

$$u: Y \ni y \longrightarrow (\Psi \circ q_Y^{-1})_{q(y)} \in \mathcal{O}_{\tilde{E},A^*}$$

is continuous, $q = \pi \circ u$, $j_\Phi = u \circ k$. Hence $u(Y) \subset \Omega_\Phi$, and the proof of (i) is concluded.

(ii) Given any a.e. (Y,j,B) of (Ω,A) it is enough to observe that $j^{*-1}: (A,\tau) \longrightarrow (B,\tau_c)$ is continuous (because j^* is an algebraic isomorphism and $\tau = \tau_{max}$) and next to repeat the proof of Theorem 5.1.

9.6 Let $E = \mathbb{R}^n$, and let $A \subset \mathcal{A}(\Omega)$ be a Frechet space when endowed with the topology τ_c.

Let A^* denote the topological dual of (A,τ_c) with the topology of pointwise convergence. Then by Proposition 8.3 the evaluation function $\Phi: \Omega \longrightarrow A^*$ is $\mathbb{R}$-analytic and there exists an open set $\tilde{\Omega} \subset \mathbb{C}^n$ such that $\Omega \subset \tilde{\Omega}$, $\forall f \in A \exists \tilde{f} \in \mathcal{O}(\tilde{\Omega})$ with $f = \tilde{f}|\Omega$, and there exists $\tilde{\Phi} \in \mathcal{O}(\tilde{\Omega},A^*)$ with $\Phi = \tilde{\Phi}|\Omega$.

Let (Y,j,B) be any a.e. of (Ω,A). Then (B,τ_c) is a Frechet space. Indeed, if $\{g_n\} \subset B$ is a Cauchy sequence then there exists $g \in \mathcal{O}(Y)$ such that $g_n \longrightarrow g$ uniformly on compact subsets of Y. So $g_n \circ j \longrightarrow g \circ j$ uniformly on compact subsets of Ω. Therefore $g \circ j \in A$, and so $g \in B$.

The mapping $j^*: B \longrightarrow A$ is, by Banach theorem, a topological isomorphism.

Therefore we are lead to the following

9.7 PROPOSITION *If the assumptions of 9.6 are satisfied then the statements (i) and (ii) of Theorem 9.5 hold true.*

10. A THEOREM ON NATURAL FRECHET SPACES

Let E be a complex t.v.s. admitting a countable basis $\mathcal{W} = \{\omega_n\}$ of open sets, Z a Banach space, $(X,p) \in \mathcal{R}(E)$ and $A \subset \mathcal{O}(X,Z)$ - a natural Frechet space (i.e. the topology of A is stronger than the topology of pointwise convergence, and consequently, stronger than the topology τ_c).

10.1 THEOREM *If* (X, id_X, A) *is a m.a.e. of* (X,A), *then there exists* $f \in A$ *such that* (X, id_X, f) *is a m.a.e. of* (X,f).

PROOF 1º If ω is any connected open subset of E and x is a point of X such that $p(x) \in \omega$, then the set

$$A_{\omega,x} := \{f \in A: \exists \tilde{f} \in \mathcal{O}(\omega, Z),\ \tilde{f} \text{ is bounded and } (\tilde{f})_{p(x)} = (f \circ p_x^{-1})_{p(x)}\}$$

is either a set of I category or $A_{\omega,x} = A$.

Indeed, given any continuous seminorm q on A define a seminorm $\tilde{q}$ on $A_{\omega,x}$ by the formula

$$\tilde{q}(f) := q(f) + \sup_{y \in \omega} ||\tilde{f}(y)||, \quad f \in A_{\omega,x}$$

The vector space $A_{\omega,x}$ with the topology $\tilde{\tau}$ defined by seminorms $\tilde{q}$ is Frechet. Indeed, if $\{f_n\}$ is a Cauchy sequence in the space

$A_{\omega,x}$, then it is also a Cauchy sequence in the space A. So there exists $f \in A$, such that $f_n \longrightarrow f$ (in A). In particular $f_n(y) \longrightarrow f(y)$ for every y in a neighbourhood of x.

The sequence $\{\tilde{f}_n\}$ is uniformly convergent in ω to a bounded analytic function $\tilde{f}$. But $\tilde{f}_n(y) = (f_n \circ p_x^{-1})(y)$ in a neighbourhood of $p(x)$, $n \geq 1$. Therefore $\tilde{f}(y) = (f \circ p_x^{-1})(y)$ in a neighbourhood of $p(x)$. Hence $(\tilde{f})_{p(x)} = (f \circ p_x^{-1})_{p(x)}$, and $f \in A_{\omega,x}$.

It is obvious that the imbedding

$$A_{\omega,x} \ni f \longrightarrow f \in A$$

is continuous. By a well known Banach theorem we obtain that $A_{\omega,x} = A$ or $A_{\omega,x}$ is of the first category.

2º Given $x,y \in X$, $x \neq y$, $p(x) = p(y)$, define

$$A(x,y) = \{f \in A: (f \circ p_x^{-1})_{p(x)} = (f \circ p_y^{-1})_{p(y)}\}.$$

We claim that $A(x,y)$ is a closed subspace of A and $A(x,y) \neq A$, so that $A(x,y)$ is nowhere dense.

Let $(f_n) \subset A(x,y)$ tend to $f \in A$. Since

$$(f_n \circ p_x^{-1})_{p(x)} = (f_n \circ p_y^{-1})_{p(y)}, \text{ so}$$

$$(f_n \circ p_x^{-1})(z) = (f_n \circ p_y^{-1})(z), \; n \geq 1$$

for every z in an open neighbourhood ω of $p(x)$. Therefore $(f \circ p_x^{-1})(z) = (f \circ p_y^{-1})(z)$ for $y \in \omega$. Hence $f \in A(x,y)$.

Let $(X_f, j_f, \hat{f})$ denote the canonical m.a.e. of (X,f), where $j_f(x) = (f \circ p_x^{-1})_{p(x)}$. Because the triple (X, id_X, A) is a m.a.e. of (X,A), we may identify any point $x \in X$ with the family $(j_f(x))_{f \in A}$. Therefore if $x,y \in X$ and $x \neq y$ then there exists $f \in A$ such that $j_f(x) \neq j_f(y)$ (see 3.11). So $A(x,y) \neq A$ and $A(x,y)$ is nowhere dense.

3° For any $\omega \in \mathcal{W}$ choose a point z_ω in ω and define $Q = \bigcup_{\omega \in \mathcal{W}} p^{-1}(z_\omega)$. By the Poincaré-Volterra theorem Q is a countable dense subset of X. Define

$$\Delta := \{(x,y) \in Q^2 : x \neq y,\ p(x) = p(y)\}$$

and

$$\Lambda := \{(\omega, x) \in \mathcal{W} \times Q : A_{\omega,x} \neq A\}.$$

Then by 1° and 2° the set $\mathcal{P} := \bigcup_{(\omega,x) \in \Lambda} A_{\omega,x} \cup \bigcup_{(x,y) \in \Delta} A(x,y)$ is of the first category. So the set $A \setminus \mathcal{P}$ is not empty (it is of the II category).

4° We claim that for every $f \in A \setminus \mathcal{P}$ the triple (X, id_X, f) is a m.a.e. of (X,f), i.e. the mapping

$$j_f : X \ni x \longrightarrow (f \circ p_x^{-1})_{p(x)} \in X_f$$

is bijective.

a) j_f is an injection. This is a direct consequence of the fact that $f \notin A(x,y)$ for every $(x,y) \in \Delta$. Indeed, if $j_f(a) = j_f(b)$, $a \neq b$, $p(a) = p(b)$, then $f \circ p_a^{-1} = f \circ p_b^{-1}$ in a neighbourhood $\omega \in \mathcal{W}$ of $p(a)$. We may assume that there are schlicht connected neighbourhoods ω', ω'' of a and b, respectively, such that $\omega \subset p(\omega') \cap p(\omega'')$. Put $x = p_a^{-1}(z_\omega)$, $y = p_b^{-1}(z_\omega)$. Then $(x,y) \in \Delta$ and $(f \circ p_x^{-1})_{p(x)} = (f \circ p_a^{-1})_{p(x)} = (f \circ p_b^{-1})_{p(y)} = (f \circ p_y^{-1})_{p(y)}$. This, however, contradicts the definition of f. Thus $j_f(a) \neq j_f(b)$.

b) j_f is onto. j_f being injective, we may consider X as a subdomain of X_f. If $X \neq X_f$, there would exist a point $x_0 \in X_f$ belonging to the boundary of X, a schlicht neighbourhood U of x_0 such that $\omega = p_f(U) \in \mathcal{W}$ and $\hat{f}$ is bounded on U, and a point $z \in Q \cap U$. The domain X being a m.a.e. with respect to A, we have $A_{\omega,x} \neq A$. Further we have $(\hat{f} \circ (p_f)_x^{-1})_{p(x)} = (f \circ p_x^{-1})_{p(x)}$

and $\tilde{f} := \hat{f} \circ (p_f)_x^{-1}$ is analytic and bounded on ω. Thus $\tilde{f} \in A_{\omega,x}$, what contradicts the definition of f. The proof is concluded.

10.2 COROLLARY *Let* $(X,p) \in \mathcal{R}(E)$, *where* E *is the same as in Theorem* 10.1. *Let* $F = \{f_1,\ldots f_s\} \subset \mathcal{O}(X)$ *be a finite system of holomorphic functions on* X *such that* (X,id_X,F) *is a m.a.e. of* (X,F). *Then there exist numbers* $\lambda_1,\ldots \lambda_s \in \mathbb{C}$ *such that* (X,id_X,f) *is a m.a.e. of* (X,f), *where* $f = \lambda_1 f_1 +\ldots+ \lambda_s f_s$.

PROOF By putting $A = \{z_1 f_1 +\ldots+ z_s f_s : z_i \in \mathbb{C},\ f_i \in F\}$.

10.3 COROLLARY *If* $\Omega \subset \mathbb{R}^n$ *is a domain,* $A \subset \mathcal{A}(\Omega)$ *is a Frechet space when endowed with the topology* τ_c, *and if* $(X,j,\hat{A})$ *is a maximal a.e. of* (Ω,A), *then there exists* $f \in A$ *such that* $(X,j,j^{*-1}f)$ *is a m.a.e. of* (Ω,f).

In particular, if Ω is a domain in $\mathbb{R}^n$ and $(X,j,\hat{H})$ denotes the m.a.e. of $(\Omega,H(\Omega))$, where $H(\Omega)$ is the Frechet space of complex-valued harmonic functions in Ω, then there exists $\hat{F} \in \hat{H}$ such that $(X,j,\hat{f})$is a m.a.e. of $(\Omega,\hat{f} \circ j)$.

In other words, the harmonic envelope of holomorphy of any domain $\Omega \subset \mathbb{R}^n$ is a natural domain of existence of a harmonic function $f \in H(\Omega)$ (see [4], [5], [6]).

REFERENCES

|1| J.BOCHNAK and J.SICIAK, Analytic functions in topological vector spaces, Studia Math., 39(1)(1971), 59-112.

|2| G.COEURÉ, Analytic Functions and Manifolds in Infinite Dimensional Spaces,Nort Holland/American Elsevier,1974.

[3] R. C. GUNNING and H. ROSSI, Analytic functions of several complex variables, Englewood Cliffs, N. J.: Prentice-Hall 1965.

[4] M. JARNICKI, Analytic continuation of pluriharmonic functions, Zeszyty Naukowe UJ, Prace Mat. 18 (to appear).

[5] C. O. KISELMAN, Prolongement des solutions d'une équation aux derivées partielles à coefficient constants, Bull. Soc. Math. France 97(4) (1969), 329-356.

[6] P. LELONG, Prolongement analytique et singularités complexes des fonctions harmoniques, Bull. Soc. Math. Belg., 7(2)(1955), 10-23.

[7] E. LIGOCKA and J. SICIAK, Weak analytic continuation, Bull. de l'Acad. Polon. des Sci., 20(6) (1972), 461-466.

[8] R. NARASIMHAN, Several Complex Variables, University of Chicago Press, Chicago and London, 1971.

[9] H. H. SCHAEFFER, Topological vector spaces, Macmillan Company, N. Y., London, 1966.

[10] M.SCHOTTENLOHER, Uber analytische Fortsetzung in Banach raumen, Math. Ann. 199 (1972), 313-336.

[11] M. SCHOTTENLOHER, Riemann domains: Basic results and open questions, Proc. in Infinite Dimensional Holomorphy 1973, Springer Lecture Notes 364 (1974), 196-212.

[12] M.SCHOTTENLOHER, Das Leviproblem in Unendlichdimensionalen Raumen mit Schauderzerlegung, Habilitationsschrift, Munchen 1974.

Mathematics Institute,
University Jagielloński,
ul. Reymonta 4,
Kraków, Poland.

Infinite Dimensional Holomorphy and Applications, Matos (ed.)

POLYNOMIAL APPROXIMATION ON COMPACT SETS

By *MARTIN SCHOTTENLOHER**

In This note a simple proof of the following result is given: A pseudoconvex, finitely Runge open set U in a locally convex Hausdorff space with the approximation property is polynomially convex. This generalizes results of Dineen [4] and Noverraz [13]. We also prove an approximation theorem of the Runge type for such domains.

Finally, in the last section of this note, we discuss a strong form of the Oka-Weil approximation theorem, which is useful in the study of the Nachbin topology τ_ω on the space $\mathcal{H}(U)$ of holomorphic functions on U (see Mujica [11]).

1. NOTATIONS AND PRELIMINARIES

Let E be a locally convex Hausdorff space over $\mathbb{C}$ (for short: lcs), and let cs(E) denote the set of continuous seminorms on E. For $\alpha \in cs(E)$, $x \in E$ and $r > o$ the "α-ball" about

---

* Research supported by the Brazilian-German Cooperation Agreement (Conselho Nacional de Pesquisas - Gesellschaft für Mathematik und Datenverarbeitung).

x with the radius r is

$$B^{\alpha}(x,r) = \{y \in E \mid \alpha(x-y) < r\} .$$

The "α-boundary distance" $d_U^{\alpha} : U \longrightarrow [o,\infty]$ for an open set $U \subset E$ is defined by

$$d_U^{\alpha}(x) = \sup \{r > o \mid B^{\alpha}(x,r) \subset U\} , \; x \in U.$$

For $B \subset U$ we put $d_U^{\alpha}(B) = \inf \{d_U^{\alpha}(x) \mid x \in B\}$. Another distance function $\delta_U : U \times E \longrightarrow [o,\infty]$ is given by

$$\delta_U(x,a) = \sup \{r > o \mid x + \lambda a \in U \text{ for all } \lambda \in C, |\lambda| < r\}, \; (x,a) \in U \times E.$$

d_U^{α} is continuous, while δ_U is in general only lower semicontinuous.

An open set $U \subset E$ is called pseudoconvex if $-\log \delta_U$ is plurisubharmonic, i.e. if the restrictions of $-\log \delta_U$ to complex lines in $U \times E$ are subharmonic. Let $\mathcal{P}(U)$ (resp. $\mathcal{P}_c(U)$) denote the set of plurisubharmonic (resp. continuous plurisubharmonic) functions on U, and let $\mathcal{H}(U)$ denote the vector space of holomorphic functions on U. For $Q \subset \mathcal{P}(U)$ and $K \subset U$ the "Q-convex hull" of K is defined by

$$\hat{K}_Q = \{x \in U \mid v(x) \leq \sup_{y \in K} v(y) \text{ for all } v \in Q\},$$

and for $A \subset \mathcal{H}(U)$ the "A-convex hull" of K is

$$\hat{K}_A = \{x \in U \mid |f(x)| \leq ||f||_K \text{ for all } f \in A\},$$

whereby $||f||_K = \sup \{|f(y)| \mid y \in K\}$. From the characterization of pseudoconvex sets in finite dimensional spaces one can deduce (cf. [17]):

PROPOSITION 1 *For an open set $U \subset E$ the following properties are equivalent*

1º *U is pseudoconvex.*

2º $-\log d_U^{\alpha}$ *is plurisubharmonic on* $\{x \in U \mid d_U^{\alpha}(x) > o\}$ *for every* $\alpha \in cs(E)$.

3º *For every compact* $K \subset U$ *there is* $\alpha \in cs(E)$ *with* $d_U^\alpha(\hat{K}_{\mathcal{P}(U)}) > 0$.

4º $\hat{K}_{\mathcal{P}(U)}$ *is precompact in* U *for every compact* $K \subset U$.*Here,* $L \subset U$ *is called precompact in* U *if* L *is precompact and if there exists* $\alpha \in cs(E)$ *with* $d_U^\alpha(L) > 0$.

It is an open question whether the above equivalences hold if $\mathcal{P}(U)$ is replaced by $\mathcal{P}_c(U)$. A partial answer is given in section 2.

U is called holomorphically convex if $\hat{K}_{\mathcal{H}(U)}$ is precompact in U for every compact $K \subset U$. A holomorphically convex open set $U \subset E$ is pseudoconvex. The converse is true for $E = \mathbb{C}^n$ (Levi problem), for $\mathbb{C}^{(\mathbb{N})}$ [7] and $\mathbb{C}^A$ [2], and for certain separable spaces E with a basis (cf. [8], [6], [17]). It is an open question whether the converse holds in general. A partial answer is given in the next section.

$U \subset E$ is polynomially convex if $\hat{K}_\pi$ is precompact in U for all compact $K \subset U$, whereby $\pi \subset \mathcal{H}(U)$ denotes the space of all continuous polynomials from E to $\mathbb{C}$. Since π is dense in $\mathcal{H}(E)$ with respect to the compact open topology on $\mathcal{H}(E)$, we have $\hat{K}_\pi = \hat{K}_{\mathcal{H}(E)}$. $\hat{K}_\pi$ is contained in

$$\tilde{K} = \{x \in E \mid |p(x)| \leq \|p\|_K \text{ for all } p \in \pi\}.$$

We don't know, whether for a polynomially convex U, $\tilde{K} = \hat{K}_\pi$ is true in general. In the case of a Fréchet space E with the approximation property this will be proved in section 3.

Closely related with polynomial convexity is the notion of a Runge open set U E. U is called Runge if π is dense in $\mathcal{H}(U)$ with respect to the compact open topology. This is the same as to say that $\mathcal{H}(E)$ is dense in $\mathcal{H}(U)$. Finally, U is called finitely Runge (resp. finitely polynomially convex) if for all finite dimensional vector subspaces F of E, $U \cap F$ is Runge (resp. polynomially convex) in F.

2. POLYNOMIAL CONVEXITY

Throughout this section let E be a lcs with the approximation property, i.e. for every compact $K \subset E$, every $\alpha \in cs(E)$ and every $\varepsilon > o$ there exists a continuous linear map $\phi: E \longrightarrow E$ with $\dim_{\mathbb{C}} \phi(E) < \infty$ and $\alpha(x - \phi(x)) < \varepsilon$ for all $x \in K$.

PROPOSITION 2 *Let U be a pseudoconvex, finitely Runge open set in E. Then for every compact* $K \subset U$:

$$\hat{K}_{\mathcal{P}(U)} = \hat{K}_{\mathcal{P}_c(U)} = \hat{K}_{\mathcal{H}(U)} = \hat{K}_{\mathcal{H}(E)} .$$

PROOF Evidently $\hat{K}_{\mathcal{P}(U)} \subset \hat{K}_{\mathcal{P}_c(U)} \subset \hat{K}_{\mathcal{H}(U)} \subset \hat{K}_{\mathcal{H}(E)}$.

In order to show $\hat{K}_{\mathcal{H}(E)} \subset \hat{K}_{\mathcal{P}(U)}$ let $x_o \in U$, $x_o \notin \hat{K}_{\mathcal{P}(U)}$.

Then there is $v \in \mathcal{P}(U)$ with

$$v(x_o) > \sup_{y \in K} v(y).$$

Due to the semicontinuity of v there are $\alpha \in cs(E)$, $\eta > o$ and $s > o$ such that for all $x \in K$

$$B^\alpha(x,2s) \subset U \text{ and } v(y) < v(x_o) - \eta \text{ for } y \in B^\alpha(x,2s).$$

Because E has the approximation property there exists a continuous linear $\phi: E \longrightarrow E$ with $\dim_{\mathbb{C}} \phi(E) < \infty$ and $\alpha(\phi(x) - x) < s$ für $x \in K$. Now $\psi = \phi + x_o - \phi(x_o)$ satisfies $\alpha(\psi(x) - x) < 2s$ for $x \in K$, hence

$$\psi(K) \subset U \cap F \text{ , where } F = sp\ (\psi(F)), \text{ and}$$

$$v \circ \psi(x) < v(x_o) - \eta \text{ for all } x \in K.$$

It follows for $w = v|_{U \cap F} \in \mathcal{P}(U \cap F)$:

$$\sup_{y \in \psi(K)} w(y) \leq v(x_o) - \eta < v(x_o) = w(x_o) .$$

Consequently, $x_o \notin \widehat{\psi(K)}_{\mathcal{P}(U \cap F)}$. Since $U \cap F$ is pseudoconvex and Runge, there exists a polynomial $g : F \longrightarrow \mathbb{C}$ with $|g(x_o)| > ||g||_{\psi(K)}$ (cf. [9, p. 53]). Now $g \circ \psi \in \pi$ and

$$|g \circ \psi\ (x_o)| = g(x_o)| > ||g||_{\psi(K)} = ||g \circ \psi||_K ,$$

hence $x_o \notin \hat{K}_{\mathcal{H}(E)}$.

From the propositions 1 and 2 we deduce:

COROLLARY 1 *A pseudoconvex, finitely Runge open set* $U \subset E$ *is holomorphically convex.*

REMARK Since for domains $\Omega \subset \mathbb{C}^n$ the main step in solving the Levi problem is to show that pseudoconvexity implies holomorphic convexity, corollary 1 is a certain contribution to the solution of the Levi problem. However, in infinite dimensional spaces E a holomorphically convex domain need not be a domain of holomorphy. In fact, Josefson [10] gives an example of a pseudoconvex domain Ω in $E = c_o(A)$, A uncountable, which is not a domain of holomorphy. Moreover, since this domain can be defined by a global plurisubharmonic function $v \in \mathcal{P}(E)$, it is finitely Runge and hence holomorphically convex.

But in certain infinite dimensional spaces, for example in Silva spaces [15], every holomorphically convex domain is a domain of holomorphy or even a domain of existence.

THEOREM 1 *For an open set* $U \subset E$*, the following properties are equivalent:*

1º U *is pseudoconvex and finitely Runge.*

2º U *is holomorphically convex and Runge.*

3º U *is polynomially convex.*

4º U *is finitely polynomially convex.*

PROOF "1º $\Longrightarrow$ 2º". According to corollary 1, U is holomorphically convex. Therefore, it suffices to show that a finitely Runge set is a Runge set. This was proved in [1] in a more general context. In our situation the proof of [1] is as follows: Let $f \in \mathcal{H}(U)$, $K \subset U$ compact, $\varepsilon > o$. Because of the continuity of f there are $\alpha \in cs(E)$ and $s > o$ such that for all $x \in K$

$$B^{\alpha}(x,s) \subset U \text{ and } |f(x) - f(y)| < \frac{\varepsilon}{2} \text{ for } y \in B^{\alpha}(x,s).$$

There exists a continuous linear $\phi: E \longrightarrow E$ with $\dim_{\mathbb{C}} \phi(E) < \infty$ and $\alpha(\phi(x) - x) < s$ for all $x \in K$. Hence

$$\phi(K) \subset U \cap F, \text{ where } F = \phi(E), \text{ and}$$

$$||f - f \circ \phi||_K \leq \frac{\varepsilon}{2} .$$

Since $U \cap F$ is finitely Runge there is a polynomial $g: F \longrightarrow \mathbb{C}$ with $||f_{U \cap F} - g||_{\phi(K)} < \frac{\varepsilon}{2}$. Now $g \circ \phi \in \pi$ and

$$||f - g \circ \phi||_K \leq ||f - f \circ \phi||_K + ||f \circ \phi - g \circ \phi||_K < \varepsilon.$$

"2º $\Longrightarrow$ 3º". Let $K \subset U$ be compact. Evidently $\hat{K}_{\mathcal{H}(U)} \subset \hat{K}_{\mathcal{H}(E)}$. If $x_o \in U$, $x_o \notin \hat{K}_{\mathcal{H}(U)}$, there is $f \in \mathcal{H}(U)$ with $|f(x_o)| > ||f||_K$. Since U is Runge, there exists a polynomial $p \in \pi$ with

$$||f - p||_{K \cup \{x_o\}} < \frac{1}{2} (|f(x_o)| - ||f||_K), \text{ hence } x_o \notin \hat{K}_{\mathcal{H}(E)}.$$

Now $\hat{K}_{\mathcal{H}(E)} = \hat{K}_{\mathcal{H}(U)}$ is precompact in U since U is holomorphically convex.

"3º $\Longrightarrow$ 4º" is trivial and "4º $\Longrightarrow$ 1º" follows from the finite dimensional results [9, p. 53].

REMARK Applying a result of Noverraz [14, th. 3] we see that for a Fréchet space E with the approximation property the following is true: For any dense vector subspace $F \subset E$ the intersection of all pseudoconvex domains $\Omega \subset E$, with $F \subset \Omega$, is equal to the intersection of all domains of existence $\Omega \subset E$, with $F \subset \Omega$. In other words, the pseudoconvex completion $F_{\mathcal{P}}$ of F agrees with the holomorphic completion $F_{\mathcal{O}}$ of F.

3. POLYNOMIAL APPROXIMATION

In this section let E be a metrizable lcs with the approximation property. Moreover, let E be holomorphically complete, i.e. $E = E_{\mathcal{O}}$ (cf. for example [14]). We need the following characterization [14, Prop. 10] : A metrizable lcs E is holomorphically complete if for every non-convergent Cauchy sequence (x_n) in E there exists $f \in \mathcal{H}(E)$ with $\sup |f(x_n)| = \infty$.

THEOREM 2 *Let U be a pseudoconvex, finitely Runge open set in E, and let $K = \hat{K}_{\mathcal{P}(U)}$ be a compact subset of U. Then every function f which is holomorphic in a neighborhood of K can be approximated uniformly on K by continuous polynomials on E.*

Note that for any compact $K \subset U$, $\hat{K}_{\mathcal{H}(E)}$ is compact according to the above characterization of holomorphically complete spaces. Hence $\hat{K}_{\mathcal{P}(U)}$ is compact since $\hat{K}_{\mathcal{P}(U)} = \hat{K}_{\mathcal{H}(E)}$ (Proposition 2).

PROOF OF THEOREM 2 Let $\varepsilon > 0$. There are $\alpha \in cs(E)$ and $s > 0$ such that f is holomorphic on $W = K + B^{\alpha}(0,s)$ U, and for all $x \in K$, $|f(x) - f(y)| < \frac{\varepsilon}{2}$ if $y \in B^{\alpha}(x,s)$. We first show

(*) There exists a finite rank linear operator $\phi : E \longrightarrow E$ with $\widehat{\phi(K)}_{\mathcal{H}(\phi(E))} \subset W$ and $\alpha(\phi(x) - x) < s$ for all $x \in K$.

Let (α_n) be an increasing sequence of continuous seminorms on E, $\alpha \leq \alpha_n$, which generates the topology of E. Since E

has the approximation property there are continuous linear maps $\phi_n : E \longrightarrow E$, $\dim_{\mathbb{C}} \phi_n(E) < \infty$, such that $\alpha_n(\phi_n(x) - x) < \frac{s}{n}$ for all $x \in K$. Assume that $\widehat{\phi_n(K)} \not\subset W$ for all $n \in \mathbb{N}$. Then there are $x_n \in \widehat{\phi_n(K)} \setminus W$. Put $E_o = \operatorname{sp}(K \cup \bigcup\{\phi_n(E) \mid n \in \mathbb{N}\})$ and let E_1 denote the completion of E_o. For all $g \in \mathscr{H}(E_1)$, $|g(x_n)| \leq$ $\leq \|g\|_{\phi_n(K)}$, since $\phi_n(E)$ is a finite dimensional, hence complemented subspace of E_1. According to our choice of (α_n) and (ϕ_n),

$$\|g\|_{\phi_n(K)} = \|g \circ \phi_n\|_K \longrightarrow \|g\|_K \quad \text{for} \quad n \longrightarrow \infty.$$

Hence, (x_n) is bounding in E_1 and, since E_1 is a separable Fréchet space, (x_n) has a convergent subsequence $x_{nj} \longrightarrow x_o$. (This is a straightforward generalization of the result that every bounding set in a separable Banach space is relatively compact; for example the proof in [16] can be transferred directly). Now, (x_{nj}) is a bounding Cauchy sequence in E:

$$|g(x_{nj})| \leq \|g\|_{\phi_{nj}(K)} \longrightarrow \|g\|_K \text{ for all } g \in \mathscr{H}(E).$$

Thus $x_o \in E$, because E is holomorphically complete. Finally, for all $g \in \mathscr{H}(E)$

$$|g(x_o)| = \lim_{j\to\infty} |g(x_{nj})| \leq \lim_{j\to\infty} \|g \circ \phi_{nj}\|_K = \|g\|_K.$$

This is a contradiction to $x_n \notin W$ for all $n \in \mathbb{N}$. We thus have proved (*).

Now $f|_{W \cap F}$ is holomorphic in $W \cap F$, whereby $F = \phi(E)$, hence in a neighborhood of $L = \widehat{\phi(K)}_{\mathscr{H}(F)} \subset W$. Let $g: F \longrightarrow \mathbb{C}$ be a polynomial with $\|f - g\|_L < \frac{\varepsilon}{2}$ [9, p. 55]. Then $g \circ \phi \in \pi$ with

$$\|f - g \circ \phi\|_K \leq \|f - f \circ \phi\|_K + \|f \circ \phi - g \circ \phi\|_K < \varepsilon.$$

COROLLARY 2 (Runge) *Let $K = \tilde{K}$ be compact in E. Then every function f which is holomorphic in a neighborhood of K can be* *approximated uniformly on* K *by continuous polynomials.*

COROLLARY 3 *The properties* 1° - 4° *in theorem* 1 *are equivalent to*

5° $\tilde{K} \subset U$ *for all compact* $K \subset U$.

PROOF Evidently "5° $\Longrightarrow$ 3°", since $\hat{K}_{\mathcal{H}(E)} \subset \tilde{K}$. To show "3° $\Longrightarrow$ 5°" let $K \subset U$ be compact. Then $\tilde{K}$ and $\hat{K}_{\mathcal{H}(E)}$ are compact since E is holomorphically complete and since U is polynomially convex. Also, $K_o = \tilde{K} \cap (E \setminus U)$ is compact. Now a holomorphic function f can be defined in a neighborhood of $\tilde{K} = \hat{K}_{\mathcal{H}(E)} \cup K_o$ by putting f equal to 0 in a neighborhood of $\hat{K}_{\mathcal{H}(E)}$ and f equal to 1 in a neighborhood of K_o. Corollary 2 implies the existence of a continuous polynomial $p: E \longrightarrow \mathbb{C}$ with $||f - p||_{\tilde{K}} < \frac{1}{2}$. Thus, $||p||_K < \frac{1}{2} < |p(x)|$ for all $x \in K_o$. It follows $K_o = \emptyset$, i.e. $\tilde{K} = \hat{K}_{\mathcal{H}(E)} \subset U$.

4. Theorem 2 can be formulated in the following way. For a compact $L \subset U$ let $\mathcal{H}(L)$ denote the space of germs of holomorphic functions on L. Then, if U is a pseudoconvex, finitely Runge open set in a holomorphically complete, metrizable lcs with the approximation property, the following holds:

(1) *To every compact* $K \subset U$ *there corresponds a compact* $L \subset U$, *containing* K, *such that the image under the "restriction map"*

$$\mathcal{H}(U) \longrightarrow \mathcal{H}(L)$$

is dense in $\mathcal{H}(L)$ *with respect to the sup norm topology on* $\mathcal{H}(L)$.

(Take $L = \hat{K}_{\mathcal{H}(U)}$). The same approximation result can be shown for a pseudoconvex domain U spread over a Fréchet or Silva space with a finite dimensional Schauder decomposition [17] (again with $L = \hat{K}_{\mathcal{H}(U)}$). A straightforward reasoning shows that in the case of $E = \mathbb{C}^n$, a sharper version (1) holds for a pseudoconvex domain U (again with $L = \hat{K}$):

(2) *To every compact* $K \subset U$ *there corresponds a compact* $L \subset U$, *containing* K, *such that the image under the "restriction map"*

$$\mathcal{H}(U) \longrightarrow \mathcal{H}(L)$$

is sequentially dense with respect to the natural inductive limit topology on $\mathcal{H}(L)$.

The inductive limit topology on $\mathcal{H}(L)$ is defined by

$$\mathcal{H}(L) = \lim_{\substack{\longrightarrow \\ V \in \mathcal{N}(L)}} \text{ind } \mathcal{H}^{\infty}(V),$$

where $\mathcal{N}(L)$ is a base of open neighborhoods of L and $\mathcal{H}^{\infty}(V)$ is the Banach space of bounded holomorphic functions on V with the sup norm.

PROOF OF (2) FOR A PSEUDOCONVEX DOMAIN $U \subset \mathbb{C}^n$: Let $K \subset U$ be compact and $L = \hat{K}_{\mathcal{H}(U)}$. Let $f \in \mathcal{H}(L)$. There exists $s > o$ such that f is holomorphic on $L(s) = L + B(o,s) \subset U$. Assume that $\widehat{L(t)}_{\mathcal{H}(U)} \not\subset L(s)$ for all $o < t < s$. Then there are $x_n \in \widehat{L(\frac{1}{n})} \setminus L(s)$, hence $|g(x_n)| \leq \|g\|_{L(\frac{1}{n})}$ for all $g \in \mathcal{H}(U)$. It follows $x_{nj} \longrightarrow x_o \in L$ for a subsequence (x_{nj}) of (x_n). Contradiction. Now let $\widehat{L(t)} \subset L(s)$ for $o < t < s$. According to the approximation theorem of Oka-Weil [9, p. 91], there exist $f_m \in \mathcal{H}(U)$ with $\|f_m - f\|_{L(t)} \longrightarrow o$, i.e. $f_m \longrightarrow f$ in $\mathcal{H}^{\infty}(L(t))$. Since the restriction map $\mathcal{H}^{\infty}(L(t)) \longrightarrow \mathcal{H}(L)$ is continuous, it follows that the germs of f_m converge to f in $\mathcal{H}(L)$.

We don't know whether or not (2) holds for all open

subsets of $\mathbb{C}^n$. There exist domains in $\mathbb{C}^n$ satisfying (2) but not being pseudoconvex. For example, if Ω is pseudoconvex and $K \subset \Omega$ is compact such that $U = \Omega \setminus K$ connected, then U satisfies (2).

Property (2) is particularly interesting in the infinite dimensional case: If an open set U in a normed space E satisfies (2), then the Nachbin topology τ_ω [12] is obtained as the projective limit of all $\mathcal{H}(K)$, $K \subset U$ compact:

$$(\mathcal{H}(U), \tau_\omega) = \lim_{\substack{K \subset U \\ \text{cpt.}}} \text{proj} \ \mathcal{H}(K).$$

This is discussed in [11] and [3].

There are only few examples of open sets in infinite dimensional spaces for which it is known that (2) is satisfied, e.g. for a balanced open set in an arbitray lcs E and for certain Reinhardt open sets in a Banach space with an unconditional basis [11]. Unfortunately, the methods presented above as well as that of [17] do not provide more examples. Also, the above proof of (2) for a pseudoconvex $U \subset \mathbb{C}_n$ cannot be transferred to domains in infinite dimensional normed spaces. However, it can be transferred to the case of a pseudoconvex domain U in an arbitrary product $\mathbb{C}^\Lambda$ of lines, because such a domain is the product $\Omega \times C^{\Lambda'}$ of a pseudoconvex Ω in a certain $\mathbb{C}^n$ and the space $C^{\Lambda'}$, $\Lambda' = \Lambda \setminus \{1,...n\}$ [2].

ACKNOWLEDGEMENT: I want to thank M.C. Matos for helpful comments.

REFERENCES

[1] R. ARON - M. SCHOTTENLOHER, Compact holomorphic mappings on Banach spaces and the approximation property. To appear in J. Funct. Analysis. (Announcement in: Bull. Amer. Math. Soc. 80 (1974), 1245 - 1249).

[2] V. AURICH, Characterization of domains of holomorphy over an arbitrary product of complex lines. Diplomarbeit. München 1973.

[3] S. B. CHAE, Holomorphic germs on Banach spaces. Ann. Inst. Fourier 21 (1971), 107 - 141.

[4] S. DINEEN, Holomorphic functions on locally convex vector spaces II. Ann. Inst. Fourier 23 (1973), 155 - 185.

[5] S. DINEEN, A growth property of pseudoconvex domains in locally convex topological vector spaces. Preprint.

[6] S. DINEEN - Ph. NOVERRAZ - M. SCHOTTENLOHER, Le problème de Levi dans certains espaces vectoriels topologiques localement convexes. To appear in Bull. Math. France.

[7] L. GRUMAN, The Levi problem in certain infinite dimensional vector spaces. Ill. J. Math. 18 (1974), 20 - 26.

[8] L. GRUMAN - C. KISELMAN, Le problème de Levi dans les espaces de Banach à base. C. R. Acad. Sci. Paris 274 (1972), 1296 - 1299.

[9] L. HORMANDER, "An Introduction to Complex Analysis in Several Variables". Van Nostrand, Princeton 1966.

[10] B. JOSEFSON, A counterexample to the Levi problem. In: "Proceedings on Infinite Dimensional Holomorphy". Springer Lecture Notes 264 (1974) 168 - 177.

[11] J. MUJICA, Spaces of germs of holomorphic functions. Thesis, Univ. of Rochester 1974. To appear in Adv. in Math.. See also this proceedings.

[12] L. NACHBIN, "Topology on spaces of holomorphic mappings". Springer-Verlag, New York 1969.

[13] PH. NOVERRAZ, Sur la pseudo-convexité et la convexité polynomiale en dimension infinie. Ann. Inst. Fourier 23 (1973), 113 - 134.

[14] PH. NOVERRAZ, Pseudo-convex completion of locally convex topological vector spaces. Math. Ann. 208 (1974), 59 69.

[15] PH. NOVERRAZ, Pseudo-convexité et base de Schauder dans les e.l.s. In: Sém. Lelong 73/74. Springer Lecture Notes 474 (1975), 63 - 82.

[16] M. SCHOTTENLOHER, Bounding sets in Banach spaces and regular classes of analytic functions. In: Functional Analysis and Applications (Symposium Recife, 1972), Springer Lecture Notes 284 (1974), 109 - 122.

[17] M. SCHOTTENLOHER, Das Leviproblem in unendlichdimensionalen Räumen mit Schauderzerlegung. Habilitationsschrift. München 1974. The Levi problem for domains spread over locally convex spaces with a finite dimensional Schauder decomposition. To appear in Ann. Inst. Fourier.

Mathematisches Institut
der Universität Munchen
Theresienstr. 39
D 8 München 2

Infinite Dimensional Holomorphy and Applications, Matos (ed.)

$\tau_\omega = \tau_o$ for Domains in $C^{\mathbb{N}}$

by MARTIN SCHOTTENLOHER

Let $\mathcal{H}(U)$ be the space of holomorphic functions on a domain $U \subset E$, where E is a Fréchet space over C, and let τ_o (resp. τ_ω) denote the compact open topology (resp. ported topology of Nachbin [6]) on $\mathcal{H}(U)$. It is clear that $\tau_\omega \neq \tau_o$ if E is not a Montel space: Consider the semi-norm $f \to \sup\{|Df(a).x| \; |x \in B\}$, $f \in \mathcal{H}(U)$, where $a \in E$ and $B \subset E$ is a bounded, closed, non-compact subset of E. However, for infinite dimensional Fréchet-Montel spaces it seems to be unknown whether $\tau_\omega = \tau_o$ or $\tau_\omega \neq \tau_o$, except for a result in [3], where $(\mathcal{H}(C^{\mathbb{N}}), \tau_\omega) = (\mathcal{H}(C^{\mathbb{N}}), \tau_o)$ is shown. The question of whether $\tau_\omega = \tau_o$ or $\tau_\omega \neq \tau_o$ holds is of some interest for Fréchet-Schwartz and Fréchet nuclear spaces (see [4]).

In this short note we want to extend Barroso's result [3] to the following:

Proposition: $(\mathcal{H}(U), \tau_\omega) = (\mathcal{H}(U), \tau_o)$ *for any domain* $U \subset C^{\mathbb{N}}$.

PROOF: Let $j : U \to U'$ be a simultaneous analytic continuation (s.a.c.) of $\mathcal{H}(U)$ with the properties

1° $j^* : (\mathcal{H}(U'), \tau_o) \longrightarrow (\mathcal{H}(U), \tau_o)$, $g \to g \circ j$, $g \in \mathcal{H}(U')$, is an open map(i.e. j is a "normal" s.a.c. of $\mathcal{H}(U)$ in the sense of [1]).

2° j is maximal with respect to 1° (i.e. if $i:U \to U''$

is another s.a.c. of $\mathcal{H}(U)$ such $i^* : (\mathcal{H}(U''),\tau_o) \to (\mathcal{H}(U),\tau_o)$ is open there exists an s.a.c. $k : U'' \to U'$ of $\mathcal{H}(U'')$ with $j = k \circ i$). Such a maximal normal s.a.c. of $\mathcal{H}(U)$ exists, and it can be constructed using the set of τ_o-continuous homomorphisms from $\mathcal{H}(U)$ to C (cf. [1]). Because of this construction it follows that U' is holomorphically convex. Therefore, U' is isomorphic to a product $\Omega \times C^{\mathbb{N}}$, where Ω is a Stein domain spread over some C^m, $m \in \mathbb{N}$ (cf. [5] for a schlicht domain $U' \subset E$, and [2] for the general case). Now it suffices to show that $(\mathcal{H}(\Omega \times C^{\mathbb{N}}),\tau_\omega) = (\mathcal{H}(\Omega \times C^{\mathbb{N}}), \tau_o)$, since then $\tau_\omega = \tau_o$ on $\mathcal{H}(U)$ according to the following diagram of continuous bijections ($\to$) and homeomorphisms ($\cong$) :

$$\begin{array}{ccccc} (\mathcal{H}(\Omega \times C^{\mathbb{N}}),\tau_\omega) & \cong & (\mathcal{H}(U'),\tau_\omega) & \to & (\mathcal{H}(U),\tau_\omega) \\ \| \wr & & & & \downarrow \\ (\mathcal{H}(\Omega \times C^{\mathbb{N}}),\tau_o) & \cong & (\mathcal{H}(U'),\tau_o) & \cong & (\mathcal{H}(U),\tau_o). \end{array}$$

$\tau_\omega = \tau_o$ on $\mathcal{H}(\Omega \times C^{\mathbb{N}})$ can be proven in a similar way to [3]. Let p be a τ_ω-continuous semi-norm on $\mathcal{H}(\Omega \times C^{\mathbb{N}})$ ported by the compact set $K \subset \Omega \times C^{\mathbb{N}}$, and let L be a compact neighborhood of $pr_1(K) \subset \Omega$, where pr_1 denotes the canonical projection $\Omega \times C^{\mathbb{N}} \to \Omega$. Choose constants $r_i < \infty$ with $\sup\{|z_i| \mid z = (z_1,z_2,\ldots) \in pr_2(K)\} < r_i$ for $i \in \mathbb{N}$, and set $W_j := \{z \in C^{\mathbb{N}} \mid |z_i| < r_i \text{ for } i = 1,2,\ldots,j\} \subset C^{\mathbb{N}}$. Since p is ported by $K \subset L \times W_j$ there exist constants c_j with $p(f) \leq c_j \|f\|_{L \times W_j}$ for all $f \in \mathcal{H}(\Omega \times C^{\mathbb{N}})$ ($\|f\|_X := \sup\{|f(x)| \mid x \in X\}$). Hence, for $s_j := c_j r_j$ and $v = (v_1,v_2,\ldots,v_j) \in \mathbb{N}^j$ we have $p(az^v) \leq c_j \|az^v\|_{L \times W_j} \leq c_j \|a\|_L r^v \leq \|a\|_L s^v$ for all $a \in \mathcal{H}(\Omega)$ ($z^v := z_1^{v_1} z_2^{v_2} \ldots z_j^{v_j}$). Now set $t_j := (2j)^2 s_j$ and $M := \{z \in C^{\mathbb{N}} \mid |z_j| \leq t_j \text{ for all } j \in \mathbb{N}\}$. Since $f \in \mathcal{H}(\Omega \times C^{\mathbb{N}})$ "depends only on a finite number of variables" (cf. [5] , [2]) there exists $j \in \mathbb{N}$ such that f has a power series expansion of the form

$$f(x,z) = \sum_{\nu \in \mathbb{N}^j} a_\nu(x) z^\nu , \quad (x,z) \in \Omega \times C^{\mathbb{N}} , \text{ with } a_\nu \in \mathcal{H}(\Omega).$$

By the Cauchy inequalities $|a_\nu(x)|t^\nu \leq \|f\|_{\{x\} \times M}$, $\nu \in \mathbb{N}^j$.
Hence,

$$p(f) \leq \Sigma\, p(a_\nu z^\nu) \leq \Sigma\, \|a_\nu\|_L s^\nu \leq \Sigma\, \|f\|_{L\times M}(t^\nu)^{-1}s^\nu \leq \tfrac{\pi}{2}\|f\|_{L\times M}.$$

Since $L \times M$ is compact it follows that p is τ_o-continuous.

REMARK: The proposition also holds for domains U spread over $C^{\mathbb{N}}$.

ADDED IN PROOF: A different proof of the above results was given by Barroso and Nachbin (to appear elsewhere).

REFERENCES

1. H. ALEXANDER, Analytic functions on Banach spaces. Thesis, University of California, Berkeley (1968).

2. V. AURICH, Characterization of domains of holomorphy over an arbitrary product of complex lines. Diplomarbeit, Universität München (1973).

3. J. A. BARROSO, Topologias nos espaços de aplicações holomorfas entre espaços localmente convexos. Anais de Acad. Bras. de Ciências 43 (1971), 527-546.

4. K. D. BIERSTEDT - R. MEISE, Nuclearity and the Schwartz property in the theory of holomorphic functions on metrizable locally convex spaces. Preprint.

5. A. HIRSCHOWITZ, Remarques sur les ouverts d'holomorphie d'un produit dénombrable de droites. Ann. Inst. Fourier 19 (1969), 219 - 229.

6. L. NACHBIN, Topology on spaces of holomorphic mappings. Ergebnisse der Mathematik 47 (1969), Springer, New York - Heidelberg.

Martin Schottenloher (May 76)
8024 Kreuzpullach 5
Germany

Infinite Dimensional Holomorphy and Applications, Matos (ed.)

HOLOMORPHY OF COMPOSITION

By *JAMES O. STEVENSON* (*)

ABSTRACT

In this paper we study the holomorphy, with respect to various locally convex topologies, of the composition map $\phi : (f,g) \in X \times Y \mapsto g \circ f \in Z$ of holomorphic functions between Banach spaces E and F, and F and G. If X includes all the constant functions, then ϕ will not be holomorphic when Y is the space of entire functions from F to G. Positive results generally require the topology on Y to be that of uniform convergence on certain subsets of F and the topology on X to be the same type as that on Z, such as compact - open, τ_σ, τ_ω.

AMS (MOS) subject classifications (1970). Primary 46E10, 58B10. Key words and phrases. Infinite dimensional holomorphy, composition, G - holomorphy, amply bounded, holomorphic convexity.

(*) This work is based on part of the author's doctoral dissertation at the University of Rochester under the supervision of Leopoldo Nachbin, and was supported in part by an NSF Traineeship.

I. INTRODUCTION

We wish to consider the following two problems for E, F, G Banach spaces over the complex field $\mathbb{C}$ and $H(E;F)$, $H(F;G)$, $H(E;G)$ the corresponding spaces of holomorphic functions between them (we follow the definitions and notation given in [8]): (1) For what vector subspaces $X \subset H(E;F)$, $Y \subset H(F;G)$, $Z \subset H(E;G)$ and corresponding locally convex topologies τ_X, τ_Y, τ_Z will the composition $\phi : (f,g) \in (X,\tau_X) \times (Y,\tau_Y) \mapsto g \circ f \in (Z, \tau_Z)$ be holomorphic? (2) Investigate the holomorphy of

$$\phi : H(U;V) \times H(V;W) \to H(U;W)$$

for $U \subset E$, $V \subset F$, $W \subset G$ open. We are driven to consider non-normable locally convex topologies on X, Y, Z since if ϕ is holomorphic, then it will be separately continuous, and so in particular the evaluation $f \in (H(F;\mathbb{C}),\tau) \mapsto f(x) \in \mathbb{C}$ will be continuous. But combining results from Alexander [1], Dineen [5], and Josefson [11], we then infer that τ is not even first countable when F is a Banach space. (Note: in the sequel we shall often write "iff" for "if and only if").

2. PRELIMINARIES

We introduce here some basic definitions and results (see Nachbin [10]). Throughout, $\mathbb{R}$ will denote the reals, $\mathbb{C}$ the complexes, and $\mathbb{N}$ the nonnegative integers. In this section let X and Y be complex locally convex topological vector spaces (LCS's) and $W \subset X$ an open, nonempty subset. For $m = 1,2,\ldots$ let $L_s(^mX;Y)$ represent the vector space of continuous, symmetric, m-linear maps from X^m to Y, and $P(^mX;Y)$ the vector space of continuous m - homogeneous polynomials from X to Y. Then $A \in L_s(^mX;Y) \mapsto \hat{A} \in P(^mX;Y)$ is a linear bijection where $\hat{A}(x) = Ax^m = A(x,\ldots,x)$. (Let $L_s(^0X;Y) = P(^0X;Y) = Y$ and $Ax^0 = A \in Y$).

DEFINITION 2.1 $f : W \to Y$ *is said to be holomorphic on* W *if for every* $x_0 \in W$ *there is a sequence* $A_m \in L_s(^mX;Y)$, $m \in \mathbb{N}$, *such that for every continuous seminorm* q *on* Y *there is an open subset* $V \subset W$ *containing* x_0 *for which*

$$\lim_{M\to\infty} q\left[f(x) - \sum_{m=0}^{M} A_m(x - x_0)^m\right] = 0$$

uniformly for $x \in V$. *Let* $H(W;Y)$ *represent the vector space of holomorphic maps from* W *to* Y. *If* Y *is Hausdorff* (T_2), *then for each* f *and* x_0 *the sequence* (A_m) *is unique and we write*

$$d^mf(x_0) = m!\, A_m,\quad \hat{d}^mf(x_0) = m!\, \hat{A}_m,$$

which represent the derivatives of order m *of* f *at* x_0. *The Taylor series of* f *at* x_0 *is written*

$$\sum_{m=0}^{\infty} \frac{1}{m!}\, d^mf(x_0)(x - x_0)^m = \sum_{m=0}^{\infty} \frac{1}{m!}\, \hat{d}^mf(x_0)(x - x_0).$$

DEFINITION 2.2 $f : W \to Y$ *is said to be* G - *holomorphic if (provided* X *is* T_2) *for every* $x_0 \in W$ *and* $x \in X$, *the map* $z \in V \mapsto f(x_0 + zx) \in Y$ *is holomorphic where* $V = \{z \in \mathbb{C} : x_0 + zx \in W\}$. $H_G(W;Y)$ *will denote the vector space of* G - *holomorphic functions from* W *to* Y.

DEFINITION 2.3 $f : W \to Y$ *is said to be amply bounded if for every continuous seminorm* q *on* Y, $q \circ f$ *is locally bounded on* W, *that is, for every* $x \in W$ *there is an open subset* $V \subset W$ *containing* x *on which* $q \circ f$ *is bounded.* $AB(W;Y)$ *will denote the vector space of amply bounded functions from* W *to* Y. *Notice if* f *is either continuous or locally bounded, then it is amply bounded. We have* $H_G(W;Y) \cap AB(W;Y) = H(W;Y)$.

DEFINITION 2.4 $f : W \to Y$ *is said to be weakly holomorphic if for every* ψ *in the dual space* Y' *of* Y, $\psi \circ f : W \to \mathbb{C}$ *is holomorphic.*

3. TOPOLOGIES

We shall now introduce the locally convex topologies we shall be considering in the sequel and discuss some of their relationships. Let E and F be complex Banach spaces and $U \subset E$ a nonempty, open subset. See Nachbin [8]. (Throughout our discussion if f is a bounded map from a set A into a Banach space, then $|f|_A$ denotes $\sup_{x \in A} \|f(x)\|$).

DEFINITION 3.1 *The locally convex topology on* $H(U;F)$ *generated by the seminorms* $f \mapsto |f|_K$, $K \subset U$ *compact, is called the compact-open topology or topology of uniform convergence on compact subsets and is denoted* τ_o. *The topology* τ_∞ *on* $H(U;F)$ *is defined by the family of seminorms* $f \mapsto |\hat{d}^m f|_K$ *for* $K \subset U$ *compact and* $m \in \mathbb{N}$.

It follows that $\tau_o \leq \tau_\infty$, and $\tau_o = \tau_\infty$ iff E is finite dimensional or $F = 0$ (Alexander [1]).

DEFINITION 3.2 *A seminorm* p *on* $H(U;F)$ *is said to be ported by the compact subset* $K \subset U$ *if for every open* $V \subset U$ *containing* K, *there is a real number* $c > 0$ *such that* $p(f) \leq c|f|_V$ *for all* $f \in H(U;F)$. *The topology* τ_ω *on* $H(U;F)$ *is defined by all seminorms each ported by a compact subset of* U .

Suppose $x_o \in U$ and $K \subset U$ is compact and x_o - balanced (that is, $K - x_o$ is balanced). Then for every sequence $\varepsilon = (\varepsilon_m)$ in c_o^+ (the space of sequences of positive real numbers tending to 0),

$$p_{\varepsilon,K}(f) = \sum_{m=0}^{\infty} \sup_{x \in K+\varepsilon_m B_1} \left\| \frac{1}{m!} \hat{d}^m f(x_o)(x - x_o) \right\|$$

defines a seminorm on $H(U;F)$ where B_1 is the open unit ball in E . If U is x_o - balanced, then the topology τ_ω on $H(U;F)$ is defined by the seminorms $p_{\varepsilon,K}$ for all ε and K as above (cf. Remark 4.2 in Aron [2]).

Again we have $\tau_\infty \leq \tau_\omega$ and $\tau_\infty = \tau_\omega$ iff E is finite dimensional or $F = 0$.

DEFINITION 3.3 *The topology* τ_σ *on* $H(U;F)$ *is defined by all seminorms of the form*

$$p_{\alpha,K}(f) = \sum_{m=0}^{\infty} \alpha_m^m \left| \frac{1}{m!} \hat{d}^m f \right|_K$$

where $\alpha = (\alpha_m) \in c_o^+$ *and* $K \subset U$ *is compact.*

If follows that $\tau_\infty \leq \tau_\sigma \leq \tau_\omega$, and $\tau_\infty = \tau_\sigma$ iff E is finite dimensional or $F = 0$.

DEFINITION 3.4 *Let* $K \subset E$ *be compact and let* $h(K;F)$ *be the union of* $H(U;F)$ *for all open* $U \supset K$. *We define an equivalence relation* $\sim$ *on* $h(K;F)$ *as follows:* $f \sim g$, *where* $f \in H(U;F)$ *and* $g \in H(V;F)$, *if there is an open subset* $W \subset E$ *such that* $K \subset W \subset U \cap V$ *and for all* $x \in W$, $f(x) = g(x)$. *An equivalence class will be called a germ of holomorphic mappings around* K *and will be denoted* $[f]$ *where* f *is any representative.* *Let*

$H(K;F) = h(K;F)/\sim$. *Then* $H(K;F)$ *has a unique vector space structure such that the maps*

$$f \in H(U;F) \mapsto [f] \in H(K;F)$$

are linear for all open $U \supset K$. *We give* $H(K;F)$ *the finest locally convex topology for which all the above maps are continuous with respect to* τ_ω *on each* $H(U;F)$. *This is an inductive limit topology and we may designate this by*

$$H(K;F) = \lim_{\substack{\rightarrow \\ U \supset K}} (H(U;F), \tau_\omega).$$

We shall denote this topology on $H(K;F)$ *also by* τ_ω. *(We get the same topology if instead of* $(H(U;F),\tau_\omega)$ *we take the inductive limit of the (Banach) spaces of bounded holomorphic functions on* U, $HB(U;F)$, *with sup norm* $|\cdot|_U$*).*

Let $U \subset E$ *be open and nonempty. We define the topology* τ_π *on* $H(U;F)$ *to be the coarsest locally convex topology on* $H(U;F)$ *such that for all compact* $K \subset U$, *the maps*

$$f \in H(U;F) \mapsto [f] \in H(K;F)$$

are continuous with respect to τ_ω *on each* $H(K;F)$. *This is a projective limit topology and we can write*

$$(H(U;F), \tau_\pi) = \lim_{\substack{\leftarrow \\ K \subset U}} (H(K;F), \tau_\omega).$$

We have $\tau_\sigma \leq \tau_\pi \leq \tau_\omega$. If U is x_o - balanced, then $\tau_\pi = \tau_\omega$ (cf. [3]).

DEFINITION 3.5 *Let* L *be a vector subspace of* $H(U;F)$. *The locally convex topology* $\tau_\lambda(L)$ *is defined as follows. For every open cover* I *of* U, *let* L_I *be the vector subspace of* L *consisting of all* $f \in L$ *such that* f *is bounded on every* $V \in I$. *We define the natural topology on* L_I *by the family of seminorms*

$f \mapsto |f|_V$ *for* $V \in I$. *Then* $\tau_\lambda(L)$ *is defined to be the finest locally convex topology on* L *such that the inclusions* $L_I \hookrightarrow L$ *are continuous for all open covers* I *of* U. *In fact, the collection of open covers is directed by refinements and* $(L, \tau_\lambda(L))$ *becomes the inductive limit of the* L_I. *In terms of seminorms,* $\tau_\lambda(L)$ *is defined by all seminorms ported by all open covers of* U, *where* p *is ported by the open cover* I *of* U *if there are* $c > 0$ *and a finite union* V *of sets in* I *such that* $p(f) \leq c|f|_V$ *for all* $f \in L_I$.

The locally convex topology $\tau_\delta(L)$ *is defined the same way except that only countable open covers* I *are used.*

If $L = H(U;F)$, we set $\tau_\lambda(L) = \tau_\lambda$, $\tau_\delta(L) = \tau_\delta$. Notice that $\tau_\lambda|L \leq \tau_\lambda(L)$ and $\tau_\delta|L \leq \tau_\delta(L)$, that is, the inclusion $(L, \tau_\lambda(L)) \hookrightarrow (H(U;F), \tau_\lambda)$ is continuous and similarly for τ_δ.

We have the following significant properties for $\tau_\delta(L)$. It is a bornological topology on L, and since U is metrizable, it is in fact the bornological locally convex topology associated with $\tau_0|L$. If $\tau_\omega(L)$ is the topology defined by all seminorms p on L each ported by a compact subset $K \subset U$ (that is, for every open $V \subset U$ containing K there is $c > 0$ such that $p(f) \leq c|f|_V$ for all $f \in L$), then $\tau_\omega(L) \leq \tau_\lambda(L) \leq \tau_\delta(L)$.

DEFINITION 3.6 *A subset* $A \subset U$ *is called a bounding subset of* U *if every complex-valued holomorphic function on* U *is bounded on* A. *This is equivalent to requiring every* $f \in H(U;F)$ *to be bounded on* A *for every* $F \neq 0$. *By replacing compact subsets of* U *with bounding subsets in the definitions of* τ_0, τ_π, *and* τ_ω, *we obtain the corresponding topologies* τ_{0B}, $\tau_{\pi B}$, *and* $\tau_{\omega B}$.

Hence $\tau_{oB} \leq \tau_{\pi B} \leq \tau_{\omega B}$.

Since every compact subset of U is bounding, we have $\tau_o \leq \tau_{oB}$, $\tau_\pi \leq \tau_{\pi B}$, $\tau_\omega \leq \tau_{\omega B}$. Hirschowitz [6] showed that τ_ω and $\tau_{\omega B}$ have the same bounded subsets, so that $\tau_\omega \leq \tau_{\omega B} \leq \tau_\delta$.

DEFINITION 3.7 *The space of functions of bounded type,* $H_b(U;F)$, *consists of all* $f \in H(U;F)$ *which are bounded on every* $A \subset U$ *which is bounded in* E *and* $d(A,\partial U) > 0$, *where* $d(A,\partial U)$ *is the distance from* A *to the boundary of* U, ∂U *(such* A *are said to be* U - *bounded). The natural topology* τ_{ob} *on* $H_b(U;F)$ *is defined by the seminorms* $f \mapsto |f|_A$ *where* $A \subset U$ *is* U - *bounded.*

$(H_b(U;F), \tau_{ob})$ is a Frechet space. We also have the useful result that for every $F \neq 0$, $H_b(U;F) \neq H(U;F)$. Indeed, Josefson [11] showed that in the dual space E' of E every bounded sequence has a weak* convergent subsequence. Dineen [5] used this property to show $H_b(E;\mathbb{C}) \neq H(E;\mathbb{C})$, and so $H_b(U;F) \neq H(U;F)$ for all $F \neq 0$. One consequence of this result is the following (where $H_b = H_b(U;F)$).

PROPOSITION 3.1 *In general* $\tau_\delta(H_b) \leq \tau_{ob}$, *and if* U *is* x_o - *balanced, then* $\tau_\delta(H_b) \neq \tau_{ob}$.

PROOF: Let p be a seminorm on $H_b(U;F)$ ported by all countable open covers of U . Consider the countable open cover $I = (U_n)_{n \in \mathbb{N}}$ where $U_n = \{x \in U : d(x, \partial U) > \frac{1}{n+1}, \|x\| < n+1\}$. Therefore, each U_n is U - bounded and $U_n \subset U_{n+1}$. Hence, there are a $c > 0$ and $U_n \in I$ such that $p(f) \leq c|f|_{U_n}$ for all f bounded on every $U_k \in I$, and so in this case for all $f \in H_b(U;F)$.

Thus p is τ_{ob}-continuous, so that $\tau_\delta(H_b) \le \tau_{ob}$.

Now suppose that $\tau_\delta(H_b) = \tau_{ob}$ (and therefore τ_{ob} is the bornological topology associated with $\tau_o(H_b)$). Since $H(U;F) \neq H_b(U;F)$, choose $g \in H \setminus H_b$. Then g has a Taylor series at x_o, $\sum_{k=0}^{\infty} P_k(x - x_o)$ where $P_k \in P(^kE;F)$ $(k \in \mathbb{N})$, which converges uniformly on compact subsets of U. Let $S_n(x) = \sum_{k=0}^{n} P_k(x - x_o)$. Then $S_n \in H_b(U;F)$, and since S_n converges to g uniformly on compact sets, we have $\{S_n : n \in \mathbb{N}\}$ is τ_o-bounded, and so τ_{ob}-bounded also (since $\tau_\delta(H_b) = \tau_{ob}$). Therefore, if $A \subset U$ is U-bounded, then there is a $c > 0$ such that for all n, $|S_n|_A \le c$. But this means $|g| \le c$ also, which yields the contradiction $g \in H_b(U;F)$. Hence, $\tau_\delta(H_b) \neq \tau_{ob}$.

In summary, then since $\tau_\delta | H_b \le \tau_\delta(H_b)$, we have the following diagram of continuous inclusions, where in general all but j are bijections.

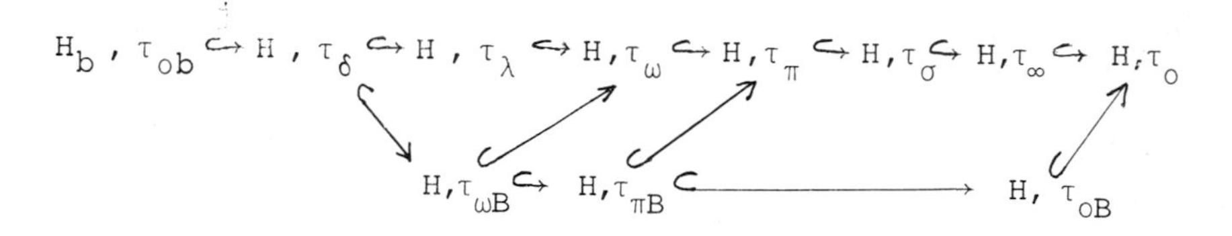

4. BASIC SETTING FOR THE PROBLEM

In order to establish the most general setting in which to investigate our two problems (short of manifolds), we wish to consider whether $H(U;V) \subset H(U;F)$ and $H_b(U;V) \subset H_b(U;F)$ are vector subspaces or open subsets, where $U \subset E$ and $V \subset F$ are open and nonempty. Since they both contain the constant functions, they will be vector subspaces iff $V = F$. Hence, we assume $V \neq F$ and turn to the question of whether they are open subsets.

LEMMA 4.1 *If* $U \subset E$ *is open and nonempty, then there is an open cover* I *of* U *such that for every* $\{U_1,\ldots,U_n\} \subset I$ *with* $V = \cup U_i$, *there exists an* $x_0 \in U \setminus V$ *for which given any* $\varepsilon > 0$ *there is a* $g \in H(U;\mathbb{C})_I$ *satisfying* $g(x_0) = 1$ *and* $|g|_V < \varepsilon$.

(Recall $H(U;\mathbb{C})_I = \{f \in H(U;\mathbb{C}) : f$ bounded on each $W \in I\}$).

PROOF: First we assert there is an $f \in H(U;\mathbb{C})$ which is unbounded. Indeed, if $U = E$, then arguing as in the classical case using the Cauchy inequalities, every nonconstant entire function is unbounded. If $U \neq E$ and if there were no such unbounded f, then U would be an open bounding subset of E which would imply that $H_b(U;\mathbb{C}) = H(U;\mathbb{C})$ (cf. Dineen [5]), an impossibility.

Now for each $x \in U$ there is an open $V(x) \subset U$ containing x such that

$$f(y) = \sum_{m=0}^{\infty} \frac{1}{m!} \hat{d}^m f(x)(y - x)$$

uniformly on $V(x)$. In particular, f is bounded on $V(x)$. Then $I = \{V(x) : x \in U\}$ is an open cover of U on each member of which f is bounded.

Let V be the union of any finite subset of I, then $|f|_V < \infty$. But since f is not bounded on U, there is an $x_0 \in U \setminus V$ such that $|f(x_0)| > |f|_V$. Let $h(x) = f(x) / f(x_0)$. Then

$$h \in H(U;\mathbb{C})_I, \ h(x_0) = 1, \quad |h|_V < 1 .$$

Therefore, for every $\varepsilon > 0$, there is an $n \in \mathbb{N}$ such that $|h^n|_V = |h|_V^n < \varepsilon$, and so $g(x) = |h(x)|^n$ satisfies the conclusion to the lemma.

PROPOSITION 4.1 *If* $U \subset E$, $V \subsetneq F$ *are open and nonempty, then* $H(U;V)$ *is not open in* $(H(U;F), \tau_\lambda)$.

PROOF: Let $y_o \in V$ and take $f_o(x) = y_o$ for all $x \in U$. Then $f_o \in H(U;V)$, but we shall show it is not a τ_λ - interior point.

The sets $N_{p,\varepsilon} = \{f \in H(U;F) : p(f) < \varepsilon\}$, for $\varepsilon > 0$ and p a seminorm ported by all open covers of U, form a base of τ_λ - neighborhoods of zero. Let I be the open cover in Lemma 4.1. Then there exist a $c > 0$ and finite union W of subsets of I such that $p(f) \leq c|f|_W$ for all $f \in H(U;F)_I$. Let $y_1 \in F \setminus V$ and choose g as in Lemma 4.1 so that $g(x_o) = 1$ and

$$|g|_W < \varepsilon / (c\| y_1 - y_o\|) .$$

Define $f(x) = g(x)(y_1 - y_o)$. Then $f \in N_{p,\varepsilon}$ and $(f_o + f)(x_o) = y_1 \notin V$ implies that $f_o + N_{p,\varepsilon} \not\subset H(U;V)$.

We shall now see $H_b(U;V)$ is not open in $(H_b(U;F), \tau_{ob})$ under certain restrictions on U.

DEFINITION 4.1 *For* $A \subset U$ *and* $\mathcal{F} \subset H(U) = H(U;\mathbb{C})$, *we define the* $\mathcal{F}$ *- convex hull of* A *to be*

$$\hat{A}_{\mathcal{F}} = \{x \in U : |f(x)| \leq |f|_A \text{ for all } f \in \mathcal{F}\}.$$

U *(open, nonempty) is said to be* $H_b(U)$ *- convex if for every* U *- bounded subset* A *of* U, $\hat{A}_{H_b(U)}$ *is* U *- bounded. If* U *is convex (in particular, all of* E *), then* U *is* $H_b(U)$ *- convex. Also* U *is* $H_b(U)$ *- convex if it is a domain of* H_b *- holomorphy (cf. Dineen [4]).*

LEMMA 4.2 *If for every* U *- bounded subset* A *of the nonempty* *open set* U, $\hat{A}_{H_b(U)} \neq U$, *then for each* U-*bounded subset* $A \subset U$ *there is an* $x_0 \in U \setminus A$ *such that for all* $\varepsilon > 0$ *we have a* $g \in H_b(U)$ *with* $g(x_0) = 1$ *and* $|g|_A < \varepsilon$.

PROOF: By hypothesis is $A \subset U$ is U-bounded, then there is an $x_0 \in U \setminus A$ and $f \in H_b(U)$ such that $|f(x_0)| > |f|_A$. The rest of the proof follows as in the proof of Lemma 4.1.

PROPOSITION 4.2 *If* $U \subset E$, $V \subsetneq F$ *are open and nonempty and if* U *is* $H_b(U)$-*convex, then* $H_b(U;V)$ *is not open in* $(H_b(U;F), \tau_{ob})$.

PROOF: The argument is analogous to that of Proposition 4.1 using Lemma 4.2 in place of Lemma 4.1.

Hence, the most reasonable setting in which to discuss the holomorphy of the composition function ϕ from $X \times Y$ into Z is to take $X \subset H(U;F)$, $Y \subset H(F;G)$ and $Z \subset H(U;G)$ as vector subspaces.

5. G-HOLOMORPHY OF ϕ

As indicated in section 2, we shall investigate the holomorphy of ϕ by examining separately when it is G-holomorphic and amply bounded. We may reduce the problem by using a theorem of Nachbin [11] which implies that if M is a LCS, W is an open subset of M, and $\tau_1(N) \leq \tau_2(N)$ are two locally convex topologies on a vector space N such that the $\tau_1(N)$ — closure of every $\tau_2(N)$-bounded set is $\tau_2(N)$-bounded (designated condition (A)), then

$$H_G(W;N_1) \cap AB(W;N_2) = H(W;N_2)$$

where $N_i = (N,\tau_i(N))$ for $i = 1,2$. Condition (A) is implied by (B): every $\tau_1(N)$ - bounded subset of N is $\tau_2(N)$ - bounded, or (C): $\tau_2(N)$ is $\tau_1(N)$ - locally closed (that is, $\tau_2(N)$ has a base of neighborhoods of zero which are $\tau_1(N)$ - closed).

Set $W = M = (X,\tau_X) \times (Y,\tau_Y)$ where $X \subset H(U;F)$ and $Y \subset H(F;G)$ are vector subspaces, and let $N_1 = (H(U;G),\tau_o)$. Then condition (B) holds for all the topologies $\tau_2(N)$ listed in section 3 since they satisfy $\tau_o \leq \tau_2(N) \leq \tau_\delta$ and so have the same bounded sets as $\tau_1(N) = \tau_o$.

Now let $N_1 = (H_b(U;F),\tau_o)$, and $\tau_2(N) = \tau_{ob}$. In this case, Proposition 3.1 shows condition (B) fails in general.The next result shows condition (C) will hold, however.

PROPOSITION 5.1 τ_{ob} *on* $H_b(U;F)$ *is* τ_o *- locally closed.*

PROOF: It suffices to show if A is a U - bounded subset of U, then $V = \{f \in H_b(U;F) : |f|_A \leq 1\}$ is τ_o - closed. Let $(f_d)_{d \in D}$ be a net in V converging uniformly o compact subsets to $f \in H_b(U;F)$. Then for every $\varepsilon > 0$, $K \subset U$ compact, there is a $d_o \in D$ such that for all $d \geq d_o$, $|f_d - f|_K < \varepsilon$. Take $K = \{x\}$ where $x \in A$. Then for every $\varepsilon > 0$, $|f(x)| \leq 1 + \varepsilon$, so that $f \in V$.

Hence, we need only study the holomorphy of ϕ when the topology on the range space is τ_o. We shall make use of the following equivalent condition for G - holomorphy.

PROPOSITION 5.2 *Let* X *and* Y *be complex,*T_2, LCS'*s and* W *a nonempty, open subset of* X . *If* Y *is sequentially complete,then the following are equivalent.*

(i) $f : W \subset X \to Y$ *is* G - *holomorphic*.

(ii) *There is a function* $L : W \times X \to Y$ *such that for all* $x_o \in W$ *and* $x \in X$,

$$\frac{f(x_o + zx) - f(x_o)}{z} - L(x_o, x) \to 0$$

in Y *as* $z \to 0$ *in* $\mathbb{C}$.

PROOF: (i) => (ii). Let $x_o \in W$, $x \in X$, and $V = \{z \in \mathbb{C} : x_o + zx \in W\}$. Then from Definition 2.2, $g = f \circ h : V \to Y$ is holomorphic where $h(z) = x_o + zx$. By the Cauchy inequalities (cf. Nachbin [9]) we have for every continuous seminorm q on Y ,

$$q\left[(g(z) - g(0) - dg(0)z) / z \right] \leq \frac{|z|}{\rho(\rho - |z|)} \sup_{|t| \leq \rho} q[g(t)]$$

where $\{t \in \mathbb{C} : |t| \leq \rho\} \subset V$. This yields the desired result , setting $L(x_o, x) = dg(0)$.

(ii) => (i). For every $\psi \in Y'$, we have for $z_o \in V$,

$$\frac{\psi \circ f \circ h(z_o + z) - \psi \circ f \circ h(z_o)}{z} - \psi\left[L(x_o + z_o x, x)\right] \to 0$$

as $z \to 0$. Hence $f \circ h$ is weakly holomorphic on V . But Y sequentially complete implies that $f \circ h$ is holomorphic (cf. Nachbin [9]). Thus f is G - holomorphic.

The requirement that Y be sequentially complete and T_2 is no restriction in our case, since $(H(U;G), \tau_o)$ and $(H_b(U;G), \tau_{ob})$ are complete and T_2 (because E is metrizable and G is complete and T_2).

Now suppose $X \subset H(U;F)$, $Y \subset H(F;G)$, and $Z \subset H(F;G)$ are vector subspaces (such that ϕ is defined). Take f_o, $f \in X$, g_o, $g \in Y$, and $z \in \mathbb{C}$, and set

$w_z(f_o, g_o, f, g) = \phi\left[(f_o, g_o) + z(f, g)\right] - \phi(f_o, g_o) - zL\left[(f_o, g_o), (f, g)\right]$

where $L : (X \times Y) \times (X \times Y) \to Z$ is given by

$$L[(f_o,g_o),(f,g)](x) = [dg_o(f_o(x))](f(x)) + g[f_o(x)]$$

for $x \in U$. We may write $w_z = u_z + v_z$ where

$$u_z(x) = \phi(f_o + zf, g_o)(x) - \phi(f_o,g_o)(x) - z[dg_o(f_o(x))](f(x))$$

and

$$v_z(x) = z[\phi(f_o + zf, g) - \phi(f_o,g_o)](x).$$

Now

$$u_z(x) = \sum_{m=2}^{\infty} z^m \frac{1}{m!} \hat{d}^m g_o(y_o)(y)$$

where $y_o = f_o(x)$ and $y = f(x)$. Hence, for $r > 0$ we have

$$\left\| \frac{1}{z} u_z(x) \right\|_G \leq \frac{1}{|z|} \sum_{m=2}^{\infty} \frac{|z|^m}{r^m} \sup_{|t|=r} \| g_o(y_o + ty) \|_G$$

$$= M_x(g_o) \frac{|z|}{r(r-|z|)}, \quad 0 < |z| < r,$$

where $M_x(g_o) = \sup_{|t|=r} \| g_o(f_o(x) + tf(x)) \|_G$, $x \in U$. Also

$$v_z(x) = z \sum_{m=1}^{\infty} \frac{1}{m!} z^m [\hat{d}^m g(y_o))](y),$$

so that

$$\left\| \frac{1}{z} v_z(x) \right\|_G \leq M_x(g) \frac{|z|}{r-|z|}, \quad 0 < |z| < r.$$

Hence, for $0 < |z| < r$, $x \in U$, we have

$$\left\| \frac{1}{z} w_z(f_o,g_o,f,g)(x) \right\|_G \leq \frac{|z|}{r-|z|} \left[\frac{1}{r} M_x(g_o) + M_x(g)\right] \quad (*)$$

where $M_x(g) = \sup_{|t|=r} \| g(f_o(x) + tf(x)) \|_G$.

PROPOSITION 5.3 *Let* $X \subset H(U;F)$ *and* $Y \subset H(F;G)$ *be vector subspaces. Then* $\phi : (X,\tau_X) \times (Y,\tau_Y) \to (H(U;G),\tau_o)$ *is G-holomorphic for any locally convex Hausdorff topologies* τ_X, τ_Y.

PROOF: τ_0 is generated by the seminorms $|\cdot|_K$ where $K \subset U$ is conpact. From equation (*) above we have

$$\sup_{x\in K} \left\|\frac{1}{z} w_z(f_0,g_0,f,g)(x)\right\| \leq \frac{|z|}{r-|z|}\left[\frac{1}{r} \sup_{x\in K} M_x(g_0) + \sup_{x\in K} M_x(g)\right].$$

Since $K_0 = \bigcup_{|t|=r} f_0(K) + tf(K)$ is compact,

$$\sup_{x\in K} M_x(g) \leq |g|_{K_0} < \infty .$$

Hence, $\left|\frac{1}{z} w_z(f_0,g_0,f,g)\right|_K \to 0$ as $z \to 0$, so that ϕ is G - holomorphic by Proposition 5.2.

6. AMPLE BOUNDEDNESS OF ϕ

Let $X \subset H(U;F)$, $Y \subset H(F;G)$, and $Z \subset H(U;G)$ be vector subspaces for which ϕ is defined, where $U \subset E$ is open and nonempty, and let $\mathcal{M}$ be a collection of subsets of U. Define $X_{\mathcal{M}}$ to be the LCS of all $f \in X$ bounded on each set in $\mathcal{M}$ with topology defined by the family of sup seminorms $(|\cdot|_W)_{W\in\mathcal{M}}$ and likewise for Z (cf. similar notation in Definition 3.5). Then $\phi : X_{\mathcal{M}} \times (Y,\tau_Y) \to Z_{\mathcal{M}}$ is amply bounded iff for every $W \in \mathcal{M}$, $|\phi(\cdot)|_W$ is locally bounded, that is, for every $(f_0,g_0) \in X_{\mathcal{M}} \times Y$, there are neighborhoods of zero M in $X_{\mathcal{M}}$ and N in $(Y;\tau_Y)$ such that

$$\sup_{f\in M, g\in N} |\phi(f_0 + f, g_0 + g)|_W < \infty .$$

N.B. From now on assume X contains the set $\tilde{F}$ of all constant functions on U to F (then so does $X_{\mathcal{M}}$).

LEMMA 6.1 *For all* f_0, $f \in X$, $g \in Y$, $W \in \mathcal{M}$, *and* $\varepsilon > 0$,

$$\sup_{\substack{f=\text{constant}\\ |f|_E<\varepsilon}} |g(f_o+f)|_W = \sup_{|f|_W<\varepsilon} |g\circ(f_o+f)|_W = |g|_{B_\varepsilon(f_o(W))}$$

where $B_\varepsilon(A) = A + \varepsilon B_1$.

PROOF: The inequalities $\leq$ are clear. Now take $z_o \in B_\varepsilon(f_o(W))$, that is, $z_o = y_o + f_o(x_o)$ where $\|y_o\| < \varepsilon$ and $x_o \in W$. Define $f(x) = y_o$ for all $x \in E$. Then $|f|_E < \varepsilon$ and so

$$\sup_{\substack{f=\text{constant}\\ |f|_E<\varepsilon}} |g\circ(f_o+f)|_W \geq \|g(z_o)\|.$$

LEMMA 6.2 $\phi : X_{\mathcal{M}} \times (Y,\tau_Y) \to Z_{\mathcal{M}}$ *is amply bounded iff for every* $W \in \mathcal{M}$, $f_o \in X_{\mathcal{M}}$, *there is an* $\varepsilon > 0$ *such that* $(Y,\tau_Y) \subset H(F;G)_{\{W'\}}$ *continuously where* $W' = B_\varepsilon(f_o(W))$.

PROOF: First we show the condition is sufficient for ϕ to be amply bounded. Let $f_o \in X_{\mathcal{M}}$, $g_o \in Y$ and let $W \in \mathcal{M}$. Then $N = \{g \in Y : |g|_{W'} < \eta\}$ is a τ_Y-neighborhood of zero. Since g_o is bounded on W', we have

$$\sup_{|f|_W<\varepsilon, g\in N} |(g_o+g)\circ(f_o+f)|_W \leq \sup_{g\in N} |g_o+g|_{W'} < \infty,$$

so that ϕ is amply bounded.

Conversely, if ϕ is amply bounded, then for $f_o \in X_{\mathcal{M}}$, $g_o = 0 \in Y$, and $W \in \mathcal{M}$, we have an $\varepsilon > 0$, $V \in \mathcal{M}$, and τ_Y-neighborhood of zero N such that, with the aid of Lemma 6.1 and letting $W' = B_\varepsilon(f_o(W))$, we have

$$\sup_{g\in N} |g|_{W'} = \sup_{\substack{f=\text{constant}\\ |f|_E<\varepsilon, g\in N}} |g\circ(f_o+f)|_W \leq \sup_{|f|_V<\varepsilon, g\in N} |(g_o+g)\circ(f_o+f)|_W < \infty.$$

This means there is an $\eta > 0$ such that $N \subset \eta N_o$ where $N_o = \{g \in Y : |g|_{W'} < 1\}$. Therefore N_o is a τ_Y-neighborhood of zero and so in particular is absorbing. Hence, each g in Y is absorbed by N_o and so is bounded on W'. Therefore $Y \subset H(F;G)_{\{W'\}}$, and the inclusion is continuous (since N_o is a τ_Y-neighborhood of zero).

We can combine the lemmas to yield the following result, letting $J_\varepsilon = J_\varepsilon(X_{\mathcal{M}}, \mathcal{M}) = \{B_{\varepsilon(f,W)}(f(W)) : f \in X_{\mathcal{M}}, W \in \mathcal{M}\}$ where $\varepsilon : X_{\mathcal{M}} \times \mathcal{M} \to \mathbb{R}^+$. (Note J_ε is an open cover of F since X contains all the constant functions).

PROPOSITION 6.1 *The following are equivalent.*

(i) $\phi : X_{\mathcal{M}} \times (Y, \tau_Y) \to Z_{\mathcal{M}}$ *is amply bounded.*

(ii) *There is an open cover* J_ε *of* F *such that* $Y \subset H(F;G)_{J_\varepsilon}$ *continuously (so that* $\tau_Y \geq \tau_\lambda(Y)$).

(iii) $(Y, \tau_Y) = Y_{\mathcal{M}'}$, *where for every* $f_o \in X_{\mathcal{M}}$ *and* $W \in \mathcal{M}$, *there is an* $\varepsilon > 0$ *and* $W' \in \mathcal{M}'$ *such that* $B_\varepsilon(f_o(W)) \subset W'$.

Now let Z have the topology defined by all seminorms on Z ported by sets in $\mathcal{M}$, that is, p is ported by $A \in \mathcal{M}$ if for every open subset W of U containing A, there is a $c > 0$ such that $p(h) \leq c|h|_W$ for all $h \in Z$. In general, we denote τ_Z by $\tau_{\omega\mathcal{M}}$. But if $\mathcal{M}$ is the collection of compact (resp. bounding) subsets of U, we write the usual τ_ω (resp. $\tau_{\omega B}$).

PROPOSITION 6.2 *Let* $\mathcal{M}$ *and* $\mathcal{M}_1$ *be collections of subsets of* U *and* $\mathcal{M}_2$ *a collection of subsets of* F *such that for every* $f_o \in X_{\mathcal{M}_1}$, $A \in \mathcal{M}$, *we have a* $W' \in \mathcal{M}_1$, *an open set* $W \subset U$ *with* $A \subset W \subset W'$, *an* $\varepsilon > 0$, *and* $V \in \mathcal{M}_2$ *such that* $B_\varepsilon(f_o(W)) \subset V$.

Then $\phi : X_{\mathcal{M}_1} \times Y_{\mathcal{M}_2} \to (Z, \tau_{\omega \mathcal{M}})$ *is amply bounded.*

PROOF: Let p be ported by $A \in \mathcal{M}$. For every $f_0 \in X_{\mathcal{M}_1}$, choose W', W, ε, and V as in the statement of the theorem. Now take $c > 0$ such that $p(h) \leq c|h|_W$ for all $h \in Z_{\mathcal{M}}$. Then for every $g_0 \in Y_{\mathcal{M}_2}$,

$$\sup_{|f|_W < \varepsilon} |\phi(f_0 + f, g_0 + g)|_W \leq |g_0|_V + |g|_V ,$$

so that

$$\sup_{|f|_{W'} < \varepsilon, |g|_V < 1} p[\phi(f_0 + f, g_0 + g)] \leq c(|g_0|_V + 1) < \infty .$$

The topology defined by the family of seminorms $(|\cdot|_W)_{W \in \mathcal{M}}$ is called the topology of uniform convergence on sets in $\mathcal{M}$. We shall be most interested in the cases where $\mathcal{M}$ is

(a) one of the following collections of subsets of U:

$\mathcal{K}$ = set of compact subsets of U,

$\mathcal{B}$ = set of bounding subsets of U,

$\mathcal{L}$ = set of U-bounded subsets of U,

I = open cover of U,

I_0 = countable open cover of U,

$I_\delta(\mathcal{M}') = \{B_{\delta(W)}(W) \cap U : W \in \mathcal{M}'\}$ where $\mathcal{M}'$ is one of the above collections and $\delta : \mathcal{M}' \to \mathbb{R}^+$, and

(b) one of the following open covers of F:

$J_\varepsilon(X, \mathcal{M}') = \{B_{\varepsilon(f,W)}(f(W)) : f \in X, W \in \mathcal{M}'\}$ where X is a vector subspace of $H(U;F)$ containing the constant functions, $\mathcal{M}'$ is one of the collections in (a), and $\varepsilon : X \times \mathcal{M}' \to \mathbb{R}^+$.

We now discuss by cases the ample boundedness (and therefore the holomorphy) of ϕ for the topologies given in §3, using Proposition 6.1 and 6.2.

But first we mention one general negative result. Identify F with the vector subspace $\tilde{F} \subset X$ of all constant functions from U to F. Then F and $\tilde{F}_{\mathcal{M}}$ are homeomorphic as well. The same holds for G. This yields the following.

PROPOSITION 6.3 *The evaluation map* $\omega: (y,g) \in F \times (H(F;G),\tau_Y) \mapsto g(y) \in G$ *is not locally bounded (= amply bounded) for any* LCS *topology* τ_Y.

PROOF: Assume the contrary. Now ω is just $\phi : \tilde{F}_{\mathcal{M}} \times (H(F;G),\tau_Y) \to \tilde{G}_{\mathcal{M}}$. Thus if ϕ is amply bounded, then from Proposition 6.1 we have $H(F;G) \subset H(F;G)_J$ for some open cover $J = J_\varepsilon$ of F. Therefore, an open set $V \in J$ is a bounding subset of F, so that $H(F;\mathbb{C}) = H_b(F;\mathbb{C})$ (Dineen [5]), which is impossible.

COROLLARY *Let* (X,τ_X) *and* (Z,τ_Z) *be* LCS'*s containing* F *and* G *respectively as locally convex subspaces and such that* $\phi : (X,\tau_X) \times (H(F;G),\tau_Y) \to (Z,\tau_Z)$ *is defined. Then* ϕ *is not amply bounded.*

From the discussion preceding the Proposition we have that every topology of uniform convergence on some collection of sets when restricted to F agrees with the Banach space topology on F. Thus in particular, τ_o and τ_{ob} agree on F with its own topology, and therefore so do all topologies between τ_o and τ_{ob}. Likewise for G. Hence all the topologies we are studying yield F and G as locally convex subspaces, so that the Corollary applies. Any positive results for our problem, then, can only come taking the second space Y as proper subspace of H(F;G).

CASE 1 τ_o, τ_ω ($\mathcal{M} = \mathcal{K}$ = {compact subsets of U})

From Proposition 6.1 we get (since $X_{\mathcal{K}} = (X, \tau_o(X))$

PROPOSITION 6.4 *If* $X \subset H(U;F)$ *and* $Y \subset H(F;G)$ *are vector subspaces and* X *contains the constant functions, then* $\phi: (X, \tau_o(X)) \times (Y, \tau_Y) \to (H(U;G), \tau_o)$ *is amply bounded iff there is a* $J = J_\varepsilon(X\,\mathcal{K})$ *such that* $Y \subset H(F;G)_J$ *continuously (and so* $\tau_Y \geq \tau_\lambda(Y)$*).*

Now if J is any open cover of F, then for every $f_o \in X$ and $K \in \mathcal{K}$, we have $f_o(X) \subset V_1 \cup \ldots \cup V_n$ for some $V_k \in J$. Hence $B_\varepsilon(f_o(K)) \subset V_1 \cup \ldots \cup V_n$ for some $\varepsilon > 0$. Thus there is a function $\varepsilon : X \times \mathcal{K} \to \mathbb{R}^+$ such that $H(F;G)_J \subset H(F;G)_{J_\varepsilon(X,\mathcal{K})}$ continuously, from which we obtain the following.

COROLLARY $\phi : (X, \tau_o(X)) \times (Y, \tau_Y) \to (H(U;G), \tau_o)$ *is amply bounded iff there is an open cover* J *of* F *such that* $Y \subset H(F;G)_J$ *continuously.*

PROPOSITION 6.5 $\phi : H(U;F)_I \times H(F;G)_J \to (H(U;G), \tau_\omega)$ *is amply bounded for all* $I = I_\delta(\mathcal{K})$ *and* $J = J_\varepsilon(H_I, \mathcal{K})$.

PROOF: We shall use Proposition 6.2 with $\mathcal{M} = \mathcal{K}$, $\mathcal{M}_1 = I$ and $\mathcal{M}_2 = J$. Let $f_o \in H_I$ and $K \in \mathcal{K}$, and take $V = B_{\varepsilon(f_o,K)}(f_o(K)) \in J$. Then for $r = \frac{1}{2}\varepsilon(f_o,K)$, there is an open set $W \subset U$ such that $K \subset W \subset W' = B_{\delta(K)}(K) \cap U \in I$ and $f_o(W) \subset B_r(f_o(K))$. Hence $B_r(f_o(W)) \subset V$, and Proposition 6.2 applies.

CASE 2 τ_{oB}, $\tau_{\omega B}$ ($\mathcal{M} = \mathcal{B}$ = {bounding subsets of U}).

Again by Proposition 6.1 we have

PROPOSITION 6.6 *If* $X \subset H(U;F)$ *and* $Y \subset H(F;G)$ *are vector subspaces and* X *contains the constant functions, then* $\phi: (X, \tau_{oB}(X)) \times (Y, \tau_Y) \to (H(U;G), \tau_{oB})$ *is amply bounded iff there is a* $J = J_\varepsilon(X, \mathcal{B})$ *such that* $Y \subset H(F;G)_J$ *continuously (and so* $\tau_Y \geq \tau_\lambda(Y)$*).*

PROPOSITION 6.7 $\phi : H(U;F)_I \times (H_b(F;G), \tau_{ob}) \to (H(U;G), \tau_{\omega B})$ *is amply bounded for all* $I = I_\delta(\mathcal{B})$.

PROOF: Again we use Proposition 6.2, letting $\mathcal{M} = \mathcal{B}$, $\mathcal{M}_1 = I$, and $\mathcal{M}_2$ = {bounded subsets of F}. For each $f_o \in H_I$ and $A \in \mathcal{B}$, take $W = B_{\delta(A)}(A) \cap U \in I$. Then W is open and $f_o(W)$ is bounded in F, so that $V = B_r(f_o(W)) \in \mathcal{M}_2$ for any $r > 0$.

CASE 3 τ_π, $\tau_{\omega B}$

There seem to be no other reasonable subspaces to take for X and Y than those given for τ_ω and $\tau_{\omega B}$. Hence the results in this case are the same as those for τ_ω, $\tau_{\omega B}$ since $\tau_\pi \leq \tau_\omega$ and $\tau_{\pi B} \leq \tau_{\omega B}$.

CASE 4 τ_λ, τ_δ ($\mathcal{M}$ = open cover (countable)).

If I is an open cover of U, then from Proposition 6.1 we have for any $J = J_\varepsilon(H_I, I)$ that $\phi : H(U;F)_I \times H(F;G)_J \to H(U;G)_I$ is amply bounded. Since $H(U;G)_I \hookrightarrow (H(U;G), \tau_\lambda)$ is continuous, we have

PROPOSITION 6.8 *For every open cover (resp. countable open cover)* I *of* U *and* $J = J_\varepsilon(H_I, I)$, $\phi : H(U;F)_I \times H(F;G)_J \to (H(U;G), \tau)$

is amply bounded when $\tau = \tau_\lambda$ *(resp.* $\tau = \tau_\delta$*).*

CASE 5 τ_{ob} ($\mathcal{M} = \mathcal{L}$ = {U-bounded subsets of U}).

From Proposition 6.1 we get

PROPOSITION 6.9 *If* $X \subset H_b(U;F)$ *and* $Y \subset H(F;G)$ *are vector subspaces and* X *contains the constant functions, then* $\phi : (X, \tau_{ob}(X)) \times (Y, \tau_Y) \to (H_b(U;G), \tau_{ob})$ *is amply bounded iff there is a* $J = J_\varepsilon(X, \mathcal{L})$ *such that* $Y \subset H(F;G)_J$ *continuously (where* $\tau_Y \geq \tau_\lambda(Y)$*).*

CASE 6 τ_∞, τ_σ, and (U x_o - balanced) τ_ω

To treat this final case we need the Taylor series expansion of gof about the point $x_o \in U$. This is given by (cf. Nachbin [7])

$$gof(x) = gof(x_o) + \sum_{m=1}^{\infty} \Big[\sum_{n=1}^{m} \sum_{|j|=m} \frac{1}{n!} d^n g(f(x_o)) \frac{1}{j!} d^j f(x_o)(x - x_o)^j \Big]$$

where $j = (j_1, \ldots, j_n) \in \mathbb{N}^n$ $(j_k \geq 1)$, $|j| = j_1 + \ldots + j_n$, $j! = j_1! \ldots j_n!$

and

$$d^j f(x_o)(x - x_o)^j = (d^{j_1} f(x_o)(x - x_o)^{j_1}, \ldots, d^{j_n} f(x_o)(x - x_o)^{j_n}) \in F^n$$

so that

$$\frac{1}{m!} d^m(gof)(x_o)(x - x_o)^m = \sum_{n=1}^{m} \sum_{|j|=m} \frac{1}{n!} d^n g(f(x_o)) \frac{1}{j!} d^j f(x_o)(x - x_o)^j$$

and

$$\frac{1}{m!} \hat{d}^m(gof)(x_o)(x - x_o) = \sum_{n=1}^{m} \sum_{|j|=m} \frac{1}{n!} d^n g(f(x_o)) \frac{1}{j!} \hat{d}^j f(x_o)(x - x_o).$$

Hence,

$$\Big\| \frac{1}{m!} \hat{d}^m(gof)(x_o) \Big\| \leq \sum_{n=1}^{m} \sum_{|j|=m} \Big\| \frac{1}{n!} d^n g(f(x_o)) \Big\| \frac{1}{j!} \|\hat{d}^{j_1} f(x_o)\| \ldots \|\hat{d}^{j_n} f(x_o)\|.$$

PROPOSITION 6.10 $\phi : (H(U;F),\tau_\infty) \times H(F;G)_J \to (H(U;G),\tau_\infty)$ *is amply bounded for every* $J = J_\varepsilon(H,\mathcal{K})$.

PROOF: For $K \in \mathcal{K}$, $m \in \mathbb{N}^+$, (f_o,g_o) and $(f,g) \in H(U;F) \times H(F;G)_J$, we have

$$\left|\frac{1}{m!}\hat{d}^m\phi(f_o+f,g_o+g)\right|_K \le \sum_{n=1}^{m} \sum_{|j|=m} \left|\frac{1}{n!} d^n(g_o+g)\right|_{(f_o+f)(K)} a_{j_1}\cdots a_{j_n}$$

where $a_n = \left|\frac{1}{n!}\hat{d}^n(f_o+f)\right|_K$. Let $M = \sup_{1\le i\le m}\left|\frac{1}{i!}\hat{d}^i f_o\right|_K$ and set $N_m = \{f \in H(U;F) : \sup_{0\le i\le m}\left|\frac{1}{i!}\hat{d}^i f\right|_K < \delta\}$. Then for $f \in N_m$ and $\rho > 0$, we get

$$\left|\frac{1}{m!}\hat{d}^m\phi(f_o+f,g_o+g)\right|_K \le \sum_{n=1}^{m} m^n(M+\delta)^n \left|\frac{1}{n!}d^n(g_o+g)\right|_{B_\delta(fo(K))}$$

$$\le \sum_{n=1}^{m} \frac{m^n(M+\delta)^n}{(\rho/2)^n}\frac{n^n}{n!}|g_o+g|_{B_{\rho+\delta}(fo(K))}$$

$$= c|g_o+g|_V,$$

where $V = B_{\rho+\delta}(f_o(K))$. Therefore if we choose ρ and δ such that $\rho+\delta < \varepsilon(f_o,K)$, then the right hand side is bounded by a fixed number for all $f \in N_m$ (a τ_∞ - neighborhood of zero) and g such that $|g|_V < \eta$.

We shall consider next the topology τ_σ on $H(U;G)$. Recall that it is defined by all seminorms of the form

$$q_{\alpha,K}(h) = \sum_{m=0}^{\infty} \alpha_m^m \left|\frac{1}{m!}\hat{d}^m h\right|_K$$

for $\alpha = (\alpha_m) \in c_o^+$ and $K \in \mathcal{K}$. In fact, we may assume $\alpha_n \ge \alpha_{n+1}$; since for any $\alpha \in c_o^+$ there is a $\beta \in c_o^+$ with $\beta_n \ge \beta_{n+1}$ (namely, $\beta_m = \sup_{n\ge m}\{\alpha_n\}$) such that $q_{\alpha,K} \le q_{\beta,K}$.

PROPOSITION 6.11 $\phi : H(U;F),\tau_\sigma) \times (H_b(F;G),\tau_{ob}) \to (H(U;G),\tau_\sigma)$

is amply bounded.

PROOF: Let (f_o, g_o) and $(f,g) \in H(U;F) \times H_b(F;G)$, $K \in \mathcal{K}$ and $\alpha = (\alpha_m) \in c_o^+$ with $\alpha_m \geq \alpha_{m+1}$. Then

$$q_{\alpha,K}(\phi(f_o + f, g_o + g)) = \sum_{m=0}^{\infty} \alpha_m^m \left| \frac{1}{m!} \hat{d}^m(g_o + g) \circ (f_o + f) \right|_K$$

$$\leq c_o + \sum_{m=1}^{\infty} \alpha_m^m \left[\sum_{n=1}^{m} \sum_{|j|=m} \left| \frac{1}{n!} d^n(g_o + g) \right|_{(f_o+f)(K)} a_{j_1} \cdots a_{j_n} \right]$$

(where $c_o = |g_o + g|_{(f_o + f)(K)}$ and $a_n = \frac{1}{n!} \hat{d}^n(f_o + f)|_K$)

$$\leq c_o + \sum_{n=1}^{\infty} \left| \frac{1}{n!} d^n(g_o + g) \right|_{(f_o + f)(K)} \left[\sum_{m=n}^{\infty} \sum_{|j|=m} \alpha_m^m a_{j_1} \cdots a_{j_n} \right].$$

Suppose $M = p_{\alpha,K}(f_o)$ and $p_{\alpha,K}(f) < \delta$ where f_o and $f \in (H(U;F), \tau_\sigma)$. Then

$$\sum_{m=1}^{\infty} \alpha_m{}^m a_m = p_{\alpha,K}(f_o + f) \leq M + \delta .$$

Hence,

$$\left[\sum_{m=1}^{\infty} \alpha_m a_m \right]^n = \sum_{m=n}^{\infty} \sum_{|j|=m} \alpha_{j_1}{}^{j_1} a_{j_1} \cdots \alpha_{j_n}{}^{j_n} a_{j_n} \leq (M + \delta)^n$$

But $\alpha_m \geq \alpha_{m+1}$ means $\alpha_m^m \leq \alpha_{j_1}^{j_1} \cdots \alpha_{j_n}^{j_n}$ for $|j| = m$. Hence, for $M = p_{\alpha,K}(f_o)$, $p_{\alpha,K}(f) < \delta$, and $V = B_{\rho+\delta}(f_o(K))$, we have

$$q_{\alpha,K}((g_o + g) \circ (f_o + f)) \leq \sum_{n=0}^{\infty} (M + \delta)^n \left| \frac{1}{n!} d^n(g_o + g) \right|_{B_\delta(f_o(K))}$$

$$\leq \sum_{n=0}^{\infty} \frac{(M + \delta)^n}{(\rho / 2)^n} \frac{n^n}{n!} |g_o + g|_V$$

$$\leq \sum_{n=0}^{\infty} \left[\frac{2e(M + \delta)}{\rho} \right]^n |g_o + g|_V$$

If $\rho > 2e(M + \delta)$, then the right hand side is bounded by a fixed number for all $f \in H(U;F)$ such that $p_{\alpha,K}(f) < \delta$ and $g \in H_b(F;G)$ such that $|g|_V < \eta$.

PROPOSITION 6.12 *If* U *is* x_0*-balanced, then* $\phi : (H(U;F),\tau_\omega) \times (H_b(F;G),\tau_{ob}) \to (H(U;G),\tau_\omega)$ *is amply bounded.*

PROOF: Let (f_0,g_0) and $(f,g) \in H(U;F) \times H_b(F;G)$, $\varepsilon = (\varepsilon_n)$ in c_0^+ (with $\varepsilon_n \geq \varepsilon_{n+1}$), $K \in \mathcal{K}$, and set $c_0 = |(g_0+g)\circ(f_0+f)(x_0)|$ and $a_n = \sup_{x \in K+\varepsilon_n B_1} \|\frac{1}{n!}\hat{d}^n(f_0+f)(x_0)(x-x_0)\|$, $n \in \mathbb{N}^+$. Then

$$q_{\varepsilon,K}(\phi(f_0+f,g_0+g)) = \sum_{m=0}^{\infty} \sup_{x \in K+\varepsilon_m B_1} \|\frac{1}{m!}\hat{d}^m(g_0+g)\circ(f_0+f)(x_0)(x-x_0)\|$$

$$\leq c_0 + \sum_{m=1}^{\infty}\sum_{n=1}^{m}\sum_{|j|=m} \|\frac{1}{n!}[d^n(g_0+g)](f_0+f)(x_0)\|\, a_{j_1}\cdots a_{j_n}$$

$$= c_0 + \sum_{n=1} \|\frac{1}{n!}[d^n(g_0+g)](f_0+f)(x_0)\| \Big[\sum_{m=n}^{\infty}\sum_{|j|=m} a_{j_1}\cdots a_{j_n}\Big].$$

Since f_0, $f \in (H(U;F),\tau_\omega)$, take $M = p_{\varepsilon,K}(f_0)$ and $p_{\varepsilon,K}(f) < \delta$. Then

$$\sum_{m=1}^{\infty} a_m \leq p_{\varepsilon,K}(f_0+f) \leq M+\delta,$$

so that

$$\Big[\sum_{m=1}^{\infty} a_m\Big]^n = \sum_{m=n}^{\infty}\sum_{|j|=m} a_{j_1}\cdots a_{j_n} \leq (M+\delta)^n.$$

Hence,

$$q_{\varepsilon,K}((g_0+g)\circ(f_0+f)) \leq \sum_{n=0}^{\infty}(M+\delta)^n \left|\frac{1}{n!}d^n(g_0+g)\right|_{B_\delta(f_0(x_0))}$$

$$\leq \sum_{n=0}^{\infty}\left[\frac{2e(M+\delta)}{\rho}\right]^n |g_0+g|_{B_{\rho+\delta}(f_0(x_0))} < \infty$$

if $\rho > 2e(M+\delta)$, since g_0 and $g \in H_b(F;G)$. Thus ϕ is amply bounded.

BIBLIOGRAPHY

[1] ALEXANDER, H., - "Analytic Functions on Banach Spaces", Thesis, University of California, Berkeley, California (1968).

[2] ARON, R., - "Topological Properties of the Space of Holomorphic Mappings", Thesis, University of Rochester, New York (1970).

[3] CHAE, S. B., - "Holomorphic germs on Banach spaces", Ann. Inst. Fourier, 21 (1971), 107 - 144.

[4] DINEEN, S., - "The Cartan - Thullen theorem for Banach spaces", Ann. Scuola Norm. Sup. Pisa, 24 (1970), 667 - 676.

[5] DINEEN, S., - "Unbounded holomorphic functions on a Banach space", J. Lon. Math. Soc., 4 (1972), 461 - 465.

[6] HIRSCHOWITZ, A., - "Bornologie des espaces de fonctions analytiques en dimension infinite", Sēminaire P. Lelong, 1970, Springer - Verlag, Bd. 205 (1971).

[7] NACHBIN, L., - "Lectures on the Theory of Distributions", University of Rochester (1963). Reproduced in textos de Matemãtica, 15, Universidade Federal de Pernambuco, Recife, Brazil (1964).

[8] NACHBIN, L., - "Concerning Spaces of Holomorphic Mappings, Dept. of Math. Rutgers University, New Brunswick, New Jersey (1970)

[9] NACHBIN, L., - "Course in Infinite Dimensional holomorphy", Rio de Janeiro (1971).

[10] NACHBIN, L., - "Limites et perturbation des applications holomorphes", Colloq. sur les Fonctions Analytiques de Plusieurs Variables Complexes, Paris, 1972 C.N.R.S., Agora Mathematica, Gauthier Villars (to appear).

[11] JOSEFSON, B., - "Weak sequential convergence in the dual of a Banach space does not imply norm convergence", Bull. Amer. Math. Soc., 81 (1975), 166 - 168.

University of Arkansas
Fayetteville, AR 72701

Infinite Dimensional Holomorphy and Applications, Matos (ed.)

THE NUCLEARITY OF $\mathcal{O}(U)$*

by *L. WAELBROECK*

P. Boland has proved that the compact - open topology is a nuclear topology on the algebra of entire functions on the dual of a Fréchet nuclear space. This result has been generalised independently by P. Boland and the author after we discussed the matter at the complex analysis meeting in Cracow, in 1974. It turns out that the compact open topology is also a nuclear topology on the algebra $\mathcal{O}(U)$ of holomorphic functions on an open subset of the dual of a Fréchet nuclear space.

The proofs of Boland and the present author are different. Each uses devices that will probably be useful elsewhere in infinite dimensional complex analysis. It appears reasonable that both proofs get published. This paper will contain the present author's proof. P. Boland's proof will appear elsewhere [2].

The interessed reader is also referred to P.Boland's earlier paper [1], where Boland shows not only that $\mathcal{O}(E)$ is nuclear when E is dual Fréchet nuclear, but also that all reasonable topologies coincide on $\mathcal{O}(U)$ if U is open in a dual Fréchet

---

(*) To Professor G. Köthe, on his seventieth anniversary.

nuclear space.

1 - Let us first state the most general form of the main theorem in this paper.

$(E, \mathcal{T})$ will be a locally convex space, $\mathcal{B}$ will be a nuclear boundedness on E which is finer that the topological boundedness (cf [3], p. 69, or think of a duality between E and a nuclear space $(F, \mathcal{S})$ where E is the dual of F, both $\mathcal{T}$ and $\mathcal{S}$ being stronger than the weak topologies $\sigma(E, F)$, $\sigma(F, E)$; will be the set of equicontinuous subsets of $E = F^{x}$). If U is an open subset of E, $\mathcal{B}_U$ will be the set of elements of $\mathcal{B}$ which are relatively compact in U.

THEOREM. *Uniform convergence on the elements of $\mathcal{B}_U$ is a nuclear topology on the algebra $\mathcal{O}(U)$ of holomorphic functions on U.*

The theorem is easier to state when E is dual nuclear and sequentially complete. We can then let $\mathcal{B}$ be the set of all bounded subsets of E, $\mathcal{B}_U$ will be the relatively compact subsets of U, uniform convergence on the elements of $\mathcal{B}_U$ will be the compact open topology.

COROLLARY 1. *Let U be an open subset of a dual nuclear sequentially complete space. The compact open topology is nuclear on $\mathcal{O}(U)$.*

Fréchet nuclear spaces and their duals (dual Fréchet nuclear, or nuclear Silva spaces) are dual nuclear and complete.

COROLLARY 2. *Let U be an open subset of a Fréchet nuclear space, or of a Silva nuclear space. The compact open topology is nuclear on* $\mathcal{O}(U)$.

The main theorem can also be applied to spaces which are not dual nuclear. All that we need is a nuclear boundedness which is compatible with the duality.

A bounded subset of a locally convex space is rapidly decreasing if it is contained in the closed absolutely convex hull of a rapidly decreasing sequence. The rapidly decreasing boundedness on a locally convex space is nuclear when the closed absolutely convex bounded subsets are completant. (B is completant when B is absolutely convex, does not contain any one-dimensional subspace, and is such that its Minkowski functional is a Banach space norm on the vector space E_B absorbed by B).

COROLLARY 3. *Let* U *be open in a locally convex space whose closed absolutely convex bounded sets are completant. The topology of uniform convergence on the relatively compact, rapidly decreasing subsets of* U *is nuclear on* $\mathcal{O}(U)$.

2. DEFINITION 1. *Let* X *be a compact subset of a locally convex space* E. *We shall call* $\mathcal{Q}(X)$ *the Banach algebra (in the uniform norm) of continuous functions on* X *whose restriction to the* E_o*-interior of* $X \cap E_o$ *is holomorphic for all finite dimensional affine subspaces* E_o *of* E.

The reader may complain that our definition of $\mathcal{Q}(X)$ is not reasonable. It would be more reasonable to require holomorphy on the E_o-interior of $X \cap E_o$ for all finite dimensional analytic sets E_o. In this paper, fortunately, we will have no reason to consider $\mathcal{Q}(X)$ except when $X = X_1 + X_2$, where X_1 is compact in a finite-dimensional subspace and equal to the closure of its relative interior, and X_2 is compact and convex. For such compact X, all reasonable definitions of $\mathcal{A}(X)$ coincide. And by the way, amateurs of uniform algebras can have lots of fun playing around with such "simple" $\mathcal{A}(X)$.

Let D be the closed unit disc. $D^{\mathbb{N}} \subset \mathbb{C}^{\mathbb{N}}$ is compact. So is $\prod_{n \in \mathbb{N}} \lambda_n D$, if the λ_n are positive real numbers. Restriction maps $\mathcal{A}(D^{\mathbb{N}})$ into $\mathcal{A}(\prod_{n \in \mathbb{N}} \lambda_n D)$ if $0 < \lambda_n \leq 1$ for all n.

PROPOSITION 1. *This restriction is nuclear if* $0 < \lambda_n < 1$ *for all* n *and* $\Sigma \lambda_n < \infty$.

We consider $f \in \mathcal{A}(D^{\mathbb{N}})$ and write, at least formally its Taylor series,

$$f(z) = \Sigma_{k \in \mathbb{N}^{(\mathbb{N})}} \frac{1}{k!} \frac{\partial^k f}{\partial z^k}(0)\, z^k$$

The notations are standard, $\mathbb{N}^{(\mathbb{N})}$ is the set of sequences (k_n) of integers such that $k_n = 0$ for n large. If $k = (k_n)$ and $k_n = 0$ when $n > N$,

$$k! = k_o! \dots k_N!$$

$$\frac{\partial^k}{\partial z^k} = \frac{\partial^{k_o + \dots + k_N}}{\partial z_o^{k_o} \dots \partial z_N^{k_N}}$$

$$z^k = z_o^{k_o} \dots z_N^{k_N}$$

$$\lambda^k = \lambda_o^{k_o} \dots \lambda_N^{k_N}$$

The Cauchy evaluation shows that the linear form on $\mathcal{A}(D^{\mathbb{N}})$

$$f \to \frac{1}{k!} \frac{\partial^k f}{\partial z^k}(0)$$

has norm unity. On the other hand, z^k is an element if $\mathcal{Q}(D^{\mathbb{N}})$ whose restriction to $\prod \lambda_n D$ has norm λ^k.

The restriction to $\prod \lambda_n D$ of the formal Taylor series we wrote for f will be a nuclear mapping of $\mathcal{Q}(D^{\mathbb{N}})$ into $\mathcal{Q}(\prod \lambda_n D)$ if the series

$$\Sigma_{k \in \mathbb{N}^{(\mathbb{N})}} \lambda^k$$

converges. And the series does converge

$$\Sigma_{k \in \mathbb{N}^{(\mathbb{N})}} \lambda^k = \Sigma_{k \in \mathbb{N}^{(\mathbb{N})}} \prod_{n \in \mathbb{N}} \lambda_n^{k_n}$$

$$= \prod_{n \in \mathbb{N}} \Sigma_{k_n = 0}^{\infty} \lambda_n^{k_n}$$

$$= \prod \frac{1}{1 - \lambda_n}$$

Our assumptions ensure that the product converges. And the mapping $\mathcal{Q}(D^{\mathbb{N}}) \to \mathcal{Q}(\prod \lambda_n D)$ that we obtain is of course the restriction mapping.

3 - Let next S_1 and T_1 be compact subsets of $\mathbb{C}^n$, each equal to the closure of its interior, T_1 being furthermore in the interior of S_1. We can consider the sets $S = S_1 \times D^{\mathbb{N}}$ and $T = T_1 \times \prod \lambda_n D$. As previously, we assume that $0 < \lambda_n < 1$ for all n and that $\Sigma \lambda_n < \infty$.

PROPOSITION 2. *In these circumstances, the restriction mapping* $\mathcal{Q}(S) \to \mathcal{Q}(T)$ *is nuclear.*

We note that $\mathcal{Q}(S_1 \times D^{\mathbb{N}})$ and $\mathcal{Q}(T_1 \times \prod \lambda_n D)$ are respectively the slice product of $\mathcal{Q}(S_1)$ and of $\mathcal{Q}(D^{\mathbb{N}})$, and of $\mathcal{Q}(T_1)$ and $\mathcal{Q}(\prod \lambda_n D)$. The restriction mapping $\mathcal{Q}(S) \to \mathcal{Q}(T)$ is the slice product of the restriction mapping $\mathcal{Q}(S_1) \to \mathcal{Q}(T_1)$ and of the restriction mapping $\mathcal{A}(D^{\mathbb{N}}) \to \mathcal{A}(\prod \lambda_n D)$. Both factor mappings are nuclear. So is their slice product.

Some readers may not know what a slice product is. If E and F are Banach space their slice product $E \phi F$ can be identified with the linear mappings $E^* \to F$ whose restriction to the unit ball of E^* is weak - star continuous, or with the linear mappings $F^* \to E$ whose restriction to the unit ball of F^* is weak - star continuous, or yet with the bilinear forms on $E^* \times F^*$ with a weak - star continuous restriction to the product of the unit balls.

If $E \subseteq C(X)$ and $F \subseteq C(Y)$ are closed subspaces with the induced norm, X and Y being compact spaces. $E \phi F$ can be identified with the space of continuous functions on $X \times Y$ which belong separately to E and to F, i.e. with the space of $u \in C(X \times Y)$ such that $u(\cdot, x) \in E$ for all $y \in Y$ while $u(x, \cdot) \in F$ for all $x \in X$.

Nuclear mappings $E \to E_1$, $F \to F_1$ can be factored

$$E \to c_0 \to \ell_1 \to E_1$$

$$F \to c_0 \to \ell_1 \to F_1$$

where the mappings $c_0 \to \ell_1$ are obtained by multiplying the sequences coordinatewise by summable sequences, (μ_n) and (ν_n) respectively. Multiplication by $(\mu_n \nu_m)$ induces a nuclear mapping $c_0(\mathbb{N} \times \mathbb{N}) \to \ell_1(\mathbb{N} \times \mathbb{N})$. We consider then the commutative diagram

$$\begin{array}{ccccccc} E \hat{\otimes} F & \to & c_0 \hat{\otimes} c_0 & \to & \ell_1 \hat{\otimes} \ell_1 & \to & E_1 \hat{\otimes} F_1 \\ \downarrow & & & \nearrow & \downarrow & & \downarrow \\ E \phi F & \to & c_0 \phi c_0 & \to & \ell_1 \phi \ell_1 & \to & E_1 \phi F_1 \end{array}$$

where the diagonal arrow $c_0 \phi c_0 \to \ell_1 \hat{\otimes} \ell_1$ comes form identifying these spaces respectively with $c_o(\mathbb{N} \times \mathbb{N})$ and $\ell_1(\mathbb{N} \times \mathbb{N})$ and coordinatewise multiplication with $(\mu_n \nu_m)$. This diagonal arrow is nuclear. It follows that many other mappings in this diagram are, among others the mapping $E \phi F \to E_1 \phi F_1$.

4 - Let now E be a locally convex space. Let S_1 and T_1 be two compact subsets of a finite dimensional subspace E_o of E, each equal to the closure of its interior (with respect to E_o), T being also contained in the interior of S relative to E_o. Let also (x_n) be a rapidly decreasing sequence of elements of E. We let

$$S = S_1 + \Sigma D x_n$$

$$T = T_1 + \Sigma \lambda_n D x_n$$

where as usual, (λ_n) is a sequence of real numbers such that $0 < \lambda_n < 1$ for all n and $\Sigma \lambda_n < \infty$.

PROPOSITION 3. *For such choices of* S *and* T, *the restriction mapping* $\mathcal{A}(S) \to \mathcal{A}(T)$ *is subnuclear.*

Recall that a linear map of Banach spaces $G \to H$ is subnuclear if an isometry $H \to H_1$ exists such that the composition $G \to H_1$ is nuclear. The composition of two subnuclear mappings is nuclear.

When $f \in \mathcal{A}(S)$, $s_1 \in S_1$, $t \in D^{\mathbb{N}}$, we let

$$u_1 f(s_1, t) = f(s_1 + \Sigma t_n x_n)$$

similarly, when $f \in \mathcal{A}(T)$, $s_1 \in T_1$, $t \in \prod \lambda_n D$, we let

$$u_2 f(s_1, t) = f(s_1 + \Sigma t_n x_n)$$

The mappings $u_1 : \mathcal{A}(S) \to \mathcal{A}(S_1 \times D^{\mathbb{N}})$, $u_2 : \mathcal{A}(T) \to \mathcal{A}(T_1 \times \prod \lambda_n D)$

are isometric inbeddings. We have the commutative diagram

$$\begin{array}{ccc} \mathcal{A}(S) & \to & \mathcal{A}(S_1 \times D^{\mathbb{N}}) \\ \downarrow & & \downarrow \\ \mathcal{A}(T) & \to & \mathcal{A}(T_1 \times \prod \lambda_n D) \end{array}$$

In this diagram, the horizontal arrows are isometric inbeddings. The vertical arrows are restriction mappings, the right - hand vertical arrow is nuclear. The left - hand vertical arrow is therefore subnuclear.

5 - To prove our main theorem, we must still prove

PROPOSITION 4. *Let* $X \in \mathcal{B}_U$. *It is possible to find compact subsets* S_1 *and* T_1 *of a finite dimensional subspaces* E_o *, each equal to the closure of its interior,* T_1 *contained in the interior of* S_1 *relative to* E_o *, to find a rapidly decreasing sequence of vectors* x_n *and a sequence of constants* λ_n *with* $0 < \lambda_n < 1, \Sigma \lambda_n < \infty$, *such that*

$$X \subseteq T \subseteq S \subseteq U$$

if $T = T_1 + \Sigma \lambda_n D x_n$, $S = S_1 + \Sigma D x_n$.

As in the main theorem, in the statement of proposition 4, U is an open subset of a locally convex space, $\mathcal{B}$ is a nuclear boundedness on that space, whose elements are topologically bounded, and $\mathcal{B}_U$ is the set os elements of $\mathcal{B}$ which are relatively compact in U .

The theorem of Komura and Komura [4] proves the existence of a sequence x_n of elements of E , which is rapidly decreasing for the boundedness $\mathcal{B}$ and whose closed absolutely convex hull contains X . The elements of $\mathcal{B}$ being topologically bounded, the sequence is rapidly decreasing for the vector space topology.

X is relatively compact in U, and U is open. We can find an open, absolutely convex neighbourhood of the origin in E, say V, such that $U \supseteq X + V$. We choose $\alpha > 0$ sufficiently small, $\alpha = 10^{-1}$ will do, and n_o such that $(1+n^4)x_n \in \alpha V$ when $n > n_o$. We let E_o be the vector space generated by $(x_o,\dots,x_{n_o})$ and

$$X' = \{\Sigma_o^{n_o} s_n x_n \mid \Sigma_o^{\infty} |s_n| \leq 1,\ \Sigma_o^{\infty} s_n x_n \in X\}$$

We note that X' is compact and contained in E_o. Also for every $x \in X$ there is an $x' \in X'$ and for every $x' \in X'$ there is an $x \in X$ such that $x - x'$ is in the closed absolutely convex hull of $\{x_n \mid n > n_o\}$.

Since $x_n \in \alpha V$ when $n > n_o$, and V is open absolutely convex,

$$X' \subseteq X + \alpha V .$$

We choose T_1 and S_1, compact subsets of E_o, each equal to the closure of its interior, T_1 in the interior of S_1, such that

$$X' \subseteq T_1 \subseteq S_1 \subseteq X + \alpha V .$$

On the other hand, letting

$$X'' = \{\Sigma_{n_o+1}^{\infty} s_n x_n \mid \forall n : |s_n| \leq 1\} = \Sigma_{n_o+1}^{\infty} D x .$$

we see that X'' contains the closed absolutely convex hull of the set of x_n, $n > n_o$, hence that

$$X \subseteq X' + X'' \subseteq T_1 + X'' = T_1 + \Sigma_{n_o+1}^{\infty} D x_n$$

To prove proposition 4, it will be sufficient to show that

$$S_1 + \Sigma_{n_o+1}^{\infty} (1 + n^2) D x_n \subseteq U$$

but

$$(1 + n^2)\, D x_n \subseteq \frac{1 + n^2}{1 + n^4} \quad \alpha V$$

hence

$$\Sigma\, (1 + n^2)\, D x_n \subseteq \left(\sum \frac{1 + n^2}{1 + n^4}\right) \alpha V$$

and

$$S_1 + \Sigma\, (1 + n^2)\, D x_n \subseteq X + \alpha V + \left(\sum \frac{1 + n^2}{1 + n^4}\right) \alpha V$$

$$\subseteq X + V \subseteq U$$

since α was chosen small enough, and

$$\alpha \left(1 + \Sigma_0^\infty \frac{1 + n^2}{1 + n^4}\right) \leq 1$$

and we know that $X + V \subseteq U$.

6 - The theorem is now essentially proved. Start with any continuous semi - norm ν on $\mathcal{O}(U)$, find a semi - norm p_X which dominates ν, where X is compact in U and

$$p_X(f) = \max_{x \in X} |f(x)|$$

Associate S and T to X as described in proposition 4. Proposition 3 shows that the restriction mapping $\mathcal{A}(S) \to \mathcal{A}(T)$ is subnuclear.

Of course, the completion of $(\mathcal{O}(U), p_S)$ is contained isometrically in $\mathcal{A}(S)$, while that of $(\mathcal{O}(U), p_T)$ is contained in $\mathcal{A}(T)$.

The natural mapping of the completion of $(\mathcal{O}(U), p_S)$ into that of $(\mathcal{O}(U), p_T)$ is therefore subnuclear. This proves the result since p_T dominates the semi - norm ν we considered initially, while p_S is a continuous semi - norm.

REFERENCES.

[1] P. J. BOLAND. Holomorphic functions on nuclear spaces. Transactions A.M.S. J. 209, 1975.

[2] P. J. BOLAND. An example of a nuclear space in infinite dimensional holomorphy. To appear, Akiv för Mathematik. 15.1, May 1977.

[3] H. HOGHE-NLEND. Théorie des homologies et applications. Springer - Verlag. Lecture Notes in Mathematics. J. 213, 1971.

[4] T. KOMURA AND Y. KOMURA. Über die Einbettung der nuklearen Räume in $(s)^A$. Math. Annalen. 162. 1966. p. 284 - 288.

Université Libre de Bruxelles
Departement de Mathematique
Faculte des Sciences
Campus Plaine, CP. 214
Bruxelles - Belgique.

INDEX OF TERMS AND CONCEPTS